Nomad's Land

FRANCE OVERSEAS: STUDIES IN EMPIRE AND DECOLONIZATION
Series editors: A. J. B. Johnston, James D. Le Sueur, and Tyler Stovall

Nomad's Land

Pastoralism and French
Environmental Policy in
the Nineteenth-Century
Mediterranean World

Andrea E. Duffy

UNIVERSITY OF NEBRASKA PRESS   LINCOLN

Portions of chapter 2 originally appeared as "Civilizing through Cork: Conservationism and *la Mission Civilisatrice* in French Colonial Algeria," *Environmental History* 23, no. 2 (2018): 270–92. Parts of chapter 6 previously appeared as "Fighting Fire with Fire: Mobile Pastoralists and French Discourse on Wildfires in Nineteenth-Century Algeria," *Resilience: A Journal of the Environmental Humanities* 3 (2015): 71–87.

Library of Congress Cataloging-in-Publication Data
Names: Duffy, Andrea E., author.
Title: Nomad's land: pastoralism and French environmental policy in the nineteenth-century Mediterranean world / Andrea E. Duffy.
Description: Lincoln: University of Nebraska Press, [2019] |
Series: France overseas: studies in empire and decolonization |
Includes bibliographical references and index.
Identifiers: ISBN 9781496219169 (epub)
LCCN 2019015630   ISBN 9781496219176 (mobi)
ISBN 9780803290976 (cloth: alk. paper)   ISBN 9781496219183 (pdf)
Subjects: LCSH: Pastoral systems—Government policy—France—
History—19th century. | Forest conservation—Government
policy—France—History—19th century. | Pastoral systems—
Mediterranean Region—History—19th century.
Classification: LCC SF55.F8 D84 2019 | DDC 636.08/450944—dc23
LC record available at https://lccn.loc.gov/2019015630

Set in Adobe Text by Mikala R. Kolander.
Designed by N. Putens.

CONTENTS

LIST OF FIGURES

Tables

ACKNOWLEDGMENTS

Writing this book has been quite a journey, and I am grateful for all of the assistance and support I received along the way. Bridget Barry, my editor at the University of Nebraska Press, was extremely helpful, attentive, and patient throughout this process. I also greatly appreciate the support of this press as a whole in giving shape to my work. In addition, I would like to thank my anonymous reviewers for taking the time to read my manuscript and to provide feedback that was constructive, considerate, and extremely valuable. Several other people read and commented on earlier versions of this book. I am particularly grateful for the suggestions of John McNeill, Gabor Ágoston, James Collins, Diana Davis, and Sam White. Although I take full responsibility for any weaknesses or limitations in this book, its strengths owe much to the insight and expertise of these scholars and friends. Investigating the history of multiple Mediterranean societies is not cheap. During the research and writing process, I benefited immensely from the generous support of various institutions, including Georgetown University, the Université de Provence, the American Society for Environmental History, and the School of Global Environmental Sustainability at Colorado State University. I also valued the presence and accessibility of archives critical to my study, as well as the willingness of personnel to assist and accommodate me. I especially appreciated the warm reception and support of staff at the Archives Nationales d'Outre-Mer (ANOM) in Aix-en-Provence, France. Many other figures deserve mention for cultivating my interest in the Mediterranean world, the environment, nomads, and mountains. I am particularly in debt to the wisdom and guidance of my late high school Latin teacher, Robert "Bobe" Simms, and to Scott "Scott Bey" Redford, the director of my study-abroad experience in Turkey. In addition, informal conversations with Sam White and Owen Miller ultimately proved instrumental in leading me to this topic. Finally, I am grateful for the enduring love and support of my family. My parents, Roger and Sylvia Wiegand, who also waded through a rough manuscript version of this book, provide an invaluable constant in my life and a perpetual source of inspiration. And this book might never have seen the light of day without the boundless patience, encouragement, and care of my loving husband, John. To everyone who helped me through this process, including many I neglected to name, I dedicate this book.

INTRODUCTION

The Nomad and the Sea

> The Turkish Mediterranean lived and breathed with the same rhythms as the Christian, [and] the whole sea shared a common destiny.
> —Fernand Braudel, *The Mediterranean and the Mediterranean World in the Age of Philip II*

The spark for this book dates to the spring of my sophomore year of college, when I participated in a study-abroad program based in Alanya, Turkey. During a two-week trip to Syria, we traveled across southwestern Anatolia via a tortuous (and torturous) bus ride that wound through the craggy passes of the Taurus range. In many places the road was barely wider than our bus and overlooked roiling waters hundreds of feet below. I learned that it was an updated version of an ancient route. It followed one of the few pregnable paths through the mountains of southwestern Anatolia, and pilgrims, merchants, officials, and other travelers had used it for millennia.

The most significant of these ancient thoroughfares is the Gülek Pass, which links Anatolia's central plateau to the coastal plains of Cilicia, near Adana, through a maze of rocky peaks. This pass, better known in the West as the Cilician Gates, has played a prominent role in history. Negotiating it proved a severe test for Alexander the Great and, later, Crusaders. Well into the twentieth century it remained a challenging obstacle to transportation between Anatolia and the Levant. As my

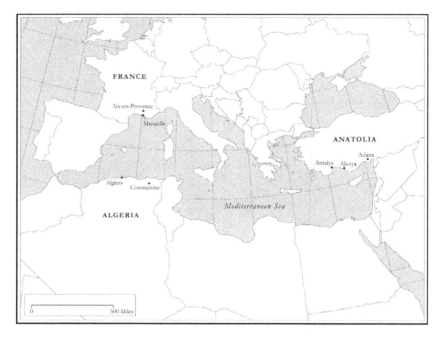

1. Provence, Algeria, and Anatolia in the Mediterranean world. Modified from *Mediterranean* (contemporary map), produced by the Cartographic Research Lab, University of Alabama, http://alabamamaps.ua.edu/contemporarymaps/world/europe/med3.jpg.

group approached this pass, we encountered a cluster of people and livestock lounging in the shade of makeshift tents. Our professor stopped to chat with them, returning a few minutes later full of excitement. "You are witnessing history," he said. "These are the last of the nomads; in a few years there will be no more."

Those words stayed with me. As I discovered then, the Cilician Gates specifically and the Taurus range in general were important landmarks not only for the people who sought to penetrate them but also for the mountain dwellers who depended on them for vital resources, shelter, and protection. For centuries nomadic tribes used this and other passes to lead herds of sheep and goats between pastures in the Taurus Mountains and sparsely populated plains along the coast of the eastern Mediterranean. Tribal leaders sometimes served as de facto rulers of these remote regions. Those who passed this way often did so with the help of these pastoralists, for many tribes raised and drove camels, the main form of transportation in the Anatolian peninsula through the late nineteenth century. Yet by the time I arrived in the region at the turn of the twenty-first century, its nomads had all but disappeared, and those who remained hardly recalled the powerful tribes that had long resisted central authority. Instead the ragtag group we met seemed to represent, as Fernand Braudel once suggested, "the relic of a tradition that is slowly disappearing."[1]

This story is not unique. Around the world, mobile pastoralists and other peripatetic peoples have witnessed the demise of their traditional lifestyle over the past two hundred years. My investigation seeks to explain why. Through case studies in Provence, northern Algeria, and southwestern Anatolia, this book chronicles the retreat of mobile pastoralism from Mediterranean coastlands in the nineteenth century. It also introduces an unlikely but critical player—the French forest regime—and it shows how the relationship between forestry and Mediterranean pastoralism fundamentally altered both institutions as well as the landscapes of the Mediterranean world.

* * *

Mediterranean history is considerably more fraught today than when Fernand Braudel first published *The Mediterranean and the Mediterranean World in the Age of Philip II* in 1949. Braudel's masterwork did much to define the field and to promote geographical, transnational, and environmental approaches to history. Yet it has not weathered the decades since its appearance without controversy, and the Mediterranean world as scholars conceptualize it today looks quite different from Braudel's original vision of a "common destiny."[2] Contemporary scholars remain divided over the suitability of the term "Mediterranean" to describe the environmental features of societies surrounding this inner sea. While many acknowledge the presence of unifying environmental characteristics, some have effectively challenged traditional ecological representations of the region. In this vein Peregrine Horden and Nicholas Purcell's survey of Mediterranean history and historiography, *The*

Corrupting Sea, emphasizes the region's "pronounced local irregularity" over its "common rhythms."[3] Over the same period scholarship has highlighted multiple "Mediterraneans," microclimates, and ecosystems within the Mediterranean region, as well as "Mediterraneanoids," or pseudo-Mediterranean systems around the world. Such developments have dampened, obscured, or dismantled many of the environmental features that once seemed to define and unify the Mediterranean region.

Others regard the Mediterranean model warily due to its history.[4] During the colonial era members of the European elite supported the Mediterranean concept based on their reading of classical authors and on their own limited, subjective interpretations of the Mediterranean region. French imperialists promoted the colonization of North Africa by emphasizing its similarity to southern France and by framing France's Mediterranean empire as a successor to Rome. By the twentieth century the presentation of the Mediterranean region as a single geographical unit had become so entrenched in European thought that it persevered long after the era of decolonization. It figured centrally in the philosophy of Fernand Braudel and other pioneers of Mediterranean history and geography, and it continues to influence perceptions of the region today. This history ties the Mediterranean concept to cultural values and perspectives that scholars no longer consider valid. Thus, some consider the Mediterranean to be a distracting, unproductive, or misleading unit of analysis. Along these lines the anthropologist João de Pina-Cabral has remarked, "The notion of the Mediterranean Basin as a 'culture area' is more useful as a means of distancing Anglo-American scholars from the populations they study than as a way of making sense of the cultural homogeneities and differences that characterize the region."[5]

Yet while contemporary critiques of the Mediterranean concept provide important checks on its use, they do not render it useless. Even critics of this concept tend to employ it extensively. Certainly this idea creates artificial boundaries and colors perceptions of the region. But while the same could be said for any nation-state, the Mediterranean provides a framework for transnational investigations. This framework is particularly useful for environmental subjects because it oversteps the political divisions among states, replacing them with broad limit-zones highlighting geographical and ecological features. The Mediterranean framework also reveals significant social, political, and cultural connections; the challenge for scholarship is to explore such commonalities without losing sight of variations within the region. As my study aims to show, the use of case studies that cut across traditional disciplinary, political, and ideological frontiers is a critical part of this endeavor.

As a work of Mediterranean history, this book reflects the legacy of Fernand Braudel, but it also updates and refines the Mediterranean concept. The story that unfolds in the following pages exposes social and environmental trends as well as divergences and eccentricities around the inner sea. Provence, northern Algeria, and southwestern Anatolia are ideally suited to this task. They lie at distinct points of the Mediterranean coast, representing its northern, southern, and eastern borders, as well as its linking of three continents. Their histories showcase the wide spectrum of political systems, languages, cultures, religions, and traditions of the Mediterranean world. Yet Provence, northern Algeria, and southwestern Anatolia all share a rich mobile pastoral tradition, and they all watched that tradition fade over the course of the nineteenth century. The modern era brought these three places together in even more significant ways. It united them through the development and dissemination of French scientific forestry and, more broadly, the French imperial mission. Provence, northern Algeria, and southwestern Anatolia provide the contours of this study because they collectively demonstrate both the rhythms and the irregularities of the Mediterranean world, and they illustrate the value of a Mediterranean lens.

This book uses a Mediterranean perspective to expose both geographic and thematic connections. In Provence, Algeria, and Anatolia, as well as other Mediterranean contexts, the main forces behind forest administration were French. Although German forestry enjoyed greater international renown in the early nineteenth century, France quickly became the unofficial leader in Mediterranean scientific forestry, a development that had much to do with its mobile pastoral tradition. France also had deeply rooted connections with the Ottoman Empire, solidified through common enemies and their shared border, the Mediterranean Sea. At the same time, France was home to some of the harshest and most vocal critics of Mediterranean pastoralism. Their view of this practice was directly influenced by gloomy prognoses about the Mediterranean environment. As French forests dwindled in the early nineteenth century, a growing number of French scientists and intellectuals began to warn of deforestation and forest degradation. They regarded the sparse, open woodlands of the Midi with particular concern, and many linked the evolution of this landscape to the ubiquitous presence of sheep and goats. From this angle Mediterranean pastoralism appeared inefficient, environmentally destructive, unsustainable, and a threat to the region's remaining forest resources. As the century advanced, French arguments against Mediterranean pastoralism came to be increasingly based on developing ideas of environmental conservation and sustainability. Such ideas also grew more widespread, entrenched, and effective in promoting antipastoral policies throughout the Mediterranean world.

The French forest regime's anxiety over deforestation, its aversion to pastoralism, and its devotion to reforestation led it to exact harsh terms on the inhabitants of Provence. In the name of forest protection the French forest administration fought transhumant pastoralism in Provence in four principal ways: by depriving communities of pastoralism's profits, by redefining the forest and protected spaces to limit or prevent pastoral use, by shrinking pastureland through afforestation and agricultural expansion, and finally by stepping up surveillance and law enforcement. In the process French foresters encountered substantial and persistent local resistance. The encounters between foresters and pastoralists in Provence played out in a variety of forms, ranging from sometimes violent conflict to compromise and accommodation. Foresters were not the only forces in the battle against mobile pastoralism. Indeed they often benefited from the support of powerful allies who were not necessarily interested in environmental conservation. Such figures ranged from policy makers to industrialists, entrepreneurs, and agriculturalists. For them, mobile pastoralism was either a nuisance or a threat, presenting obstacles and limitations to their private interests. Perhaps most critically, the French forest regime's antipastoral initiatives gained the support of the central administration, which was motivated by political as well as environmental aims. In the context of Provence the nineteenth-century forest regime greatly expanded state control of Mediterranean pastoral populations. It served as an effective form of centralization, if not internal colonization.

These trends were not unique to Provence. Private interests and state politics played an even greater role in the marginalization of mobile pastoral populations around the inner sea. The French conquest of Algeria that began in 1830 provided new opportunities to mine valuable resources, as well as an outlet for settlement, but it also created new tensions and struggles for control of colonization and resources. Within this conflict indigenous pastoralists became a convenient scapegoat, as the colonial community quickly learned to enrich itself and settle disputes by dispossessing nomads, justifying these actions through charges of environmental destruction. By the mid-nineteenth century colonial views had crystallized into an elaborate narrative blaming Arab nomads for the long-term degradation of the Algerian landscape. This narrative served to guide and legitimize colonial policies, including extensive agricultural and land management reforms, ambitious reforestation initiatives, the expropriation of tribal lands, and the overall subjugation of indigenous inhabitants.

During the same period the Ottoman state began to take note of the development of scientific forestry in other parts the world. In the wake of its defeat by Russia in the Crimean War, which earned the empire the unsolicited distinction of being termed the "sick man of Europe," and as a result of the spirit of reform characteristic of the Tanzimat era, the Ottoman government was anxious to dispel growing accusations of unchecked deforestation throughout the empire. In 1857 the sultan therefore invited French forest experts to Istanbul to institute and oversee forest administration in the empire. These forest engineers were well aware of the environmental and economic similarities between Anatolia and Provence. Indeed some had personal or professional experience in both contexts. They largely modeled their vision of Ottoman forestry on the example of France. In the process French foresters also gained insight from the Ottoman case. They noted the presence of mobile pastoral tribes within the empire and studied their relationship with forests and the state. As a result, ideas born through the development of scientific forestry in Anatolia began to filter back into France and French colonial Algeria, fueling forest initiatives around the Mediterranean.

The impact of French scientific forestry varied widely among these three locations. French foresters proved much more successful in renovating the rural landscape in France than they did in Algeria, while their environmental impact on Anatolia was virtually nonexistent. Likewise, nineteenth-century forest legislation played a critical role in the transformation of the pastoral industries of Provence and northern Algeria, but in southwestern Anatolia forest administration ultimately had little impact on the lives and livelihoods of mobile pastoral tribes. Nevertheless the inhabitants of these three regions all witnessed the dramatic decline—and in some places the total disappearance—of their mobile pastoral tradition. At the same time, the application of French scientific forestry around the Mediterranean was also shaped by the will and needs of local populations. The pastoral industries of Provence, northern Algeria, and southwestern Anatolia all emerged at the turn of the century as the product of negotiation and compromise among forest agents, communities, the environment, and the state. Throughout these regions sedentary sheep farming largely replaced mobile pastoralism, and both locals and outsiders came to view this age-old practice as a parochial folk tradition rather than a respectable and legitimate occupation.

The words and deeds of forest officials indicate that they truly believed in their cause and assumed that their efforts would serve both the environment and the common good. History and the benefit of hindsight, however, reveal a more complex picture. Beginning in the mid-nineteenth century, alongside the often noisier and more publicized concerns over deforestation, a small but growing body of research developed challenging mainstream perspectives on environmental decline.

A few decades later scientists began to reject antipastoral narratives and to defend the practice of Mediterranean mobile pastoralism. Today most specialists agree that France's nineteenth-century hysteria over deforestation was premature, and they censure early forest agents' indiscriminate demonization of sheep and goats.[6] Scientists now view mobile pastoralism—when effectively regulated—as an efficient and sustainable use of land in certain environments, including much of the Mediterranean region, and they credit goats with the ability to limit the danger of wildfires in the Mediterranean region's forests.[7] Such insights cast the nineteenth-century marginalization of mobile pastoralists as a needless and tragic exercise. Yet nineteenth-century French foresters' environmental alarm did yield important benefits: it ultimately contributed to the development of modern conservationism. If not for these individuals and their effort to preserve and expand the world's forests, our contemporary environment would be much less green.[8]

* * *

This book actively engages scholarly discussions surrounding Mediterranean societies and the environment, but it also attempts to address some major historiographical lacunae. While scientific perspectives on Mediterranean pastoralism have matured dramatically since the early nineteenth century, this subject has been sorely neglected by other academic fields. Extant works tend to approach it from primarily geographical or anthropological perspectives that lack historical analysis or contextualization.[9] French historiography has all but ignored Mediterranean pastoralism, and literature on pastoralism in Provence appears predominantly in the form of polished coffee-table books representing this practice as a romantic, bygone tradition.[10] The historiography of French forestry has seen better development, but few studies have investigated connections between forestry and pastoralism.[11] The coverage of this subject outside of France also remains spotty and unrepresentative. There is a considerable amount of scholarship on the history and traditions of Anatolian nomads, but such works have done little to elucidate the nomads' relationship with forests and forest administration.

Meanwhile, beyond a few path-breaking works, the history of Ottoman forests and forestry has received only marginal attention.[12]

The connection between forest administration and Mediterranean mobile pastoralism is perhaps best represented in the Algerian case, largely thanks to Diana K. Davis's pioneering study, *Resurrecting the Granary of Rome*.[13] In that work Davis traces the development of a narrative blaming indigenous North Africans for perceived environmental decline. During the French colonial era this narrative contributed to the appropriation of tribal land and resources, the subjugation of indigenous groups, and the rise of commercial agriculture.[14] My study builds on the work of Davis and other scholars of French colonial environmental history in several ways. First, it focuses specifically on the vehicle of forest administration and its interaction with mobile pastoralists, a distinct contingent of Algeria's indigenous population. Second, it actively engages a wide range of players, including colonists, foresters, scientists and intellectuals, indigenous Algerians, and colonial officials and administrators. The conflict, complexity, and change within these groups, as well as the evolving relations among them, are all critical elements of this story. In addition, my investigation exposes the broader significance of these developments by linking them to the establishment and application of scientific forestry in France and in Ottoman Anatolia, as well as to the decline of mobile pastoralism throughout the Mediterranean world.

Contemporary scholars have neglected the history of Mediterranean pastoralism largely for two excellent reasons, which also placed serious stumbling blocks in the way of my own investigation. The first challenge is the availability of sources. In contrast to most popular historical subjects, pastoralists left few traces. As a rule, they did not write letters or diaries, and even when they did, these documents fail to provide a clear picture of their movements. Tracking a single herd through a maze of property rights and usage agreements, complex migration routes, and administrative bureaucracies can be maddening work, when it is even possible. The magnitude of this challenge is substantiated by *Spain's Golden Fleece*, one of the few successful studies of early modern Mediterranean pastoralism. According to the authors' own testimony, they spent two full decades researching and composing this work, which focuses on the relatively well-documented wool economy of Spain.[15] The second reason is scope. Conducting Mediterranean history effectively requires in-depth knowledge of multiple languages, cultures, and historical traditions. It also poses the logistical challenge of accessing sources in archives that are dispersed, in different countries, and in some cases difficult to navigate. Researching this subject has been demanding and often frustrating, and it has sometimes felt more like a wild goose chase than an academic pursuit. It has led me into regional, national, colonial, and diplomatic archives in Turkey, France, and Britain and onto the hills, tracks, and pastures long frequented by migrant herds. Given such challenges, this book represents my best effort to illuminate the history of Mediterranean pastoralism and to reconstruct this nomads' land.

Although my study focuses on three specific Mediterranean contexts, its implications are much broader. By elucidating connections between French scientific forestry and Mediterranean pastoralism, it helps to explain why French foresters were employed in global contexts where mobile pastoralists and settled agriculturalists shared space. In addition, this book informs the global history of relations between "the desert and the sown" and the administration of subaltern groups. It reveals complex and underappreciated links between environmental policy and networks of power, and it provides new insight on mobile pastoralism's global retreat. As I show, the transformation of Mediterranean pastoralism occurred largely on environmental terms, and it involved the active participation and interplay of local and state officials, protoconservationists, entrepreneurs, and agriculturalists, as well as pastoralists themselves.

* * *

There is no magic recipe for organizing a transnational study that is both comparative and connective and that exhibits change over time. The schema of the following pages represents a compromise in that it attempts to integrate chronological and thematic organization in a way that showcases not only the significant and often surprising correlations among my three case studies but also their critical differences. As a result, it is necessarily messy in a way, skipping across time and place. Throughout the text I have attempted to temper such transitions by providing cues and context, as well as sufficient background information for nonspecialists.

This study is presented in two parts. The first part, "People, Place, and Perceptions," supplies a foundation. It describes the practice and perceptions of Mediterranean pastoralism, together with its relationship to Mediterranean forests, through the early nineteenth century. Chapter 1 presents the Mediterranean environmental context and the traditional pastoral practices of Provence, Anatolia, and Algeria. It aims to explain and justify my choice of the term "Mediterranean pastoralism." The second chapter explores French scientific, intellectual, and official perspectives on Mediterranean pastoralism and how and why such views evolved over the course of the early modern period. As it shows, French perceptions of pastoralism during this period were shaped by ideas of race and progress, as well as by burgeoning environmental perspectives. Together they fostered harsh critiques of pastoral practices in Provence, Algeria, Anatolia, and other Mediterranean contexts in the early nineteenth century. Chapter 3 exposes the role of such perceptions in the genesis of French scientific forestry.

The second part explores the transformation of Mediterranean pastoralism in the second half of the nineteenth century. Chapter 4 chronicles the application of French scientific forestry around the Mediterranean, exposing the ways in which pastoral traditions shaped and were shaped by this institution. The fifth chapter focuses on the relationship between Mediterranean pastoralism and nineteenth-century landed property legislation. This chapter demonstrates that changing interpretations and designations of landed property in the nineteenth century were important factors in both the application of French forest legislation and the alteration of Mediterranean pastoral industries. In Provence, Algeria, and Anatolia property reforms complemented forest legislation in targeting and marginalizing mobile pastoralists. The sixth chapter returns to the theme of intellectual perspectives introduced in chapter 2. It explores the role of the climate and natural disasters in nineteenth-century perceptions of and policies toward mobile pastoralists in Provence, Anatolia, and Algeria in the mid- to late nineteenth century. Even as these groups struggled to survive various environmental challenges, they were systematically held responsible for a range of natural disasters and adverse environmental phenomena. As this chapter shows, such accusations were largely influenced by the rhetoric of French forest science, together with social and political factors. The final chapter brings this story into the twentieth century. It reflects on the nineteenth-century transformation of Mediterranean environments and societies, as well as on the legacy of French scientific forestry. It shows that while nineteenth-century French environmental policy contributed to the decline of both mobile pastoralism and forests around the Mediterranean, it ultimately failed to destroy either. This book presents French scientific forestry as a powerful force against mobile pastoralism in Provence, Algeria, and Anatolia, but it also reveals that in all three cases the application of forest legislation represented a process of negotiation in which pastoralists, foresters, and administrators, as well as other interested parties, played active and decisive roles. Such developments and encounters served to reshape this traditional nomads' land into the modern Mediterranean world.

Nomad's Land

Part 1 | People, Place, and Perceptions

CHAPTER 1

Land of the Golden Fleece

Mediterranean Pastoralism in a Wider Society

This pastoral life, which appears to us so strange, holds for them so much appeal that it is extremely rare to see them abandon it.

—Christophe, comte de Villeneuve, *Voyage dans la vallée de Barcelonette*

Visit Aix-en-Provence today, and you are unlikely to encounter even a single sheep. Instead you will find the ancient *centre-ville* crammed with tourists and shops full of designer merchandise. Yet flash back to late spring a few hundred years ago, and the same squares and serpentine streets would be packed with bleating beasts. Provence's pastoral industry thrived in the medieval and early modern periods. In the late fifteenth century the region's beloved King René owned a flock of four thousand sheep.[1] In May of each year these sheep were herded into the Place des Prêcheurs in the center of Aix to be sheared, a task that engaged fifty-odd shepherds. A few days later those shepherds would depart with their newly shorn charges on a seasonal commute to the Alps of Haute-Provence. The Roi René is still remembered in Aix-en-Provence through business names as well as a prominent statue dominating Cours Mirabeau, the town's central boulevard, but his royal flock is long forgotten. Yet the Place des Prêcheurs and dozens of other central squares in this "city of a thousand fountains" still bear the traces of this springtime tradition.[2] Its historic role in Aix's economy is evident even in the fountains themselves, which typically have low, wide basins ideal for watering sheep. Pastoralism has been chiseled into Provence's past.

In much the same way nomads in southwestern Anatolia and northern Algeria long drove herds of sheep and goats between upland pastures and coastal plains. As in Provence, their trail must be sought largely in history. How did this once flourishing practice fade from these three corners of the Mediterranean, and how were their fates connected? This chapter grounds such questions by sketching out the practice of Mediterranean pastoralism in the early modern era, prior to its nineteenth-century transformation. In the process it provides an overview of Mediterranean environmental features and exposes connections between the environment and society. It reveals striking social and environmental similarities among Provence, Anatolia, and Algeria, and it shows how these three case studies together represent a Mediterranean story.

The traditional mobile pastoral economies of Provence, Anatolia, and Algeria are typically divided into distinct categories, as transhumant, seminomadic, and nomadic, respectively. In theory nomadism occurs on a longer range and engages an entire tribe or community, while transhumance is more a localized and solitary affair.[3] In addition, scholars commonly identify mobile pastoralists south and east of the Mediterranean (the Islamic sides) as nomads, while those to the north and west (the Christian sides) are considered transhumant.[4] Not only do such distinctions carry disconcerting connotations, but they obscure the many common features and complexities among traditional practices of pastoralists in these three contexts and throughout the Mediterranean world. The terminology of pastoralism is useful, but we should not think of these cases as exclusively separate and distinct. Standard distinctions among pastoral types can obscure the significant political, historical, societal, cultural, and geographical forces that bound Mediterranean pastoralists together. Because this book aims to elucidate such forces, I use the term "Mediterranean pastoralism" to deliberately blur the lines between nomadism and transhumance and to emphasize experiences shared across the Mediterranean world. What then is Mediterranean pastoralism? This term engages features shared by traditional pastoral industries around the Mediterranean, including similar environmental conditions, common history and practices, and the context of a wider agropastoral society. These key ingredients of Mediterranean pastoralism are embodied in the case studies of Provence, northern Algeria, and southwestern Anatolia.

Mediterranean Environments

The most obvious bond among Mediterranean societies is their environment. The region is typically characterized by its warm, dry summers and cool, wet winters. Visitors to far-flung parts of the Mediterranean world return with like impressions of vineyards, olive trees and orchards, and craggy limestone outcrops. Whether in Greece or Spain, Italy or Algeria, flocks of sheep traipse through similarly sunbaked coastal shrublands each spring to reach the greener mountain pastures beyond. Such images reflect real ecological commonalities, but there are also key distinctions across the Mediterranean world. The case studies of Provence, Algeria, and Anatolia effectively represent both the common elements as well as the range and variety of the Mediterranean environmental zone.

The environmental similarities of the Mediterranean region result in part from its location. The sea lies in a midlatitude position, which makes it susceptible to multiple meteorological systems. In the winter westerly wind belts descending from northern latitudes bring precipitation along with relatively mild temperatures. In the summer the arrival of subtropical high pressure systems from the south creates a hot, dry climate.[5] These climatic influences spread inland into a broader Mediterranean zone.[6] The convergence of multiple weather systems in this region also leads to the appearance of periodic high winds.[7] The Mediterranean climate, however, is far from uniform. The dramatic topography of the region, with its coastal mountain ranges, connection to multiple continents, and proximity to the Atlantic Ocean, promotes the formation of significant regional variations. Rainfall often forms in the lee of mountain ranges, which are nearly always visible along the

Mediterranean coast.[8] These high-elevation landforms can capture the precipitation of moisture-laden weather systems from the north and west, making Mediterranean mountains particularly humid.[9] The sea also contributes to the appearance of microclimates in coastal regions. There are also regional differences. Landforms closer to the Atlantic tend to experience milder temperatures, while much of the eastern Mediterranean, under the continental influences of central Europe and Asia, hosts a drier climate that is hotter in summer and colder in winter.[10] Finally, temperatures rise gradually from north to south, so that the land to the south is generally warmer than the northern shore areas. Such variations are reflected in the average temperatures and annual precipitation of Provence, northern Algeria, and south-western Anatolia.

TABLE 1. Average precipitation in Marseille, Constantine, and Antalya, showing mean number of days with precipitation and total precipitation (mm) by month

Month	Marseille		Constantine		Antalya	
Jan.	6.1	53.6	12.6	69.4	12.6	232.4
Feb.	5.1	43.5	11.6	56.0	10.8	160.7
Mar.	4.8	40.4	10.7	56.2	8.9	96.8
Apr.	6.3	57.9	10.5	58.8	6.4	46.2
May	4.9	41.2	8.3	44.7	5.3	30.0
June	3.5	25.4	5.1	19.5	2.6	9.6
July	1.4	12.6	2.6	7.1	0.6	2.2
Aug.	3.1	31.4	3.7	10.8	0.7	2.5
Sept.	4.1	60.6	6.8	35.8	1.8	12.3
Oct.	6.3	85.4	7.8	38.2	5.8	67.7
Nov.	5.2	50.6	11.3	57.7	7.6	131.9
Dec.	5.6	52.0	12.5	80.8	12.3	263.3
Total	56.4	554.6	103.5	535	75.4	1,055.6

SOURCE: World Meteorological Association, World Weather Information Service, accessed 7 October 2018, http://worldweather.wmo.int/062/c01058.htm.

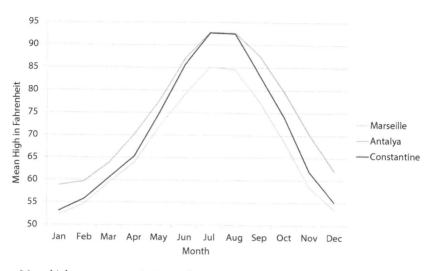

2. Mean high temperatures in Marseille, Antalya, and Constantine. Based on statistics from World Meteorological Association, World Weather Information Service, accessed 7 October 2018, http://worldweather.wmo.int/062/c01058.htm.

Such climatic elements affect Mediterranean ecosystems and societies in a host of ways. First, the Mediterranean climate limits the types of plant communities that can survive without irrigation. Mediterranean vegetation therefore tends to be well adapted to the region's signature hot, dry summers and cool, wet winters. Climate systems also influence the formation and distribution of soils. The diverse range of climatic features in the Mediterranean region has led to wide soil variability.[11] The northern basin boasts a higher percentage of soils classified as nutrient rich: 16 percent compared to just 1.3 percent in the south.[12] Yet throughout the Mediterranean world soils tend to be thin, making them susceptible to runoff and erosion, as well as nutrient poor, thus limiting the types of flora—and thus fauna—that the region can support.[13] Mediterranean vegetation has clearly adapted to the habitual presence of drought and fire. Indeed most scientists agree that fire is an essential part of the life cycle of plants regarded as typically Mediterranean.[14] The *maquis* shrublands that characterize the region are populated by olive, holm oak, kermes oak, cork oak, juniper, and Aleppo pine trees, as well as aromatic taxa such as rosemary, lavender, and thyme. Such vegetation is resistant to high temperatures, drought, and fire and can survive under a wide range of precipitation regimes, temperature extremes, and soil types.[15] In addition, much of the Mediterranean region is subject to high annual rainfall variability (cv, or coefficient of variation), which is just as central to the development of vegetation systems as is rainfall itself. In the drylands and regions of high cv that characterize much of the Mediterranean world, unirrigated agriculture is a risky venture with poor returns. For most of human history, mobile pastoralism presented an appealing alternative to agriculture throughout the Mediterranean zone because its mobility offered relative protection from climatic disturbances such as droughts, floods, and frosts.[16]

The Mediterranean environment in turn also reflects its extensive history of human settlement. Indeed human impact is so ubiquitous and deeply rooted that ecologists no longer speak of a "natural" Mediterranean environment.[17] Much of the region is covered with vegetation that is resistant to or even dependent on grazing.[18] Since the introduction of domesticated sheep and goats some ten thousand years ago, grazing has helped to prevent the invasion of open vegetation communities by trees and other woody species, thus preserving species diversity and tempering wildfires. In addition, traditional agropastoral practices facilitated the return of nutrients to the earth, which fertilized the soil and prevented its degradation.[19] Today scientists rate undergrazing to be just as great a threat to the Mediterranean environment as overgrazing, if not greater.[20] The imprint of long-term human habitation, together with other Mediterranean environmental characteristics, is clearly evident in Provence, Anatolia, and Algeria. The vegetation, topography, and past economies of these three regions expose patterns common to the Mediterranean world, as well as regional variations. As the environments of Provence, northern Algeria, and southwestern Anatolia suggest, the Mediterranean model might be best considered as a range. These three locations thus lie at different points on the Mediterranean environmental spectrum.

"Provence" is a term with historical, cultural, and geographical significance. It describes an administrative province of France in existence prior to the creation of the department system during the French Revolution. The region is roughly equivalent to the current administrative unit Provence–Alpes–Côte d'Azur, which encompasses six departments. Provence extends from the Rhône River in the west to the Var Department in the southeast, from the Mediterranean Sea in the south across the Durance River to the Luberon range and the Central Massif in the north, and northeast to the Southern Alps, covering a total of more than three million hectares. The low, coastal region of southwestern Provence is often referred to as Basse-Provence, while Haute-Provence designates the higher zone to the north and west, marked by foothills and mountains. This study focuses primarily on the Bouches-du-Rhône Department in Basse-Provence, which includes Marseille, Aix-en-Provence, and the surrounding region south of the Durance River.

Basse-Provence has been described as a land dominated by nature rather than humans, resisting domestication through its rocky hillsides and extensive wastelands.[21] It is divided among three broad environmental biomes: *garrigue/maquis*, Mediterranean forest, and plain and steppe. Garrigue refers to the sparse vegetation of rocky limestone soils, which support only a few hardy plants, such as kermes oak, thistle, and gorse.[22] It can also describe a form of degraded maquis.[23] Maquis is more vegetated; it describes a dense brushland covered in shrubs and small trees such as kermes oak.[24] The woodlands of Basse-Provence are generally open forests characterized by white and holly oak (*chêne vert*), kermes oak, and pine trees, as well as brush vegetation that blends into the surrounding shrubland. By contrast, the forests of Provence's higher zone, Haute-Provence, can become quite dense, though much of the current forest cover is "artificial," the result of extensive mountain reforestation efforts in the 1860s. Provence is also home to France's last remaining "steppe"—the Crau, a large, level, arid plain covered in flat, round pebbles called *galets*. This feature once covered more than 150 square miles and stretched from the Alpilles range to the north, to the Rhône River in the west, and to the sea in the south. The Crau has been used as pasture since at least antiquity and probably longer.[25] To this day it contains the best pastures of Provence, locally called *coussouls* or *coussous*, and produces hay of such value that it is regulated by an *appellation d'origine contrôlée* (AOC), a French quality standard normally associated with fine wine.[26] The Camargue, a marshy region just south of the Crau, also provided an important source of pasture in the past. Together the Crau and the Camargue once supported the flourishing pastoral industry of nearby Arles.

Algeria lies less than five hundred miles due south of Provence. Its largest port, Algiers, is as close to Marseille as the latter is to Paris. Prior to the French conquest in 1830, this region formed an integral part of the broader North African zone nominally under Ottoman control. It lies in the center of the Maghreb, a strip of North Africa unique for its northerly latitude and extended areas of high elevation.[27] Due to these features and to the region's relative proximity to the Atlantic Ocean, the northern Maghreb receives greater annual precipitation than other parts of the North African coast. Although Algeria comprises the largest country within this territory, more than seven-eighths of the country is occupied by the Sahara Desert.[28] In environmental terms Algeria can be divided into four topographical zones: the Tell Atlas, the Hauts-Plateaux (High Plateaus), the Saharan Atlas, and the Sahara.

The Tell is a small mountain range originating along the Moroccan border in the northwest and stretching along the Mediterranean coast to comprise an area of around fourteen million hectares. The Algerian coast itself is dominated by cliffs broken only occasionally by bays, none of which forms an ideal natural harbor.[29] Near the middle of this coastline, in northwestern Constantine, lies the hilly region of Great Kabylia, one of the last coastal parts of Algeria to resist French control. Farther inland from the coast, the Algerian landscape is adorned with layers of hills, also known as the Tell. These hills enclose the most fertile and richly cultivated plains of Algeria, sometimes called the Sahel d'Alger, not to be confused with the transitional zone in the southern Sahara also known as the Sahel. It is in this region that indigenous pastoralists once enjoyed great prosperity and have also faced great persecution.

The Tell region benefits from a relatively mild climate. The summers are dry and warm. Temperatures usually peak in the high eighties Fahrenheit, though sirocco desert winds occasionally bring hot, dry desert air and can increase this figure significantly. Winters are generally cold, with a rainy season usually beginning in late September and ending by May. When the forester Theodore Woolsey visited Algeria in the early twentieth century, he estimated the annual precipitation along the coast at between forty and forty-seven inches, while the region surrounding Algiers, Kabylia, and the northwestern part of the province of Constantine received about thirty inches on average.[30]

The Hauts-Plateaux are elevated steppe regions south of the Tell and covering a total area of more than forty-two hundred square miles. They experience very little rainfall—no more than eight inches annually—and are dominated by saline and gypseous soils of poor productivity. Traditionally these steppes were used for little other than sheep and goat grazing, and even pastoralism could succeed only when practiced sparsely and on an extensive scale. For most of the nineteenth century the unyielding soil, extremely dry climate, and remoteness of these areas thwarted colonists' efforts at cultivation, but these features did not prevent some from condemning indigenous pastoralism in even these regions. Farther south, the Saharan Atlas acts as a natural barrier between northern Algeria and the Sahara Desert. It is partially wooded, and its peaks reach nearly eight thousand feet. Until the late nineteenth century these southern regions of Algeria, or Territoires du Sud, were governed by the French military. Although parts of the Sahara were eventually incorporated into French Algeria, this region was never considered suitable for French colonization, and it continued to be populated almost exclusively by nomadic tribes. Colonial observers typically overlooked the climatic variations within these southern regions, choosing to characterize Algeria as either Mediterranean or desert.

In Algeria, as in other Mediterranean contexts, the definition of "forest" is elusive. Much of the country's northern territory is covered with underbrush comprising cistus, lentisk, heather, myrtle, arbutus, and, as in Provence, kermes oak. Certain taller species appear as well, including the Aleppo pine, maritime pine, cedar, thuya, juniper, cork oak (*chêne-liège*), and various other types of oak.[31] Of these the Aleppo pine is the most widespread, spanning from sea to desert and growing even on poor, arid soils. The province of Constantine, which makes up the eastern third of the country, boasts the greatest forest cover. In 1911 Constantine contained about thirty-four hundred square miles of national forest land, more than half of the total for Algeria's three provinces combined.[32] Yet according to many colonial era observers this was the province most affected by past forest abuse and most at risk from the ill effects of deforestation. The French scientific mission to Algeria in the years 1840–42, which represented France's first serious attempt to survey its new colony, contrasted deforestation in Constantine with relatively abundant forests in other regions.[33] This apparent discrepancy is partly explained by Constantine's ecological range. Stretching from the coast to the sub-Sahara in the south, this province represents the full range of Algeria's ecology. While the northern zone provides some of the most fertile soils and lush vegetation in all of Algeria, southern Constantine's soils are considerably less productive. Another reason for colonists' special concern for the health of Constantine's forests was their commercial potential. In Algeria the natural habitat of the cork oak was limited almost exclusively to the coastal regions of Constantine, which previously had served as pastureland for sheep.[34]

Sheep also long grazed the coastlands of southwestern Anatolia, which sits at the eastern end of the Mediterranean Sea, roughly equidistant from both Provence and Algeria. Also called Asia Minor, the landmass of Anatolia straddles the boundary between Europe and Asia and comprises modern Turkey. It describes a broad rectangle stretching more than two thousand miles from east to west and more than six hundred miles from north to south. Anatolia borders the Mediterranean Sea in the west and southwest, Europe and the Black Sea in the north, and stretches into the Central Asian Steppe in the east. Like Provence and Algeria, the Mediterranean coast of southwestern Anatolia is characterized by rocky coastlines and coastal ranges. In some places along the coast, cliffs rise directly out of the sea, but in others the protective wall of the Taurus range encloses coastal plains. Its vegetation and environmental features are also similar to those of Provence and Algeria. Much of this land is either marshy or characterized by karstic soil, making cultivation difficult.[35] While there are pockets of deep, rich soils, they are often found on hilly, uneven territory. As in other parts of the Mediterranean, persistent farmers bent this region to the plow through terracing and, more recently, irrigation and the draining of marshes.

Throughout the modern era Anatolia represented the heart of the Ottoman Empire and the chief source of its power, but it also contained some of the empire's most inaccessible frontiers. It is characterized by an extensive plateau that is bounded by mountain ranges in the north, south, and southwest, as well as rising elevation in the east. The Taurus range lining the Mediterranean coast in southwestern Anatolia is particularly pronounced. In some places mountains rise steeply from coastal plains to heights of more than ten thousand feet. Before the arrival of the railroad and other advances in transportation and communication, these ranges disrupted connections to the central plateau, and the regions beyond them remained isolated and sparsely populated.

Anatolia's southern coast is bounded by Antalya, near the southwestern corner of Anatolia, and Adana, in the easternmost corner of the Mediterranean, where Anatolia meets the Levant. Antalya was once an Ottoman administrative district, and it remains a province of modern Turkey. It is also the name of the largest city in this province. This region extends from the Gulf of Antalya in southwestern Anatolia to the western Taurus Mountains in the north and east. The mountains surrounding Antalya Province traditionally made transportation and communication between it and the inner Anatolian plateau slow, difficult, and dangerous.[36] The western Taurus range also forms a "climatic frontier" between Mediterranean and continental zones.[37] Weather is less predictable in these mountains. Annual precipitation is significantly higher, and rainfall occurs in the late spring as well as in winter. Several hundred miles to Antalya's east, on the other side of the Cilician Gates, lies Adana. The nearby Cilician Plain or "Çukurova" is now celebrated for its agricultural productivity.[38] Well into the nine-teenth century, however, much of this land was covered in swamps that became a breeding ground for mosquitoes every summer, leading most of the settled population to avoid the region for fear of malaria. This concern freed the territory for seasonal use by nomads, who avoided its hazards by visiting the plains in the winter and retreating into the nearby Taurus Mountains in the spring. In this way Mediterranean environmental elements facilitated the spread of mobile pastoralism in Anatolia, as in Algeria and Provence, and formed a critical connection among these three contexts.

Pastoral Traditions around the Mediterranean

Mobile pastoralism has been a mainstay throughout the Mediterranean region for the past two millennia, if not longer.[39] The Old Testament affirms the importance of sheep in eastern Mediterranean society in ancient times. The Israelites were themselves seminomadic pastoralists, said to have roamed the desert-like Middle East before finally settling in Palestine, "the land flowing with milk and honey," according to the Bible.[40] In ancient times a range of cultures depended on wool for clothing. In the first century CE Pliny the Younger remarked, "Sheep are likewise in great request, both in regard they serve as sacrifices to appease the Gods, and also by reason of their fleece yielding so profit-able a use."[41] As these sources suggest, mobile pastoralists in the early modern Mediterranean world were linked not only by their environ-mental context but also by similar histories and common practices. The case studies of Provence, Algeria, and Anatolia perfectly illustrate such elements of Mediterranean pastoralism. Despite their cultural, polit-ical, and ideological differences, these three societies shared a deeply rooted tradition of mobile pastoralism, which developed through time in strikingly similar ways.

The phrase "since time immemorial" haunts past and present liter-ature on pastoralism in Provence, showing the pride of place and near mystic authority ascribed to this tradition.[42] This phrase also conve-niently avoids the difficulty of tracing the roots of a practice that, despite millennia of peregrinations, left few footprints. The discovery of the remains of ancient *bergeries* (sheep pens) on the Crau affirms that pastoralism existed prior to the Roman era.[43] In addition, evidence for pastoral exploitation of the Alps exists from preliterate times.[44] Considered together, these discoveries suggest that mountain shepherds began to bring their herds to the plains for the winter in ancient times, but there is no way to prove that these shepherds were mobile or that the herds of the mountains and plains were the same. It is indeed much easier to trace the origins of pastoralism than the origins of transhumance, or *mobile* pastoralism, in Provence. Yet, as early as the Roman era observers described transhumance in Provence as an ancient tradition. In his *Natural History*, Pliny the Elder wrote, "Today, as we know, the rocky plains in the province of Narbonensis [southern France] are filled with thyme; it is nearly their only revenue, and thousands of sheep come from distant regions to graze on it."[45] This and other references suggest that the roots of transhumance run much deeper.

By the late medieval period the practice of transhumant pastoralism was firmly entrenched in Provence, and its rituals and traditions continued with little change into the modern era. From late fall to late spring Basse-Provence was annually flooded with sheep, spread out over a wide variety of territory. Not all shepherds had access to the *coussous* of the Crau and the Camargue, and these choice winter pastures were not used all the time. Instead many shepherds grazed their sheep and goats on *terres vagues et vaines*. These were *garrigues, maquis*, and *landes*—open lands left uncultivated and/or considered unsuitable for agriculture. They also by definition lacked any high forest cover.[46] *Terres gastes* furnished another, similar destination for Provence's herds. The word *gastes* comes from an Old French term meaning uncultivated, fallow, and wild.[47] *Terres gastes* were designated wastelands generally submitted to communal use. They may have contained bushes and shrubs suitable for kindling, but their vegetation was devoid of commercial value.[48] Finally, both sheep and goats grazed in public and private woodlands, especially when no other pasture was available. Beyond these zones the livestock of Provence found sustenance wherever possible, on fields, forests, parks, and roadsides. During the winter months most herds practiced micromigrations, moving, for example, from fallow fields in the fall to the prairies of the Crau and surrounding areas in the spring.

In the past most of the sheep in Provence migrated seasonally between coastal regions and the alpine pastures of Haute-Provence. This tradition is called *la grande transhumance* in French to distinguish it from pastoral migration on a smaller scale.[49] From an environmental perspective the practice of transhumance allowed inhabitants to exploit the dry landscape of Basse-Provence, which could not support significant year-round grazing, as well as to relieve pressure on the mountain pastures of Haute-Provence.

During their winter on the low plains, shepherds typically lived together with their families and others of their trade in tiny cabins constructed on the land grazed by their communal herd. If these structures were somewhat more permanent than the tents of nomadic tribes, they were not necessarily more comfortable. Christophe, comte de Villeneuve-Bargemont, who witnessed transhumant pastoralism in an 1802 tour of the Alps, described the shepherd's winter abode as a "rustic . . . sort of cottage" in which the entire family shared a single bed of straw.[50] He also claimed that shepherds wore their characteristic capes constantly in order to "brave the season's bad weather." All members of the family participated in the domestic division of labor. According to Villeneuve, "The women's work consists of preparing soup for the shepherds, twice per day and . . . they make cheeses and go to sell them as well as milk in the neighboring villages." Shepherding, like many occupations, was traditionally passed down through generations, from father to son. In addition, large families often sent children as young as nine or ten to help herd sheep during seasonal transhumance.[51] These young shepherds learned the trade and frequently pursued it.[52] The most respected shepherds, however, came from the Alps. The farmers of Basse-Provence highly valued the experience and competence of these mountain dwellers.[53]

Often the shepherds who guided herds of sheep through seasonal migration were not the beasts' owners. Most shepherds began their career as hired laborers—working with someone else's livestock. Many shepherds gradually acquired at least a part of the herd, but they still faced significant obstacles and expenses associated with owning and maintaining a herd, including finding and renting pasture and paying various tolls, fees, and local taxes.[54] Few shepherds could support themselves on their sheep alone, so most continued to work for agriculturists, adding other sheep to their herd to help meet the costs of transhumant herding.

In the early nineteenth century the large transhumant herds of Provence were administered by a hierarchy of men: the head shepherd or *bayle*, who oversaw accounting and logistics and directed the others, his assistant, and an additional shepherd for approximately every three hundred sheep.[55] Since some herds numbered in the thousands, the bayle's job could be quite demanding. Prior to the herd's departure, he had to arrange for access to pasture along the designated migration route by negotiating and settling up with landowners at each point of rest. This detail also required the careful choice of a route, and large herds were often split up to allow for sufficient pastureland. In order to prevent overcrowding, each herd followed a distinct path, and the entire trip was organized and reserved in advance. The route as well as all of the nightly stops had to be predetermined, so shepherds could pay ahead of time for the use of pastures and fields.[56] The bayle also had to provide for the herd's human travelers, who included not just the shepherds themselves but their families—women and children—as well.[57] Because everything had to be planned beforehand, a single unforeseen obstacle, such as a sick beast, bad weather, a broken bridge, or a river too high to cross, could spell disaster for pastoralists and their herds.

In addition to lambs, rams, and ewes, the migratory herd typically included goats, dogs, and mules, each assigned a specific task. Goats, recognized as being cleverer and more individualistic than sheep, helped to direct the flock, keeping them together and on the path. Originally dogs were required to ward off wolves, and these dogs were often fierce breeds and armed with spiked collars suitable to this purpose.[58] However, this role became less important following the large-scale, systematic extermination of wolves throughout France in the nine-teenth century.[59] Later shepherds showed a preference for smarter but less aggressive breeds, which began to replace goats as guides. Finally, these migrants were accompanied by mules, preferred over horses as alpine pack animals to carry provisions as well as any animals too sick to walk. Some of the best imagery of this motley mammalian crew comes from *Mirèio*, the famous mid-nineteenth-century epic poem by Provence's most beloved muse, Frédéric Mistral. He lists the mules' burden as follows:

> Food for the shepherd-folk, and flasks of wine,
> And the still bleeding hides of slaughtered kine;
> And folded garments whereon oft there lay
> Some weakly lamb, a-weary of the way.[60]

Until the late nineteenth century the seasonal migration between low and high Provence was made by foot and hoof, and it was a slow and challenging trek. Herds typically left the plains in late May.[61] To avoid overexertion, shepherds rarely allowed a herd to cover more than three miles per day. Transhumant herds frequently traveled for three to six weeks to reach their destination, which usually lay one hundred to two hundred miles from their point of departure.[62] Their path led through agricultural fields on wide, designated paths that farmers were required to leave uncultivated.[63] Along the way, sheep grazed in fallow fields, forests, and rented pastures, and they constantly nibbled on roadside vegetation.

Once the herd reached its destination, the sheep spent the summer on fertile mountain meadows, or *alpages*. These "delicious pastures," as Charles-François, baron de Ladoucette, termed them, were rented from the local population, except in cases of inverse transhumance, where alpages belonged to the owner of the herd.[64] These alpages were often highly prized and could be very expensive.[65] Here shepherds' accommodations were even more modest: primitive stone or wooden shelters called *cabanes*.[66] Although alpages and cabanes were often distant from commercial centers, a shepherd's life was hardly solitary.

As noted, they typically lived together with their families and other shepherds, constituting settlements that sometimes grew to the size of a small village. Ladoucette explains that "in the middle of prairies that extend just to the limits of all vegetation, appear the *cabanes* of shepherds, dairies, nearby chalets, entire villages uninhabited except in the summer."[67] In the fall these groups and their herds returned to the lowlands, where they attended markets to sell cheese, skins, and old sheep for slaughter.[68] Then the annual cycle would begin again.

As in Provence, mobile pastoralism in North Africa has ancient roots. Indeed the Roman name for this region, Numidia, comes from the Greek word for nomad.[69] Since antiquity and probably before, mobile pastoralism has been practiced in this region in various forms and degrees. For thousands of years it has formed the backbone of the Maghreb's economy, providing continuity in a history fraught with political, social, and religious change. This tradition is so important to the Algerian past that one historian of the region has called it the "land of sheep."[70]

When French forces arrived in the regency of Algiers in the 1830s, it was nominally part of the Ottoman Empire. By the early nineteenth century, however, much of the central Maghreb was either directly or indirectly in the grip of powerful tribal leaders. The Ottoman state maintained a handful of troops, most of them in and around Algiers. Only four hundred were stationed in the entire province of Constantine, then known as the Beylik of the East, the regency's largest and most profitable *beylik*, or province.[71] These Turkish soldiers depended on the support of local auxiliary forces, including prominent *makhzan* (government) tribes, who for their pains were exempt from taxes and allowed to tax other populations.[72] In this system Ottoman involvement was limited to urban centers in the Tell, comprising approximately thirty thousand square miles, or just one-thirtieth of the total area of modern Algeria.[73] The vast region beyond this oasis of Ottoman control was governed by confederations of nomadic tribes. The organization of such tribal units was in some ways similar to the hierarchy of transhumance in Provence, from the common shepherd to the head bayle.[74] Yet, in contrast to French pastoralists, Algerian nomads were not only central players in their regional economy but also had the upper hand.[75]

The culture and social organization of Algerian tribes also shared notable features with pastoralists across the Mediterranean. As in Provence, nomads traveled in groups of families or clans. Their caravans could be an impressive spectacle, which one witness compared to "the scenes of biblical times . . . the picturesque [image] of an entire people marching toward the Promised Land."[76] Once arrived in their summer or winter quarters, nomads pitched elaborate tents, creating a veritable, if temporary, metropolis. In the Chelif River valley in western Algeria, which prior to the French colonial era was inhabited by tribes who stayed only part of the year, one mid-nineteenth-century observer described the "cries, the bellowing of the herd, the howling of dogs" as ever-present sounds of spring and contrasted this constant noise with "the silence and solitude of this region in other parts of the year."[77]

Although nomadic pastoralists traditionally roamed throughout Algeria, the specific nature of their trade and its role in the wider regional economy varied according to environmental and political factors. As in Provence, pastoralists migrated seasonally in search of a moderate climate, access to water, and fresh pasture. This journey led them in all directions and varied greatly in length. Northern Algeria possessed the best pastureland, with its choicest pastures in what would become the province of Constantine. This territory was reserved for the richest, most powerful tribes. Of Algeria's nomadic groups, those of Constantine had the most in common with the mobile pastoralists of Provence. They migrated on a relatively small scale, leaving and returning according to a seasonal schedule identical to that of their Mediterranean neighbors. In the fall, winter, and early spring they grazed their sheep on fallow fields or other unused land in the Tell, and they benefited from the productivity of nearby mountain pastures in the summer.[78]

Other, less privileged nomads had to travel much farther to reach suitable pastures. According to one early twentieth-century scholar, some tribes' seasonal migrations covered more than four hundred miles.[79] In addition to herding their livestock, these long-distance nomads provided vital services to settled or seminomadic groups in the north. Many tribes crossed the Atlas to winter in the Sahara, taking with them foodstuffs and products for sale and bringing back dates and other southern produce on their return.[80] Indeed in Algeria the transmission of grain throughout the country was linked directly to the seasonal migration of these tribes.[81] The predominant forms of livestock also varied in different parts of Algeria. In the north, especially in the agricultural greenbelt of the Tell, cattle grazed alongside sheep, while in the heat of the far south, goats and camels predominated. In general, though, and in between these two extremes, sheep and goats were by far the most common.[82] Due to their versatility, sheep were often considered "the essential wealth of the Algerian Steppe."[83] Overall, mobile pastoralists in the regency of Algiers gained considerable power through the value of their industry, local political connections, and their very mobility, which could provide an escape from central authority.

Like other Mediterranean regions, the Anatolian peninsula has long supported both agriculture and pastoralism.[84] The history of pastoralism probably reaches back in time even further in Anatolia than in Provence. Some of the earliest evidence for the domestication of sheep and goats, dating from 6000 BCE, comes from neighboring regions.[85] The Hittites practiced pastoralism in central Anatolia from the nineteenth to twelfth centuries BCE. The gradual lowering of the water table in this region, accompanied by a shift toward a warmer, drier climate, probably made Anatolia's Mediterranean coastal regions more suitable for raising sheep and goats than cattle.[86] In any case, by the classical era sheep had become ubiquitous throughout the region.[87] They even became the stuff of legend. Aristotle told of a river in southwestern Anatolia, now known as the Menderes (Büyük Menderes Irmağı), whose water would make sheep turn a yellowish hue.[88]

The original appearance of mobile pastoralism in Anatolia is more difficult to date, but most scholars agree that this practice developed definitively in the medieval era with the arrival of Turkish tribes, who were themselves nomadic pastoralists.[89] Once the Ottoman Turks had consolidated their power in Anatolia in the early fourteenth century, they abandoned their peripatetic ways. They established a state that through centuries of conquest eventually stretched across the Mediterranean into Europe, Africa, and Asia. Throughout the Ottoman era, the imperial administration played a sort of tug-of-war with its remaining nomadic inhabitants, who would advance at times and then fall back.[90] For much of the Ottoman period, soil quality, climate, and land formation, as well as the limits of transportation and administration, helped to forge boundaries between the so-called desert and sown.[91] Agriculture was mostly limited to areas that were accessible and relatively secure, through their proximity to Istanbul, other urban centers, or transportation networks. Nomadism, in contrast, thrived on the periphery, in secluded areas outside the purview of the Ottoman state. As agriculture and the reach of Ottoman administration gradually expanded, nomads retreated ever more to these frontiers and to unoccupied land considered unsuitable for farming. Some remained in the desert plateaus of eastern Anatolia, while others were pushed west and south, into the Taurus Mountains and nearby coastal plains.[92]

The spread of nomadic tribes in southwestern Anatolia occurred in the context of general depopulation within Antalya Province. The port cities of Antalya and neighboring Alanya were once bustling centers of merchant activity. When Ibn Battuta visited in the mid-fourteenth century, Antalya was a prosperous port. He described the surrounding country as "one of the finest in the world; in it God has brought together the good things dispersed throughout other lands. Its people are the most comely of men, the cleanest in their dress, the most delicious in their food, and the kindliest folk in creation."[93] By the seventeenth century, however, Antalya's significance as a port had declined, the surrounding region faced depopulation, and corruption and piracy plagued the remaining community. It remained significant enough to appear in the mid-seventeenth-century account of Evliya Çelebi, but he complained of the disrepair and lawlessness of the city and surrounding countryside.[94] By this time much of that territory was occupied by semiautonomous nomadic tribes.

The nomads of southern Anatolia are known generally as *yörük*, which comes from the Turkish word *yürümek,* meaning "to walk."[95] A yörük tribe traditionally comprised an extended family or clan who lived and migrated together.[96] As such, the yörük were not unlike the shepherd communities of Provence. After wintering in the river valleys of the coastal plains, they left in April for summer pasturelands, called *yayla*s, in the Taurus Mountains, traveling via migration routes fixed by tradition.[97] Year after year each clan occupied the same designated space, which they claimed through ancestral use, rather than ownership in the modern, Western sense. Here they camped in communities of black felt tents among oak, juniper, grasses and shrubs, and springs vital for them and their flock.[98] The yörük of southwestern Anatolia raised sheep, goats, and sometimes camels, the preferred form of transport and beast of burden of the Near East.[99] They ate meat only rarely, while milk products from their livestock formed the bulk of their diet.[100] They passed the time tending the herd, slaughtering yearling males and lambing in the winter, and making cheese in the summer.[101] The men tanned hides and worked leather, while the women carded, spun, and dyed the wool and crafted textiles.[102]

In many respects the yörük lived in much the same way as the pastoralists of Provence, but their local politics and relationship with the central administration more closely mirrored the Algerian case. The yörük lived both outside and within the Ottoman system. They often resisted taxes and conscription, and they evinced much more affinity for the local control of their tribal leaders than for the relatively distant Ottoman sultan. In common with pastoralists in the Maghreb and to a lesser extent Provence, yörük society followed a distinct hierarchy. According to Henry John Van Lennep, a British traveler who observed yörük tribes in southern Anatolia in the mid-nineteenth century, "Their government is patriarchal, and they are divided into tribes, so many tents or families being said to be under the authority of each Sheikh, an office which is hereditary among them."[103]

The relative inaccessibility of nomadic tribes was undoubtedly a factor in Ottoman administrative policy toward them. Because they inhabited the frontiers of Ottoman authority, nomads were difficult to control. Van Lennep recalled an encounter with a nomad who proudly described his imperial commission as head of the local police, or "gendarmerie," though "he has repeatedly acknowledged that his 'gendarmes' rob whenever they can do so with the hope of not being found out."[104] Nomads did prey on villages, especially when attempts to barter failed. Tribes were also known to retreat into inaccessible mountains when tax officials drew near. The administration of nomads became even more difficult in times of war, when garrisons that might otherwise guard and protect vulnerable settled populations were called to the front.[105]

Pastoralism and Mediterranean Society

Even at the peak of their independence the yörük remained dependent on their environment and on surrounding settled society. Likewise, early modern mobile pastoralists around the Mediterranean participated in a broader agropastoral ecosystem that provided shelter and sustenance for themselves and their livestock. Forests were a critical part of this system. For pastoralists in Provence the forests provided a valuable supplement to open pasturelands. In the fall and winter sheep often grazed on forestlands while open pastures regenerated. Few shade-bearing grasses and herbs are palatable to sheep and goats, and they generally cannot reach the leaves of trees, but they will browse on small edible plants and low-hanging leaves.[106] Forests were particularly important in dry years and following a flood, when open pasture was unavailable or insufficient. Transhumant shepherds also depended on forests while traveling between summer and winter pastures.[107] They provided a shepherd and his flock with shelter, food, and bedding.

In southwestern Anatolia nomads depended on forests for shelter, fuel, and wooden tools.[108] Their livestock occasionally grazed in forests, but both sheep and goats preferred the vegetation of summer and winter pastures. The only forests that contributed measurably to their diet lay along seasonal migration routes.[109] As they passed through forests on their journey from coastal plains to upland pastures (yaylas), nomads could not prevent their herds from grazing on the vegetation that lay along their path.[110] Because migration routes were standardized, herds grazed in the same intermediate forests year after year, but these forests were used only for short periods. They thus had extensive recovery time. The yörük and their livestock did less damage to the region's forest cover than their neighbors, the *tahtacılar* (lumber workers) and the *ağaç erleri* (wood cutters). Both groups were nomadic tribes like the yörük, but they gained their livelihood through wood cutting instead of pastoralism.[111] Forests surrounding summer pastures in Anatolian mountains generally remained in fair condition throughout the Ottoman period, while depletion of forests near winter pastures in the plains was usually a result of their accessibility, and they were exploited not by nomads but by settled populations and the state.[112]

When necessary, pastoralists burned forests and brushlands to provide pasture for their animals. In the early modern era this practice was particularly widespread in the central Maghreb, where it was least regulated. Here agriculturalists also burned brushlands periodically, in a practice called *keçir*. This land was used for agriculture every four years and for pastoralism in the interval, forging another link between these two economic systems. Algerian pastoralists also grazed their herds in wooded areas and brushlands, which provided food, fodder, fuel, and welcome shelter from the sun and heat.[113] In general the environmental impact of pastoralists in Provence, Algeria, and Anatolia in the early modern era was mild, especially when compared with what was to follow.

The case studies presented here also challenge predominant narratives about relations between the so-called desert and sown. Much of the literature on mobile pastoralists emphasizes a dichotomy between them and settled agriculturalists. The biblical story in which Cain, a farmer, kills his shepherd brother Abel out of envy exposes the antiquity of this narrative. As the examples of Provence, Anatolia, and Algeria show, however, pastoral and agricultural industries often had more in common than in conflict. Relations between nomads and farmers in these three contexts varied, ranging from common practices and peaceful coexistence to rivalry, hostility, and competition. Pastoralism alone does not supply all of one's daily needs, so many pastoralists worked with settled farmers or practiced agriculture themselves. The Mediterranean environment supported such practices, as most pastoralists spent at least part of the year in regions where they or others cultivated basic food crops. At the same time, agriculturalists competed with pastoralists for rights, land, and property. Mediterranean landscapes have long hosted agriculture, as well as both mobile and sedentary pastoralism. This feature distinguishes the Mediterranean world from regions traditionally divided between "the desert and the sown." Indeed it represents one of the few global contexts in which mobile pastoralism and agriculture flourished, side by side, for so long. As the cases of Provence, Algeria, and Anatolia show, the interdependence between these groups was a hallmark of traditional Mediterranean society.

Both human and environmental factors shaped the nature and location of agricultural production in Provence. In the early modern period the cultivation of crops now considered quintessentially Mediterranean, such as grapes, olives, and orchard fruits, was far outstripped by open, uncultivated land.[114] As late as the mid-twentieth century, one visitor observed that humans occupied a "quarter of the land, the low-lying basins which are oases with harvests, olive trees, vines, and ornamental cypresses. Nature has three quarters of the land, layered rocks, reddish-brown or silver grey."[115] Throughout Provence's history environmental elements helped distinguish farmland from pasture. Provence's arid climate and intractable soils traditionally made agriculture difficult, generally restricting it to low-lying plains and valleys, while herds grazed on the rocky hillsides above.[116] Yet the geographical distinctions between pastoralism and agriculture were not always clear. Sheep provided a valuable source of fertilizer, and they frequently grazed on fallow vineyards and fields.[117] Both pastoralists and farmers, moreover, made use of communal land. In addition, agropastoralism was popular because it provided versatility and a secondary source of income.

This symbiotic relationship existed into the early nineteenth century. When the Comte de Villeneuve described rural Provençal society in the 1810s, he noted how shepherds paid their rent in kind: "In all of the places they passed through, they were able to find landowners who received them without any compensation other than the gift of some sheep or goat's milk that had been prepared before the departure."[118]

Farmers also recognized the sheep's presence as a source of fertilizer and used their encounters with shepherds to sell them fodder and produce.[119] Still, grazing violations, property rights, power rivalries, and other factors occasionally brought these two groups into conflict. Shepherds' relations with the French state were equally ambivalent in the early modern era. French administrators appreciated the services and products that pastoralism provided, but they also complained of the bureaucratic challenges of monitoring and regulating these mobile citizens.

Pastoralists and agriculturalists also coexisted in the early modern Maghreb. The population of the regency of Algiers was decidedly cosmopolitan on the eve of the French occupation. As one anthropologist has written, "The history of Algeria is not that of one people but of twenty."[120] Merchants, foreign officials, and Turkish soldiers mingled with indigenous inhabitants in the urban center of Algiers. Much of the rest of the territory was populated by communities of peasant farmers and the seasonal camps of pastoral nomads, but these too were a diverse bunch. The most commonly discussed distinction within Algeria's native population is that of Berbers and Arabs. The Berber presence has been linked to the ancient Mediterranean civilization of the Phoenicians, which peaked circa 1200–800 BCE. From this perspective Arabs are newcomers, having arrived abreast the Islamic conquests of the seventh to eleventh centuries. In practice Berbers and Arabs have long shared economic, political, and cultural ties. Members of either group might practice agriculture or pastoralism, or both and choose a sedentary or mobile lifestyle. Both groups were historically governed by tribal organizations, and both followed Islam. Given these commonalities, it makes sense to consider them, on the eve of the colonial era, to have been part of an integrated society.

In the early nineteenth century nomads comprised approximately 45 percent of the Algerian population and held much of the land and power in the region.[121] They shared this land, however, with a sizable minority of agriculturalists. The greatest concentration of farmers could be found among the Berber peasants of Great Kabylia, but substantial sedentary populations could also be found scattered throughout northern Algeria in pockets of suitable environmental conditions.[122] Powerful nomadic tribes exacted tribute from this sedentary population in the form of produce, but they also exchanged goods and services such as livestock, handicrafts, and transport with the settled population.[123] In addition, they practiced agriculture themselves, and agropastoralism formed an integral part of the pastoral tradition in Algeria, as it did in Provence.[124] In the fertile Tell region to the north, for example, wealthy pastoralists often invested in agricultural land, which they either cultivated themselves, hired hands to tend while they were away, or rented to peasant farmers.[125] The practice of agropastoralism was most pronounced in the province of Constantine, where a mild climate allowed pastoralists and agriculturalists to live and work side by side, in much the same way as in Provence.[126]

Likewise, in Anatolia nomads and peasants long shared space, and both have left their mark on the region. The mobility and relative autonomy of yörük tribes allowed them to prey on the settled population, which they were inclined to do in times of need or when dissatisfied with the terms of exchange. Yet Anatolian peasants and nomads coexisted harmoniously and symbiotically in most places, most of the time. Like their contemporaries in Provence and the Maghreb, nomads in these regions raised sheep for meat, milk, and wool. Their consumption was primarily subsistence-based, but nomads also provided the empire's supply of mutton, one of its main sources of meat, and a small portion of the wool they produced was exported.[127] As in other parts of the Mediterranean, the herders of Anatolia were known to supplement their pastoral habits with agriculture.[128] Indeed their agricultural practices have led some scholars to describe them as neither true nomads nor agriculturalists but migrants of a sort.[129] They grew some vegetables in their mountain camps during the summer months, and they cultivated small grain fields on the lower terraces of the Antalya Plain, leaving them to nature during the summer and harvesting them on their return to the coast in the fall.[130] Nevertheless the yörük generally remained dependent on nearby sedentary populations to supplement their diet.[131] This relationship could be reciprocal, since tribes produced valuable goods and services ideal for exchange, but it occasionally involved violence, when nomads had the power and will to take what they wanted by force. Encounters between nomads and sedentary inhabitants occurred both on the plains and in the surrounding mountains. In addition to furnishing the yaylas of the yörük, the Taurus range sheltered other nonpastoral nomadic tribes, as well as small settlements of peasant farmers who raised wheat, barley, and cotton and traded with the nomads.[132] In the mountains the nomadic population greatly outnumbered permanent residents.[133] Agricultural communities, moreover, grew increasingly sparse in the sixteenth to early nineteenth centuries, while the permanent population on the plains declined as well. As late as the mid-nineteenth century the western Antalya Plain was used primarily as a seasonal camp for yörük tribes.[134]

The shifting balance between pastoralism and agriculture in early modern Anatolia and other parts of the Mediterranean was due in part to climatic effects associated with the Little Ice Age. This global cooling trend developed between the fourteenth to sixteenth centuries, peaked in the seventeenth, and lasted, with a few major and many minor variations, well into the nineteenth century. As scholars have shown for contexts ranging from Europe to China to the Americas, the onset of the Little Ice Age produced frosts, droughts, famine, and epidemics that contributed to social crises.[135] In the Mediterranean region the Little Ice Age brought cooler, wetter conditions and lasting socioeconomic effects. Provence experienced frequent frosts during the height of the Little Ice Age, followed by unseasonable rain and floods during the climatic warming of the mid- to late nineteenth century. From the mid-fifteenth to the mid-nineteenth century the Rhône River froze over with relative frequency, allowing inhabitants to walk—or skate—from bank to bank. Local records indicate the occurrence of thirteen freezes in the eighteenth century alone, including three in the years 1766, 1767, and 1768.[136] Flooding was also common. During the periods from 1651 to 1720 and 1751 to 1860 the community of Arles was hit by an unparalleled number of intense floods.[137] These and other climate-related factors contributed to the outbreak of major famines throughout the Midi in the late seventeenth and early eighteenth centuries. A particularly devastating famine in 1693 cost the lives of an estimated 25 percent of the regional population, and many survivors left Provence in search of better weather or at least better government assistance.[138] In this way inclement and variable weather encouraged the expansion of Provence's pastoral industry by contributing to depopulation and by making mobile pastoralism a more feasible, adaptable alternative to settled agriculture.

Climate change also contributed to economic decline, depopulation, and the expansion of mobile pastoralism in Algeria, but its effects were most pronounced in southwestern Anatolia. The unfolding of the Little Ice Age in the sixteenth and seventeenth centuries brought wetter conditions to Anatolia's Mediterranean coastlands, making them choice breeding grounds for mosquitoes.[139] In many places farming was already restricted by the prevalence of steep hillsides and the absence of flat, arable land. Together with the heat, the prospect of malaria-ridden marshlands such as the Pamphylian Plain of Antalya and the Cilician (Çukurova) Plain of Adana made these regions nearly uninhabitable in the summer. Meanwhile, nomadic tribes that gradually had been pushed out of the central Anatolian plateau by administrative measures and the expansion of agriculture began to take advantage of the relative emptiness and isolation of these coastal plains.[140]

While these developments did not lead to the complete abandonment of the region by its sedentary residents, the Little Ice Age did contribute to a shift in the balance between the desert and sown. In the late sixteenth century both nomadic and sedentary populations were increasing, but nomads remained a small minority. In Alanya the number of settled households rose from approximately seventeen thousand to twenty thousand from the 1520s to the 1570s, an 18 percent increase. The nomadic population doubled over the same period, but it still made up only 2 percent of the total population.[141] Changing climatic conditions transformed this picture. Although reliable statistics are not available for the centuries that followed, the accounts of Turkish and foreign travelers, as well as the region's demographic transformation by the early nineteenth century, suggest an overall trend of depopulation combined with a steady influx of nomadic tribes.[142]

The environment figured heavily in the development and transformation of mobile pastoral industries around the Mediterranean, but so did politics. In Provence transhumant shepherds relied on the state for access to seasonal pastures, forests, and migration routes. In Algeria the Bedouin flourished in the absence of strong central administration and even became de facto rulers themselves, using their mobility to expand their domain and provide for settled groups such essential services as security, transportation, and trade. They also distinguished themselves from their sedentary neighbors in cultural and societal terms. In the case of Anatolia, mobility offered nomads relative freedom from central administration, which was probably as important a selling point for this lifestyle as was the prospect of extensive pasture use. For centuries the Ottoman state encouraged this arrangement by giving tribal chieftains official roles in frontier zones.[143] In all three of these cases, moreover, mobile pastoralists faced increasing competition from sedentary sheep farmers in the modern era.

The pastoral traditions of Provence, southwestern Anatolia, and northern Algeria exhibit both similarities to and differences from one another and the pastoral industries in other parts of the world. Early modern pastoralists in these three places were united not only by a shared environment but also by similar cultural and societal practices, political organization, their uses for sheep, their relationship with agriculturalists through systems of exchange and the practice of agropastoralism, their dependence on forests, and their tradition of seasonal migration between the mountains and plains of the Mediterranean hinterland. These features are descriptive not only of the three cases studied here but of many other pastoral economies around the Mediterranean Sea.[144] Together they describe a brand of pastoralism that is uniquely Mediterranean.

The nineteenth century would witness dramatic changes in all of the defining elements of Mediterranean pastoralism. The story that unfolds in the following chapters includes the indictment of mobile pastoralists by French officials and intellectuals on charges of environmental destruction, crippling restrictions on their traditional practices and movements around the Mediterranean, and additional challenges due to environmental, economic, social, and technological pressures. Even as these changes altered the meaning of Mediterranean pastoralism, however, they added a new set of common traits to this practice. They gave pastoralists in Provence, Anatolia, and Algeria a joint role in contesting state power and in shaping Mediterranean forest administration. Such patterns appear only when we—like nomads—overstep traditional boundaries and approach our subject from a transnational perspective with an inclusive conception of mobile pastoralism. As the world's remaining nomads disappear, this approach will be essential in investigating, understanding, and preserving their heritage.

CHAPTER 2

Black Sheep

The Intellectual Roots of Mediterranean Environmental Policy

> Nomadism, with its herd, tends incessantly to expand its domain, to ster-
> ilize increasingly vast regions, to overflow into surrounding cultivation, if
> one lets it. From this comes the frequency of ruin in countries inhabited
> by nomads; the destruction of urban centers, of irrigation systems, and
> insecurity favor their advance.
>
> —Augustin Bernard and Napoléon Lacroix, *L'évolution du nomadisme en Algérie*

Today we use the phrase "black sheep" to describe "the least repu-
table member of a group; a disgrace."[1] This idiom exists in multiple
languages, including French (*brebis galeuse*). In the Turkish case the
Karaca Koyunlu (a name meaning "black sheep") yörüks inhabited
the *sancak*s (districts) of Aydın and Menteşe in the fifteenth and sixteenth centuries. The phrase originated when sheep farmers viewed
black sheep as a burden, since their wool was more difficult to dye.
The color results from a rare genetic trait, so most herds had no more
than one or two black sheep. Some early modern cultures treated their
appearance as a bad sign; in England it was considered the mark of the
devil.[2] Likewise, in the eighteenth and early nineteenth centuries the
Mediterranean pastoral tradition increasingly came under fire. The
growing opposition of naturalists, intellectuals, and officials around
the inner sea effectively cast shepherds and their flocks as the black
sheep of the Mediterranean world.

Mediterranean pastoralism's antagonists attacked the practice on both social and environmental grounds. They commonly ascribed it to an earlier, more primitive stage of human development, and this characterization helped to reframe its terminology, identity, and history. At the same time, a growing number of critics expressed concern about the impact of sheep and goats on Mediterranean environments, and they wove the mobile pastoral tradition into narratives of Mediterranean environmental decline. These social and environmental perspectives gained momentum in the early nineteenth century, when they frequently appeared side by side. Together they inspired a new conception of Mediterranean pastoralism that cast mobile pastoralists as agents of deliberate environmental destruction.

This chapter chronicles the establishment and evolution of antipastoral narratives in the eighteenth- and early nineteenth-century Mediterranean world. It shows how both social and environmental perspectives served to vilify the practice and practitioners of Mediterranean pastoralism. As subsequent chapters reveal, such antipastoral narratives would have a dramatic impact on both pastoralists and the environment. Over the course of the nineteenth century state officials systematically fought Mediterranean pastoralism through restrictions, relocation, and sedentarization, but they continued to legitimize such moves through the rhetoric of civilization and environmental conservation. The marginalization of mobile pastoralists around the Mediterranean opened up new spaces for settlement, cultivation, and environmental exploitation. While serving the idols of state centralization, expansion, and capitalism, it ultimately proved detrimental to the Mediterranean pastoral tradition and the Mediterranean environment.

The Desert and the Sown

Nineteenth-century opposition to Mediterranean mobile pastoralism owed much to developments in the eighteenth century. During this period European encounters with distant peoples and places inspired new perceptions of human society, nature, and civilization. This trend was embodied in the French *philosophe* Georges-Louis Leclerc de

Buffon's immensely popular *Histoire naturelle, générale et particulière*, which suggested that humans had evolved into multiple races whose constituents shared common climate-specific temperaments.[3] Buffon's contemporary, Johann Friedrich Blumenbach, built on this foundation, dividing humanity into a hierarchical order of five distinct races and maintaining that the Caucasian (European) race was the most perfect since it most closely resembled God's original creation. At the same time, the work of Montesquieu, Adam Smith, and other Enlightenment thinkers recast human history as a story of progress, a march toward a more perfect future.[4] The idea of progress gained appeal in the nineteenth century, when developments such as industrialization, European expansion, and evolutionary thought seemed to provide evidence of human advancement. The theory of progress, together with racial theories, supported European imperialism by assigning colonized peoples, and mobile pastoral groups in particular, to an earlier, more primitive stage of civilization than the settled, industrialized societies of Europe. This perspective served to justify antipastoral policies throughout the Mediterranean world and beyond, but it would prove particularly critical in the marginalization of nomadic tribes in French colonial Algeria.

In the early nineteenth century French interests in North Africa inspired renewed interest among French intellectuals in the history, languages, religion, and culture of the Middle East and North Africa.[5] This generation of French orientalists revived the works of medieval Muslim scholars, such as the Tunisian historian-sociologist Ibn Khaldun, through translations and commentaries. Eschewing "an exact reproduction of ideas uttered in the text," their interpretations opted instead to integrate the sources with contemporary values.[6] For example, *Prolégomènes*, a translation of Ibn Khaldun's *Muqaddimah* by the prominent orientalist William de Slane, explicitly draws on and bolsters the idea of progress.[7] In a key passage Ibn Khaldun explains through the pen of Slane, "I have placed nomadism before sedentary life because (in the order of time) it has preceded all of the forms that the latter can take."[8] Later in the work Slane's translation states, "The habits and morals of nomadic life have made the Arabs a crude and savage people. . . . Such a disposition prevents the progress of civilization." The text continues, "As much as sedentary life promotes the progress of civilization, nomadism opposes it."[9] Slane even claims in his introduction to the text that "tracing the progress of civilization" formed Ibn Khaldun's main purpose in composing it.[10] Ibn Khaldun's original work, however, depicted history as cyclical. He believed that empires would rise and then inevitably descend into chaos, and he presented a relatively nuanced picture of nomadic and sedentary groups that acknowledged strengths and weaknesses within both lifestyles. In contrast, Slane's version emphasizes progress and advancement, and it represents mobile pastoralists as more primitive and even a potential threat to human civilization.

The works of Ibn Khaldun also influenced nineteenth-century French perceptions of division and conflict between the so-called desert and the sown, especially with regard to North Africa's indigenous population. In his *Muqaddimah* the medieval scholar devotes considerable time to differentiating these two groups. He associates Arabs with camel herding and the desert, describing them as "more rooted in desert life" than even other nomadic groups such as the Turks.[11] The Berbers, by contrast, "do not go deep into the desert," according to Ibn Khaldun. Instead most are settled agriculturalists, inhabiting "small communities, villages, and mountain regions."[12] At the same time, the *Muqaddimah* highlights connections and cooperation between these two groups, and it acknowledges that Berbers also practice mobile pastoralism as an alternative to agriculture. Moreover, by suggesting that nomads eventually will become sedentary, Ibn Khaldun's cyclical historiography suggests that these two societies are indelibly intertwined.

French orientalists tended to overlook such nuances in Ibn Khaldun's work. Instead they used it to support and promote the idea of a historically rooted dichotomy between North Africa's mobile and sedentary inhabitants. According to this narrative, Hilalian Arab invaders flooded into North Africa in the eleventh century, subduing the predominantly agricultural Berber, or Kabyle, population and converting the land to extensive nomadic pastoralism. This Kabyle myth, as contemporary scholars have termed it, cast Kabyles as the region's true natives and branded Arabs as relative newcomers and outsiders.[13] It also idealized Berbers as sedentary, law-abiding, industrious agriculturalists, in contrast to the destructive, nomadic Arabs. In the early decades of the occupation the Kabyle myth crystallized into a central part of the French colonial imaginary. Even Alexis de Tocqueville, an acerbic critic of French society, accepted it implicitly. In his "First Letter on Algeria," published in 1837, Tocqueville advised the new colonial administration to "focus above all on questions of civil and commercial equity" with Berbers, while emphasizing "political and religious questions" with Arabs, because their "soul is even more mobile than their dwellings."[14] Echoing Slane's translation of Ibn Khaldun, Tocqueville highlighted the "anarchy" of nomadic groups and claimed that they "have a multitude of vices and virtues that . . . belong to the stage of civilization at which they find themselves."[15]

More broadly, imperial agents sought to guide and justify the French colonization of Algeria by sharing the light of civilization. As one official explained, "It is by enlightening the populations, by civilizing them, that we wish to colonize today."[16] In the early decades of colonial rule, this charge was led by adherents of the French utopian socialist movement that had been founded by Claude Henri de Rouvroy, comte de Saint-Simon, earlier in the century. These Saint-Simonians arrived in the colony intent on molding it into an ideal society. Restyling Algerian nomads as France's own noble savages, they promoted the gospel of the *mission civilisatrice*. Their ranks included many influential members of French society, and their initiatives to uplift the indigenous population made a significant impact on French public opinion and colonial policy. Prosper Enfantin, a leader in the Saint-Simonian movement, joined the state-sponsored information-gathering expedition of 1840–42. His romanticization of indigenous tribes pervaded the expedition's official report, *Exploration scientifique de l'Algérie*.[17] Enfantin also presented his observations in an independent publication, *Colonisation de l'Algérie*. Then in 1859 Henri Duveyrier, an adherent and the son of another prominent Saint-Simonian, set off on a three-year trek through the Sahara with a member of the notorious Tuareg tribe as his guide. When he returned alive and proceeded to write *Exploration du Sahara: Les Touareg du nord*, in which he described his hosts as "hospitable, generous, kind and peaceful," many began to question their former perceptions of nomads as hostile, lawless, and dangerous.[18]

The peace-loving Saint-Simonians found unlikely allies in the military organization of the Bureaux Arabes, the offices assigned the task of overseeing indigenous affairs. The Bureaux Arabes were formed in 1830 as a branch of the French military to govern zones that had not been fully pacified. Although their main sphere of influence was in the southern military zone, the Bureaux Arabes operated throughout Algeria. In 1844 branches of the Bureaux Arabes were established in each of Algeria's three provinces as well as the southern zone.[19] The staff members of these offices spoke Arabic, they were versed in Middle Eastern history and culture, and many were self-styled "Arabophiles," or *indigenophiles*, who lived among indigenous groups and adopted their customs and dress.[20] The Bureaux Arabes, and the military administration in general, quickly gained a reputation for being more sympathetic toward the indigenous Algerians than the civil administration. This characterization was due largely to the role of the Bureaux Arabes as representatives of indigenous groups. These offices provided the principal legal avenue for native Algerians to influence French administration. Among other functions, the Bureaux Arabes held tribunals where local inhabitants could voice their concerns and complaints, and they frequently took advantage of this opportunity. Few cases actually reached the Algerian civil administration, but those that did showed the officers of the Bureaux Arabes to be strongly in support of their cause.[21]

Saint-Simonians, the Bureaux Arabes, and other "indigenophiles" all aimed to civilize Algerian nomads by encouraging them to settle.[22] Gen. Thomas-Robert Bugeaud de la Piconnerie, a leader of the Algerian conquest in the 1830s, urged the French administration to "work actively to modify" indigenous customs, and he argued that "the best way to change [the natives] is to sedentarize them."[23] Prosper Enfantin agreed. His *Colonisation de l'Algérie*, published in 1843, promotes an "agricultural" organization of indigenous tribes.[24] According to Enfantin, the best way to uplift them was to "build towns, improve agriculture, [and] create commercial relations."[25] The Bureaux Arabes, for their part, sought compromise partly out of necessity, in the interest of their own tenuous security in dangerous regions. Like the overwhelming majority of French colonists and administrators, members of the Bureaux Arabes maintained the view that sedentarization was progress, even in the desert.

French narratives of progress and civilization also influenced perspectives on mobile pastoralism beyond the French empire, in southwestern Anatolia. The Ottoman state did not always position itself against nomads.[26] In the early modern era perpetual wars and the incorporation of new territories limited the Ottomans' control over the countryside, leading them to adopt a policy of administrative flexibility.[27] During this period military conflicts robbed the Anatolian countryside of much of its male population, leaving fields uncultivated and villages unprotected. The resulting lack of security spurred many rural inhabitants to move to Istanbul and other commercial centers, contributing to steady urbanization and urban growth over the course of the eighteenth century.[28] As settlements retreated, many nomadic tribes expanded their domain, transforming unused plots into pasture. They also enjoyed nearly complete independence, thanks to the exodus of regional administrative officials. Recognizing the limits of its power, the early Ottoman state often dealt with these tribes in the only way it saw fit: by giving their leaders administrative roles in exchange for nominal promises of military service and taxation.[29] In practice this treatment increased tribal power and autonomy, as well as the insecurity of the countryside.

Early modern Ottomans recognized the benefits of their nomadic population, including military support and garrisoning, rural administration, transportation services, and the products of pastoralism.[30] Indeed nomads made up nearly one-fifth of the Ottoman population in the sixteenth century, and this figure did not decline significantly until the nineteenth century.[31] To ignore or subjugate such a significant population would have belied the Ottomans' characteristic pragmatism. Sources show that, on the contrary, they made deliberate efforts to keep the peace between their mobile and settled populations. Nomads and agriculturalists frequently brought claims of property violations to court, and Ottoman judges settled such disputes with an even hand.[32] This relationship began to change as nomadic tribes and the Ottomans' traditional flexibility toward them became more of a liability than an asset.[33]

In the late sixteenth and early seventeenth centuries the Celali Rebellions, a series of uprisings led by nomadic tribes, plagued the Ottoman countryside. The impetus for them was rural unrest as well as the unfavorable climatic factors associated with the Little Ice Age.[34] These rebellions helped to inspire state reform and centralization initiatives.[35] Members of the Ottoman elite began a process of soul-searching reform that targeted banditry, corruption, and inefficiency throughout the empire and promoted a stable lifestyle for its sedentary constituents (*reaya*).[36] Reform-minded Ottomans laid the blame for administrative weaknesses, agricultural failure, brigandage, and other problems at the feet of Anatolia's mobile pastoral population.[37] By the late seventeenth century this narrative had persuaded the administration to trade its long-standing policy of acceptance, accommodation, and compromise toward Anatolia's nomadic population for a new program promoting regulation, resettlement, and sedentarization.

The sultanate started its first formal sedentarization campaign in 1689, and this initiative was succeeded by other, similar programs over the course of the eighteenth century.[38] The sedentarization efforts of this period varied somewhat in their purpose and methods. Some merely proposed to move tribes out of the way of the settled population, while others ordered nomads to relinquish their mobile lifestyle and become agriculturalists.[39] In either case the sedentarization campaigns of the early modern era were motivated largely by concerns of security. The campaigns contained frequent references to the "banditry" of tribal members, who were presumed to "attack travelers" as well as to engage in "looting the towns and villages" and "stealing the animals and the harvest."[40] This theme continued throughout the eighteenth century, with occasional confrontations between nomads and the state, unsuccessful efforts at sedentarization, and tribal gains in the wake of rural depopulation and decentralization.[41]

Ibn Khaldun's cyclical view of history maintained a modest following among Ottoman intellectuals throughout the early modern period, though his works were not translated into Turkish until the eighteenth century.[42] Ibn Khaldun's Ottoman interpreters were primarily interested in his model of the rise and fall of empires, which they gleaned for insight into their own perceived military and economic distress.[43] In this context Ibn Khaldun's characterization of the Bedouin as more courageous, pure, and good than their sedentary counterparts helped to furnish an explanation for Ottoman decline. Thus, early modern Ottoman readings of Ibn Khaldun, far from contributing to bias against nomadic tribes, encouraged their idealization. Only in the mid- to late nineteenth century, as Europeans exercised a growing influence on the Ottoman court, did Western interpretations of Ibn Khaldun's work begin to gain currency among the Ottoman elite.[44]

European visitors to the empire shared Ottoman intellectuals' frustration with the administration of nomadic tribes. Indeed the French consul in Antalya was so fed up with government corruption that he abandoned his post in 1810, taking the French consulate with him.[45] Fifty years later little seemed to have changed. In 1857 the British consul general complained bitterly of scant law enforcement in the *vilayet* of Antalya.[46] In *Voyage dans l'Asie Mineure*, the consul general's contemporary, Baptistin Poujoulat, after explaining the Ottoman system of conscription to his readers, remarks sardonically, "This system . . . does not reveal a great advancement of civilization in this country."[47] Poujoulat's characterization of Arabs is particularly severe. He could easily be echoing the voices of French colonials in Algeria when he writes, "Arab civilization rapidly gained a place among the advanced civilizations, only to fall back into the barbarity of the desert."[48] Ottoman losses in the Crimean War of 1853–56 fueled European criticism of the state, culminating in its widespread identification as the "sick man of Europe."[49]

In this context the Ottomans' apparent inability to control its nomadic population further weakened the empire's ailing reputation. Echoing orientalist interpretations of Ibn Khaldun, European observers often cited nomadism as an example of, if not an explanation for, Ottoman administrative weakness. Travelers and diplomats alike complained of "marauding Bedouins," the "attendant evils" of nomadism, or "the lawless, predatory disposition" of certain tribes.[50] On a voyage through southern Anatolia in the 1860s, the British traveler Henry Van Lennep took the precaution of bringing along a police escort for protection, but he noted that even the police officer "skulked and hid behind us" when they passed nomadic tribes.[51] Van Lennep's compatriot Frederick Burnaby, who visited Anatolia a few decades later, was even more pointed in his criticism of Ottoman administration. His written account starkly contrasts sympathy for the Turks with condemnation of their "despotic" government.[52] He describes evidence of ubiquitous agricultural and industrial waste and decline, and his text resounds with opportunities lost. He is struck by the region's rich mineral potential, which he counts as wasted due to a lack of modern materials. To drive home his point, he quotes a local Turk as saying, "We have got nothing but paper money in Anatolia . . . all this rich metal lies buried beneath our feet."[53] He describes marble quarries unused for hundreds of years and fields of "rich soil" abandoned and left fallow. Referring to this desolate landscape, Burnaby cries, "Poor Turkey, she has descended the steps of civilization, and not ascended them like European nations." Although Burnaby's account goes on to romanticize the freedom of a people who could "pack up their chattels and migrate to the mountains" whenever the tax collector arrived, he and his contemporaries also reasoned that the state would have settled such populations if it could.[54]

As external views of the Ottoman government grew increasingly negative over the course of the nineteenth century, internal developments in Ottoman administration and society also influenced state perceptions and policy toward nomads. The rise of nationalism and nationalist unrest made the Ottoman state increasingly self-conscious about its international role and image. One result was the promulgation of national histories that emphasized Islamic identity over the empire's nomadic roots. Such narratives portrayed the founding tribe of Osman as *gazi*, or holy warriors.[55] By casting their ancestors as medieval Muslim zealots, Ottoman elites hoped to gain both support within the Islamic community as well as legitimacy for their control of the Islamic caliphate. They also used the *gazi* creation myth to present a nobler, more civilized image to Europe.[56] As the Ottoman state integrated this ideology with modernization and westernization, it increasingly characterized nomadism as primitive, describing tribal inhabitants in many of the same pejorative terms then popular in Europe.[57] For instance, nomads were subject to such epithets as "barbarous," "primitive," "raider," and "savage."[58] Ottoman officials claimed that the tribal peoples lived "in a state of nomadism and savagery" and advised that they be "gradually brought into the fold of civilization" and "liberat[ed] from the shackles of community life."[59] As the nineteenth century progressed and the administration's demonization of nomads crystallized, the drive to settle nomadic tribes became more than a matter of convenience. As Ottoman intellectuals adopted Western perspectives on nomadism, sedentarization became a symbol of Ottoman cultural advancement. In this way European biases against pastoralism provided the Ottoman state with international legitimization for its policy against nomads at a key juncture in both its internal affairs and its global reputation.

As the cases of France, Algeria, and Anatolia demonstrate, European ideals of progress and civilization fueled opposition to Mediterranean pastoralism as well as charges of Ottoman despotism. Likewise, European exposure to mobile pastoral traditions around the Mediterranean, together with narratives of race and social progress, ultimately may have helped to transform and stratify the lexicon of pastoralism. Unlike "nomad," which has been part of English usage for nearly five hundred years, "transhumance" is a relatively modern term.[60] It did not arrive in the English language until around 1911, and its roots are only somewhat deeper in French and Spanish. It first appeared in the French dictionary *Robert* in 1818.[61] Thus, the practice of transhumant pastoralism in Provence and other parts of the world predated its classification by hundreds if not thousands of years. Why did modern Europeans create a new term to describe an age-old practice? In French the word *transhumance* likely gained currency at least in part as a way to distinguish mobile pastoralism in Provence from similar practices outside the metropole, in the Near and Middle East, Africa, and Asia. As the French came into contact with other mobile pastoral systems, this terminology allowed them to maintain that mobile pastoralism at home was a modern, civilized institution, whereas abroad it was primitive and destructive, even if such industries were in practice very similar. Thus, transhumance may well have been first named when it began to be challenged.[62] In addition, the term "nomadism" has long been associated with the Orient. Some European literature even presents it as a symbol of the divide between East and West.[63] In this context the term "transhumance" provided a way for France and other European powers to preserve and reinforce their Western identity, together with their claims of superior civilization, as they expanded into the East.

Pastoralism and Environmental Decline

Nineteenth-century opposition to Mediterranean pastoralism owed much to social ideals, but it also drew on pressing environmental concerns. Like the former, the latter was largely a product of the Enlightenment. Just as this movement fueled theories of race and progress, it also inspired a narrative of long-term Mediterranean environmental decline.[64] This theory drew on classical sources, including Herodotus, Strabo, Tacitus, Pliny, Ptolemy, and others who had characterized the Mediterranean world—and North Africa in particular—as a fertile breadbasket.[65] Neoclassicist artists reinforced the image by depicting the ancient Mediterranean world as lush, fertile, and heavily forested. Contrasting these features with the relatively arid, open Mediterranean landscapes of their day, Enlightenment thinkers concluded that the region's environment had been ruined by centuries of human use and abuse.[66] The resulting declensionist narrative linked deforestation to climate change. The Comte de Buffon, whose *Histoire naturelle* attempted to explain European superiority through climatic factors, also

3. Nicolas Poussin, *Landscape with the Gathering of the Ashes of Phocion*, 1648. Courtesy of Walker Art Gallery, Liverpool.

pioneered climatology in a subsequent work, *Les époques de la nature*. Noting that temperate zones generally enjoy greater forest cover than hot deserts, Buffon's text suggests that forests cool their environment by shading it from the sun.[67] Likewise, Pierre Poivre, a naturalist and administrator of French colonial Mauritius, drew on his colonial experience in connecting forest destruction to climate change. "The rain," he claims in his description of the island in *Voyages d'un philosophe* (1768), "which in this island is the only solution and the best that the earth can receive, follows forests exactly, stopping there and no longer falling on cleared land."[68]

Deforestation became an even more prominent and troubling issue in the early nineteenth century, and interest in its impact on the climate grew apace. The Baron de Ladoucette remarked in 1820 that "the destruction of woods and thoughtless clearing have caused a change in the temperature," reasoning that greater solar reflection from rocks and rugged mountains increased the summer heat, while in the winter, gusts of wind no longer impeded by forests brought a deeper chill.[69]

In 1822 France's Ministry of the Interior conducted a national survey designed to study this phenomenon. French departments were asked to indicate the extent of their forest cover as well as their perceptions of climate change, including increased cold, dryness, flooding, violent storms, and a decline in agricultural productivity.[70] The nineteenth century also witnessed the maturation of a theory that forests promote rainfall, now called the "rain-shadow" effect. Conversely, this theory suggested that deforestation would lead to the desiccation of the soil or, in extreme cases, desertification. The rationale was that overgrazing and forest clearing dried up the soil, which in turn limited plant life so that the earth held less moisture, and thus less moisture was released into the air to return as rainfall.

Once scientists, intellectuals, officials, and others accepted that the Mediterranean environment had suffered significant decline since classical times, they began to ask why. The Enlightenment encouraged the use of reason to interpret natural phenomena, and it emphasized human factors in environmental processes. These philosophizing Lumières, as they were known, sought anthropogenic explanations for occurrences—everything from disease to extreme weather—that previously had been attributed to the will of God. Drawing on such ideas, nineteenth-century literature increasingly cast environmental phenomena as the direct consequence of destructive human actions. By midcentury even the *mistral*, the violent wind that battered Provence each spring and fall, had been recast as "the child of men, the result of their devastations."[71]

Influenced by the social and environmental perspectives of the time, French intellectual explanations for Mediterranean environmental decline increasingly focused on pastoralism. Eighteenth-century critics had aimed most of the blame for Provence's environmental transformation on tree cutting and forest clearing, after which the impact of pastoralism, if mentioned at all, was a distant second. "One thing that surprises every observer," wrote the Provençal historian Charles-François Bouche in 1775, "is that the plains of the Camargue were formerly covered with wood; today one sees only grain fields, marshes, and prairies partly covered with stagnant water."[72] Thus, Bouche scolded inhabitants for having "abandoned" their landscape, and he declared, "It is not at all necessary to look elsewhere than in the abuses and effects of forest clearing for the cause of this desertion."[73] In the nineteenth century, however, the same intellectual perspectives increasingly referenced pastoralism's role in environmental decline in Provence and throughout the Mediterranean world. One of the most influential voices on this subject was that of Étienne Laurent Joseph Hippolyte Boyer de Fonscolombe, a natural scientist from Aix-en-Provence. His widely read treatise, "On the Destruction and Reestablishment of Woods in the Departments That Composed Provence," explains how the region's landscape deteriorated over time: "All evidence proves that there were once very extensive forests in the calcareous regions that form seven-eighths of these four *départements*. The plains and valleys were covered with downy oaks; the heights with holm oaks and pines. As the population expanded, plains and valleys were cleared, where the beauty of the woods that had covered them attested to their fertility. Communal forests disappeared first. Goats and sheep that were introduced after trees were felled ate their remains; there is nothing left but brush."[74]

This and similar narratives helped to solidify the association of deforestation and pastoralism in Provence. Even Ladoucette, who as prefect of the Alpes-de-Haute-Provence was relatively sympathetic toward pastoralists, lamented "the devastation caused by goats [and] sheep" in his department.[75]

Protoconservationists vilified sheep, goats, and shepherds and derided their trade as an inefficient and unsustainable use of land, a practice that exhausted fragile Mediterranean soils. Some critics targeted transhumance in particular because it took up considerable space.[76] Critics of Mediterranean pastoralism's environmental impact reserved their most scathing denunciations for the goat. This creature was accused not only of destroying current forests but also of thwarting future regeneration by devouring roots and saplings.[77] Although they were domesticated, goats were commonly categorized with other noisome "pests," such as foxes and rabbits. A typical observer affirmed that "the mortal enemy of the tree is the goat," compared to which "the ravages of wolves are insignificant."[78] It was "the scourge of all unenclosed plantations," according to a member of the Academic Society of Aix. He continued, "It eats the shoots of all trees and saplings, without exception. It incessantly impedes their progress and after making its young subject languid and stunted for a few years, it ultimately kills it."[79] Boyer de Fonscolombe agreed, stating that "goats above all contribute to the destruction of woods, which furnish their principal nourishment," and he noted that they were equally harmful to forests and fields.[80] By the mid-nineteenth century the prominent naturalist Charles de Ribbe was so concerned about goats' environmental impact that he declared, "Only the absolute prohibition [of these animals] will prevent the total ruin of woodlands and promote their regeneration."[81]

Legislation outlawing goats or even calling for their extermination appeared throughout the modern period, but with little impact. Most French constituents considered goats' usefulness to outweigh their many unsavory characteristics. For transhumant pastoralists, their intelligence and ability to lead the herd were indispensable, and they provided a cheap and palatable source of dairy. Goats were of particular value to the poorest sector of the population, and their numbers actually rose during and following the economic hardship of the French Revolution, despite the proliferation of laws against them.[82] Yet, widespread and persistent opposition throughout France was not without effect, and it put particular strain on Provence because of the region's considerable goat population.[83]

In the early nineteenth century protoconservationists genuinely feared that mobile pastoralism would destroy the Mediterranean environment, both directly, through the impact of livestock on forests and fields, and indirectly, through erosion, flooding, and climate change. Nineteenth-century critics of this industry combined social and environmental narratives, drawing on environmental anxieties as well as new conceptions of race, progress, and civilization. But opposition to Mediterranean pastoralism targeted more than its supposed environmental side effects; it also accused pastoralists of deliberate destruction, both by nature and by trade. French stereotypes typically portrayed mobile pastoralists as lazy, corrupt, lawless, and irresponsible.[84] A late eighteenth-century tract, "On the Pastoral Economy of the Hautes-Pyrenees, of Its Vices and the Ways to Remedy Them," associated pastoralists in the Pyrenees with a colorful range of "vices" that, according to the author, wreaked environmental havoc.[85] Many considered pastoralists as lacking respect for the law and assumed that they would graze their herds illegally. "The shepherd does not have the proper respect for forests," declared one detractor, adding, "He destroys for the sake of destruction."[86] Those who kept goats were under the greatest suspicion because of restrictive grazing laws. One Provençal scholar characterized the goatherd as "a kind of poacher," leading goats into communal mountains by day and "skirting and traversing cultivated property by night."[87] Similarly, during the "plague" of these animals following the revolutionary era, the prefect of the Ardèche called the goat "the animal of the idler and the pillager," in an obvious slight toward its keepers.[88]

The practice of blaming pastoralists for environmental decline proved even more popular in French colonial Algeria, where it elided with imperial interests and social ideals. Campaigns in Algeria in the 1830s brought painful awareness of France's limited knowledge of the region. Citing this problem, the French naturalist and traveler Jean-Baptiste Bory de Saint-Vincent urged the minister of war to finance "a commission both exploratory and scientific, a single body directed toward a common purpose."[89] Scientific missions, based on those completed in Egypt and the Morea (Peloponnese) earlier in the century, were duly carried out in 1840, 1841, and 1842, ambitiously aiming to understand and classify all aspects of Algerian culture, history, and geography.[90] To this end, experts in every branch of natural science, art, and history, from zoology to portraiture, were included.[91] In the following years, the results of this expedition were published in multiple volumes under the title *Exploration scientifique de l'Algérie pendant les années 1840, 1841, 1842 publiée par ordre du gouvernement et avec le concours d'une commission académique.*[92] This document contains an extensive and detailed report on the physical geography of Algeria, and it is clear that the commission's observations are practically oriented. Algeria is divided into "productive" and "unproductive" regions.[93] The possibility of afforestation in the Sahara is considered and compared to the draining of the Landes region in France. The success and scope of this enterprise was limited by personal rivalries and politics, but it provided a precedent for other, similar endeavors, and its observations guided French administrators in the development of Algerian environmental legislation. As the French worked to subdue and consolidate their new prize, they persevered in the study of its nature and resources. The resulting reports helped to shape French concerns over deforestation and environmental decline.

French colonial Algeria also provided a fruitful context for desiccation theory. As early as 1838, a report on Algeria's forests warned that overexploitation of the cork trees surrounding La Calle could easily transform "this green and wooded country" into "a sterile and burning desert."[94] Others used desiccation theory to support the claim that the Algerian environment was once a verdant "granary of Rome," boasting a temperate climate, fertile soils, and lush vegetation.[95] According to this narrative, centuries of reckless forest destruction had desiccated this Mediterranean Eden. One source pointed to "traces of vegetation" on the southern foothills of northern Algeria as evidence that "in the not too distant past forests extended right up to the desert."[96] Indeed the Ligue du Reboisement, a powerful colonial lobby group formed in the 1880s, operated on the assumptions that "immense forests" had once covered Algeria and that continued deforestation would drastically transform Algeria's climate.[97] However, the theory of desertification was used not just to lament past destruction but also to promote the "reforestation" and reclamation of degraded lands. From this perspective the cultivation of forests, even in the desert, would encourage regeneration and positive climate change.[98] Many scientists of the era argued that the rain would follow the plow. Proponents of desiccation theory generally either singled out Arab nomads as responsible for the damage or blamed the entire North African population. In either case, they branded the indigenous population as poor environmental stewards. The use of this theory to advocate universal afforestation was particularly detrimental to mobile pastoralists in Algeria. In the colonial context the narrative of desertification legitimized the French administration's appropriation of Algerian lands, its subjugation of the indigenous population, and its manipulation of the traditional economy.

French colonial understandings of the Algerian environment also engaged orientalist sources. Slane's translation of Ibn Khaldun's history of the Arabs (*Kitāb al-'Ibar*) has them cutting down all the trees, destroying civilization, and transforming much of North Africa "into desert" during the invasions of the ninth to twelfth centuries.[99] The title of one chapter, in a telling interpretation of the original text, reads "Every country conquered by the Arabs is soon destroyed."[100] Such representations led many to theorize that significant environmental decline in North Africa had begun with the Arab invasions of the eleventh century.[101] Likewise, French colonial agents referenced the Kabyle myth in interpreting the Algerian environment. "In Algeria," explained one mid-nineteenth-century observer, "one can, without fear of error, affirm that wherever the earth appears desolated, without trees, one is in Arab territory; in contrast, wherever there is fine cultivation, beautiful trees, woods and forests, one is in Berber lands."[102]

By the mid-nineteenth century French environmental discourse on Algeria had solidified into a tale of severe, long-term environmental degradation, largely at the hands of the territory's nomadic pastoral inhabitants. An 1846 report on the state of Algeria's forests contains a typical example of this narrative: "More than any other country Algeria must have in distant times been covered with forests. One finds irrevocable indications of this in all its provinces. It is not in Algeria as in Europe, where the progress of civilization was the cause of deforestation. There, a thousand needs to satisfy either in construction or in industry; here, other uses depending on barbarity, always in accord with the interests of men."[103]

Once established, this story changed little over the course of the century. "The natives manifest a veritable hatred of trees," observed scholars Augustin Bernard and Napoléon Lacroix in the early twentieth century, adding that "passersby cut off their branches, shepherds mutilate them, and woodcutters turn them into firewood."[104] By presenting indigenous Algerians as selfish, shortsighted, and barbaric, such perspectives supported and legitimized the Algerians' subjugation by European settlers.

In contrast to France, the Ottoman administration traditionally showed little concern for nomads' ecological impact. Early modern Ottoman forest legislation was mute on the issue of pastoralism. When Anatolian nomads were charged with environmental damage, the source of the dispute generally arose from interests in property rather than preservation.[105] Indeed while peasant farmers were quick to blame nomads for ruining their land, the imperial government did not necessarily take their side. In such cases the judge's concern was with which of the two parties had rights to the land, rather than what they had done to it.[106] In some cases the Ottoman administration actively encouraged and supported nomadic livelihoods. In the early modern era the *tahtacılar* and *ağaç erleri* tribes played a central role in the extraction of forest products, including wood, timber, and charcoal. They were also involved in the trade of such products with ports around the Aegean and along the Mediterranean coast of Anatolia. As early as the fifteenth century these tribes were engaged directly by the Ottoman state.[107]

Meanwhile, European travelers to the region increasingly took time in their accounts to criticize the presence of mobile pastoralism, calling it a tragic corruption of fertile land. As early at 1823 one French agronomist listed Anatolia along with Egypt, Judea, and Greece as "once fertile countries transformed into deserts" where nothing remained but "ruins and tombs."[108] W. J. Hamilton, who visited the countryside near Bursa in the mid-nineteenth century, lamented that lands "capable of the finest cultivation" had become "a scanty pasture to the flocks of wandering tribes."[109] For such observers the cause of environmental decline was clear. Traveling in western Anatolia in the early nineteenth century, François-René de Chateaubriand recorded widespread destruction: "We began to climb through a mountainous region that would be covered with an admirable forest . . . if the Turks let anything grow. But they set fire to the young plants and mutilate the trees. These people destroy everything. They are a veritable scourge."[110] Another account described Anatolia as "the last and greatest example of the sterilization of a civilized country by nomads."[111] The archaeologist Theodore Bent echoed this charge. Calling nomads "the most inveterate enemies of the Asia Minor forests," he accused them of "lay[ing] bare whole tracts of country that they may have fodder for their flocks."[112] Such views were confirmed by nineteenth-century European diplomats posted to Anatolia; their correspondence abounds with scornful descriptions of the ravages of nomadic tribes. In a typical example a British consul posted to Kayseri described the "blackened stumps, charred and scorched trunks" that resulted from local tribes' "habit" of burning trees to increase pastureland in their *yayla*s (summer pastures).[113]

Paradoxically, the same accounts often suggested that Anatolia possessed vast forest resources.[114] The origins of this perception probably date to the early modern period, when various French travelers, including Nicolas de Nicolay, Pierre Belon, Jean-Baptiste Tavernier, Jean de Thévenot, Laurent d'Arvieux, Antoine Galland, and Joseph Pitton de Tournefort, returned from the Ottoman Empire with tales of its lush, wooded landscape. Their descriptions were popularized, exaggerated, and propagated during the period of *turquerie*, a celebration of Ottoman culture that spread throughout Europe in the seventeenth and eighteenth centuries.[115] By the nineteenth century the legend of Ottoman environmental wealth was entrenched in European thought. In this context the Ottoman state's mid-nineteenth-century announcement that it would open up its resources to international commercial exploitation generated an enthusiastic frenzy among competing European powers, but it also fueled claims that the empire was squandering its resources.[116] European encounters with the Ottoman state ultimately helped to highlight and heighten environmental anxieties at home while spreading new environmental awareness and concern to members of the Ottoman elite.

French intellectual movements of the eighteenth and early nineteenth centuries dramatically reshaped perspectives and policy on Mediterranean pastoralism. At the turn of the nineteenth century concern over deforestation was limited to a few disparate voices, and it was treated through a handful of laws that were rarely enforced. However, a growing number of legislators, officials, entrepreneurs, farmers, intellectuals, scientists, foresters, and protoconservationists began to target Mediterranean pastoralism. These critics not only blamed the practice for extensive, long-term environmental decline but also associated it with social backwardness and a lack of civilization. The growth of environmental awareness and concern among French intellectuals bred new arguments against Mediterranean pastoralism, while racial theories and the idea of progress served to cast mobile pastoralism as a more primitive form of civilization. In the nineteenth century these perspectives dovetailed with the practical and political requirements of modern state-building. At one end of the Mediterranean, France sought to justify its colonial interests in Algeria; at the other, the Ottoman administration worked to sedentarize its nomadic tribes. In this context the combination of environmental anxieties with social, political, and cultural biases against Mediterranean pastoralism created a powerful antipastoral lobby. In both cases policy makers benefited from social and environmental narratives, as well as from innovations in communication, transportation, and technology that granted them greater control over mobile populations. In some ways these developments reified the supposed age-old conflict between mobile pastoralists and settled societies—the desert and the sown—in the Mediterranean world.

CHAPTER 3

Counting Sheep

Pastoralism and the Construction of French Scientific Forestry

"Why," a prominent resident asked me, "don't you call for the plantation of woods in the more susceptible places? Why not limit the number of pastures and the number of beasts?"

—Charles-François, baron de Ladoucette, *Histoire, topographie, antiquités, usages, dialectes des Hautes-Alpes*

In the early nineteenth century Charles-François Ladoucette traveled throughout the Hautes-Alpes Department, where he served as prefect. Along the way he recorded his observations, published as *Histoire, topographie, antiquités, usages, dialectes des Hautes-Alpes*. His account features barren hilltops "devastated" by sheep and goats, as well as distraught mountain dwellers reeling from the loss of forestlands and begging for governmental action.[1] Ladoucette's descriptions reflected real environmental change. Over the course of the early modern era deforestation had transformed the French landscape quickly and dramatically. The wars of the seventeenth and eighteenth centuries, together with the French Revolution and the Napoleonic Wars, exacted a hefty toll on France's woodlands. Noble families also played a role by reaping healthy profits from wood sales. The Bourbon-Penthièvre family, for example, shifted its enterprise heavily toward the sale of wood during the course of the eighteenth century. In 1790 nearly 75 percent of their profits came from wood, compared to slightly more than 20 percent in

1737.[2] In addition, the relative scarcity of forest guards made the enforcement of forest legislation spotty at best. As a result, when Ladoucette first published his account in 1820, French forests had shrunk to about half of their extent in the mid-seventeenth century.[3]

Over the same period the appearance and institutionalization of novel administrative structures in France and throughout Europe contributed to the development of new conceptions of and roles for the state. In the 1970s the French sociologist Michel Foucault coined the term "governmentality" to describe this confluence of state power, political economy, security, bureaucratization, and the presentation and regulation of knowledge systems (*savoirs*).[4] The governmentalization of the French state, combined with the specter of deforestation, would lead to the implementation of a new and bold forestry regime throughout French lands. Modern forestry in France embodied the trends of discipline, litigation, regulation, power and knowledge, and economic administration characteristic of the governmentalized state. It represented a new "science," as well as a new science of administration. The forest regime extended the French state's presence so that it directly encountered its population, establishing a vast, hierarchical network responsible for surveillance and policing. In the process it buttressed national security and enhanced national culture. Forest agents even brought the French language into the service of the state by classifying and redefining forestlands. Through its self-identification as science, nineteenth-century French forestry produced new forms of political knowledge, expanding the French state's purview over the thoughts and actions of its constituents.

This chapter traces the construction of French scientific forestry from its roots to its spread across the Mediterranean. The formulation of scientific forestry in France was heavily influenced by the system of forest management pioneered by neighboring German states in the mid- to late eighteenth century. However, nineteenth-century French forestry was inspired at least as much by domestic perceptions of the environment cultivated through overseas expansion, the Enlightenment, and the French Revolution. In addition, Mediterranean pastoralism played a fundamental role in the development and expression of French forest administration, a feature that set French forest science apart from the German model and helped to win it recognition around the Mediterranean and around the world.

The Development of French Forest Administration

The roots of French forest administration reach at least as deep as the fourteenth century. In 1346 King Philip VI passed a law that is often considered the first French forest code.[5] The aim of this succinct document—comprising just seven pages—was to maintain a ready supply of timber for the royal naval fleet. An edict on forests promulgated in 1516, under François I, included a clause banning all livestock from forest stands too young to defend themselves against grazing, though it did not specify an age limit for protected trees.[6] A series of royal *ordonnances* during the reign of mid-sixteenth-century monarch Charles IX marked the first time different types of forests were distinguished. In addition, some were placed in reserve, and the cutting of undergrowth younger than ten years was banned.[7] Thus, early French forest laws focused on specifying seigneurial privileges, setting aside certain woodlands for state use, and imposing basic limitations on exploitation. Deforestation had not yet become a major concern for the French state, and forest grazing continued unabated. Indeed early modern monarchs harvested the forests of France voraciously. Speaking of Louis XII (r. 1498–1515), one royal attendant recalled, "The king excited fear not only among men, but also among trees."[8]

The clearest precursor to nineteenth-century French forest administration was the Edict of 1669. The brainchild of Louis XIV's finance minister, Jean-Baptiste Colbert, this expansive document echoed previous legislation in its aim to limit the use of French forests by inhabitants to provide more timber for state use. However, the edict differed from its predecessors in both its size—it totaled more than one hundred pages of provisions, stipulations, and explanations—and its scope.[9] It was the first law to place extensive limits on sheep and goat grazing throughout France. It also revolutionized French forest administration by placing previously independent forests under the control of the Chambre des Eaux et Forêts. Perhaps even more important for the future of French forest management, the Edict of 1669 represented the French state's earliest attempt to address anxieties over environmental degradation. The introduction clearly states the aim to preserve France's forests "for posterity," in other words, to maintain sustainable exploitation.[10]

In practice Louis XIV's military ambitions, combined with a lack of enforcement, prevented the Edict of 1669 from having much impact on early modern society. The forest regime's main weakness was limited staffing, which made surveillance extremely poor. Prior to the nineteenth century, forest regulations were designed to be enforced by forest guards (*gardes-champêtres*) posted sparsely throughout the realm.[11] These were veritable forest soldiers, with military rather than silvicultural training.[12] Their primary task was to prevent poaching and illegal wood cutting, but they were also assigned to enforce grazing regulations. They patrolled forests while armed and often had occasion to use their weapons.[13] Forest guards generally were obliged to work alone, far from their colleagues, and they sometimes needed protection from local residents as much as the forests did. They were, as a rule, unpopular and resented by local communities, and their attempts to enforce the law often triggered violence. Because guards were hired and paid by the communities they served, moreover, they were easily bought off or corrupted by local interests.[14] For their part, inhabitants responded to the edict's restrictions with widespread disregard. Because the government lacked the resources to enforce regulations against pastoralism, shepherds continued to graze sheep and goats in forests. In a typical case from 1804, two shepherds were caught illegally grazing a herd of 198 goats in the communal forest of Gémenos.[15] To make matters worse, this forest had been classified as deforested and was under special protection. Because forest guards were not present in sufficient numbers to prevent local inhabitants from violating forest laws, their occasional policing won them mistrust and hostility.[16] Indeed the most significant effect of the Edict of 1669 may have been to turn local populations against forestry officials and the forest administration in general because its terms were perceived by many as unjust and an infringement on customary rights.

The Edict of 1669 remained in force into the early nineteenth century, though it was supplemented periodically by legislation that did little more than restate its clauses, confirming that it was not satisfactorily enforced. Much like law enforcement in general under the ancien régime, forest administration was most effective in times of peace because forest agents doubled as soldiers and could be called to active duty. For the same reason, enforcement frequently broke down in times of war and internal conflict. Warfare also intensified the burden on national forest resources. To add to the problem, forest rights and ownership were poorly defined under the law. While it probably did reduce the extent and speed of deforestation, the Edict of 1669 ultimately increased public aversion to forest administration without effectively regulating forest use.[17]

National legislation tells only part of the story of forest management in early modern France. Local administrators and residents held their own counsel on the problem of deforestation, supplementing national laws with region-specific decrees. In the eighteenth and early nineteenth centuries the parliament of Provence implemented a number of legislative measures regarding forest use, some of which were actually stricter than national legislation.[18] Residents also tried to prevent violations of forest regulations by augmenting the number of guards stationed in their forests or by exercising vigilance themselves.[19] The French Revolution represented an important turning point for local officials and members of the Provençal elite. The narrative of revolutionary-era destruction spurred growing anxiety over deforestation in postrevolutionary Provence.[20] The mayor of Tarascon embodied this sentiment when he argued in a regional meeting in 1822, "As a result of forest clearing, floods have exercised their ravages without obstacles. . . . [We] must hurry to recreate that which has been destroyed."[21] Many blamed the denudation (*dépaissance*) of Provence's hillsides on sheep and goats. Jean-Baptiste, marquis de Montgrand, mayor of Marseille from 1815 to 1830, cited three essential steps for initiating the process of afforestation: prohibiting clearing, improving surveillance and the quality of forest guards, and, finally, restricting grazing.[22] Others tempered their accusations by acknowledging the importance of pastoralism. As the Tarascon mayor cautioned, legislation that was too strict would "destroy a very necessary type of industry."[23] Instead he promoted education and moderate grazing regulations that would protect trees under a certain age.

One of the first modern attempts to reenvision forest administration in Provence occurred under the prefecture of Antoine Claire Thibaudeau. Formerly a popular revolutionary, Thibaudeau reconciled himself to Bonapartism in the Napoleonic era, serving in the Bouches-du-Rhône for eleven years, from 1803 to Napoleon's fall in 1814.[24] The post was for him a minor demotion, and he may have taken out some of his frustration on his constituency, using what one scholar has termed his "well-known . . . administrative rigor" to pass sweeping reforms in forest management.[25] In an effort to compensate for what he considered the revolutionary era's abuses, Thibaudeau reinstated the harshest antipastoral legislation of the ancien régime. In 1804 he promulgated a new law echoing eighteenth-century legislation but with increased punishments for infractions. The 1804 ordinance left shepherds few grazing options. It banned both sheep and goats from "national, communal, or private" forests, as well as from "*landes*, moors, wastelands, or the borders of woods and forests."[26] Violations could be punished through confiscation of the beasts and a fine of three francs per head, a substantial price for small-scale pastoralists who traditionally depended on such territories.

The ordinance thus effectively banned from their traditional haunts the two types of livestock most essential to the Provençal economy. It explicitly prohibited inhabitants of the Bouches-du-Rhône from keeping goats, "except in designated lands, and under specified conditions."[27] In 1811, perhaps recognizing the impossibility of enforcing the 1804 measure, Thibaudeau passed another law regarding goat grazing. This initiative, "Arrête relative au parcours des chèvres," confined goats to wastelands (*étendues de landes* and *terres gastes*).[28] The 1811 act not only prohibited goats from entering forests but also adopted an expansive definition of forest vegetation that included the kermes oak (*Quercus coccifera*), a bush commonly found throughout *garrigue* and *maquis* scrubland. By redefining such landscapes in his legislation, Thibaudeau subtly expanded the purview of forest administration.[29] Through this seemingly minor qualification, the significance of Thibaudeau's legislation reached far beyond its time. Thibaudeau's administration was hardly successful at keeping goats out of forests, but his new definition of the Mediterranean forest had a long legacy. As we will see, the manipulation of the meaning of "forest" became a popular and pragmatic tactic for nineteenth-century French foresters throughout the Mediterranean world.

The Birth of French Forest Science

The development of French forest administration in the early modern and modern eras fueled—and was fueled by—growing environmental awareness and concern. By furnishing a foundation for environmental management, early French forest legislation such as the Edict of 1669 ultimately helped to inspire an ethic of conservation. Additional developments in the eighteenth and early nineteenth centuries, including overseas expansion, the Enlightenment, and the French Revolution, encouraged French environmental perceptions along this path.

Prior to the colonial era, the environmental transformation of Europe occurred on a relatively slow scale, over the *longue durée*. From the fifteenth to the seventeenth century, however, European states began to explore and expand overseas, motivated in part by the quest for virgin forests and other natural resources that had been depleted in Europe. Tropical islands in the Atlantic were the first to succumb to European colonial expansion, since they lay closest at hand, but the discovery of the New World and Europe's bustling trade with Asia led to the addition of island colonies in the Indian Ocean and the South Pacific in the seventeenth and eighteenth centuries. Europe's growth, the quest for new resources that helped to fuel it, and the environmental plundering that it entailed all contributed to the development and spread of concern over environmental decline.

In *Green Imperialism*, Richard Grove maintains convincingly that early European imperialists viewed tropical islands as "Edens" because their abundant forests, fresh water, and tame wildlife contrasted so sharply with the long-exploited landscapes of Europe.[30] Indeed Fernand Braudel adopted this perspective as recently as the mid-twentieth century, remarking in *Capitalism and Material Life* that "even at the end of the eighteenth century, vast areas of the earth were still a garden of Eden for animal life. Man's intrusion upon these paradises was a tragic innovation."[31] Once island colonies were secured and made to serve the state, their landscapes changed rapidly and dramatically. The precepts of mercantilism, which all early modern European imperial powers pursued to some degree, encouraged thorough exploitation of colonial resources. Entrepreneurs worked industriously to extract as much as possible, as quickly as possible. The result was what Grove terms an "environmental crisis" in the seventeenth century, when the once abundant resources of these "Edens" began to disappear.[32] Springs went dry or turned sour, forests and indigenous wildlife vanished, and goats and rats proliferated. In the British colony of St. Helena, sheep and goats apparently had become such a nuisance by 1730 that settlers petitioned the governor to order the total extermination of the current population of these beasts and to ban them on the island for a period of ten years.[33] Thus, the "tropical island edens" that Europeans first encountered were quickly transformed into devastated wastelands.[34] In this way colonialism allowed a select but influential circle of Europeans to witness firsthand the environmental impacts of civilization, which were multiplying at a greatly accelerated pace.

The French state was particularly attuned to the environmental consequences of colonialism. During the early modern period the British generally chose to make up for domestic shortages by putting greater pressure on their colonies, while the French dealt with resource depletion through management reforms.[35] Although neither approach proved truly effective in curbing early modern environmental change, both fueled lively intellectual interest in the issue. English naturalists produced a wealth of texts expressing environmental concern, and members of the eighteenth-century French intellectual elite, drawing on such sources, began to call for action. In this way environmental questions first raised in tropical island colonies found their way into Enlightenment salons.

In France the Enlightenment was characterized by an interest in the natural world, which in turn fueled the development of natural sciences such as botany, horticulture, silviculture, biology, geology, agronomy, and atmospheric studies. By focusing on the environment, these fields motivated greater and more widespread concern for its preservation. From the perspective of mid-eighteenth-century French naturalists, the Edict of 1669 appeared sorely underequipped to check the wholesale destruction of forest reserves.[36] To assure a steady supply of timber, France seemed doomed to depend on foreign sources such as Italy and the Baltic states. Spurred by the intellectual precepts of the day, French *philosophes* grew increasingly troubled by these developments.[37]

One of the most outspoken critics of French environmental legislation was Georges-Louis Leclerc, comte de Buffon. His *Histoire naturelle, générale et particulière* echoed Linnaeus in classifying and describing all known organisms and in addressing the question of forest depletion. He pursued this theme further in tracts with titles such as *Mémoire sur la conservation et le rétablissement des forêts* (Memorandum on the conservation and reestablishment of forests), which promoted the preservation of woodlands through better management.[38] French forests found an even more dedicated advocate in Henri-Louis Duhamel du Monceau, who has been credited with inspiring both French and German scientific forestry.[39] He published his multivolume magnum opus, *Traité complet des bois et forêts*, over the course of twelve years, from 1755 to 1767. As its title suggests, this work provided a "comprehensive treatment" of forests. Duhamel divided it into five volumes covering different aspects of his subject, including growth, physical characteristics, plantation, exploitation, and conservation. Duhamel presented his text as a management guide for landowners, but he was also motivated by a personal interest in forest preservation and a concern for the consequences of deforestation. In his introduction to the final volume, *Du transport, de la conservation et de la force des bois* (On the transportation, conservation, and power of woods), he rhetorically asks his readers, "Is it not obvious that a country denuded of wood would be uninhabitable?"[40] By sounding the alarm on French environmental exploitation, such tracts helped to pave the way for the birth and spread of the conservationist ethic in the nineteenth century. Although their main targets were the French state and large-scale timber operations, they raised awareness toward all possible perpetrators of environmental crimes, including pastoralists.

In the late eighteenth century one pivotal event helped to solidify concerns over deforestation by shaking France to its core, exacerbating its timber shortage, and encouraging overexploitation: the French Revolution of 1789. As one study of French forest history has remarked, "On the eve of the Revolution, the Great Ordonnance of Colbert was, in practical terms, nothing more than a beautiful text."[41] Indeed the Chambre des Eaux et Forêts had long since fallen from favor, and its administration was all but obsolete. Although seigneurial courts continued to adjudicate regional forest violations, their main concern was the poaching of wood. Forest grazing, in violation of the Edict of 1669, remained widespread. At the same time, Enlightenment figures such as Buffon and Duhamel du Monceau were inspiring other environmental advocates, and by the late eighteenth century forest management had begun to reclaim a place in French politics after an absence of more than a hundred years. This time, however, the revolution quickly erupted into a monumental and pressing concern, temporarily pushing environmental issues off the stage. This upheaval dramatically changed people's perceptions of the natural world and fostered a new sense of anxiety and urgency regarding environmental degradation.

In *The Old Regime and the French Revolution*, Alexis de Tocqueville presents the French Revolution as a hurricane barreling ruthlessly through political, cultural, and religious institutions until, through its very devastation, "a change came over men's minds."[42] Most nineteenth-century scholars accepted this characterization, though some cited an additional victim of the revolutionary storm: nature. The narrative of the day presented this period as an orgy of environmental destruction. French administrators, professional foresters, and ordinary citizens alike wept over the wastes that their woodlands had become. The forest *conservateur* Louis Tassy called the revolutionary era a "period of complete license in forest use," and he estimated that about fifteen hundred square miles of French forests disappeared between 1791 and 1802. Others set the toll as high as nearly two thousand square miles.[43] Critics of the revolutionary era's environmental impact frequently focused their blame on four related targets: forest legislation, war, property, and goats.

Given the monumental achievements of the revolution's early years—the abolition of feudalism, slavery, and monarchy; the establishment of a republic; church reform; a new calendar; and the division of France into *départements*, to name a few—it is somewhat surprising that sweeping environmental legislation did not also pass. On the contrary, the situation grew decidedly worse. The revolutionary government wasted no time in dissolving the hated and ineffective Chambre des Eaux et Forêts, the administrative body created by the Edict of 1669, but neglected to replace it with a better system. The National Convention finally produced a new forest code in September 1791, but this highly anticipated document failed to impress. It proved just as powerless as the previous forest regime to check environmental overexploitation. The Law of 1791 embodied an era of fitful but ineffective attempts at environmental reform. The failure of forest legislation during this period was certainly not due to a lack of interest—the subject remained at the forefront of political debates from the 1789 National Assembly to the 1792 National Convention.[44] Rather, this effort miscarried largely because no single governing body was strong enough, or remained in power long enough, to enforce it. Even under Napoleon I, who brought France out of the chaos of revolution in 1799, stabilized the country, and held power for fifteen years, environmental administration was supplanted by the more pressing concerns of economic rehabilitation and war. Indeed this latter issue proved a formidable obstacle to forest preservation initiatives throughout the revolutionary era. On 20 April 1792 the National Assembly voted to declare war on Austria. From this date France remained in a continuous state of war on multiple fronts for more than two decades. These conflicts took an immense toll on French forests. Desperate to bolster its naval fleet and unable to trade for timber with hostile European states, the French state systematically mined its dwindling woodlands.

The revolution increased the strain on forests not only on the grand scale of wars and politics but also through the very local, real problem of poverty. Although economic distress had helped incite the country to revolt in 1789, the revolution utterly failed to remedy this problem. Due largely to its volatile and impotent administration and to its constant wars, the French Revolution had a devastating economic impact. As more and more peasants slid below the subsistence line, they sought provisions wherever they could. Those who had nothing else relied heavily on forests, where they found firewood, edible plants and berries, and even shelter. The poor also used forests as pasture for small livestock. People generally sold or slaughtered cows and even pigs as they became too expensive to maintain. Goats, however, would eat nearly anything, and they provided a cheap source of dairy. Thus, France's population of large stock plummeted during the revolution, while the number of goats increased dramatically. Goats were particularly prominent in Provence, where the census of 1796 (*an III*) recorded a ratio of 6,193 goats to one cow in the commune of Aix-en-Provence.[45]

Some viewed this development as nothing short of an environmental catastrophe. One alarmed inhabitant of the Midi exclaimed, "Everywhere I have heard complaints about the excessive multiplication of goats.... They climb everywhere, they destroy everything. . . . We must take measures to stop the continuation of an abuse that is truly devastating."[46] Others branded these pests as a "real evil."[47] Jean Antoine Joseph Fauchet, who served as prefect of the department of Var from 1800 to 1806, bewailed the destruction of "all the young trees" in the region by forty thousand seemingly crazed goats.[48] Likewise, the nineteenth-century historian Jules Michelet vilified goats for their environmental devastation, declaring in his popular *Tableau de la France*, "The goat above all, the beast of those who possess nothing, the adventurous beast who lives off the community, the leveling animal, was the instrument of this devastating invasion, the terror of the desert."[49] This passage and others like it helped to establish the role of small ruminant livestock in the narrative of revolutionary environmental destruction.[50]

At the same time, many observers recognized the value that goats held for the country's poorest residents. Sensitive to this need, the state waffled between turning a blind eye and passing legislation to limit goat herding and grazing.[51] A series of regional and national legislative measures attempted to deal with this problem. Throughout Provence goats were banned from all communal land, and their herd size was limited even on private land. Some of France's newly created departments imposed even harsher measures.[52] Yet the state ultimately lacked the means or the heart to enforce these terms. As with many of its other policies, the result was ineffective. Thus the proliferation of goats in Provence and elsewhere outlasted the revolutionary era and remained a concern for subsequent administrations.[53]

From the perspective of nineteenth-century critics, small livestock were part of a broader story of revolutionary environmental destruction and of deforestation in particular. Most agreed that the lack of effective forest regulation during the revolution had had a catastrophic and lasting impact on the country.[54] Looking back, one scholar lamented, "The license and devastation of woods grew each day more disastrous."[55] Government officials were equally distraught. Asked to evaluate the quality of his woodlands in the mid-nineteenth century, the prefect of the Bouches-du-Rhône replied somberly, "All of these forests have suffered, more or less, from the effects of the revolutionary upheaval. To our great regret, we can no longer cite as forest any but Sainte-Beaume in comparison to the others, which are only regarded as copses."[56] Even when they extolled the power of the French Empire, nineteenth-century voices mourned the revolutionary wars' impact on forests. One early nineteenth-century treatise even used this example to argue that war had been the principal cause of deforestation throughout history and around the world, including Asia Minor, Phoenicia, Persia, and Greece.[57] Prominent scholars such as Jules Michelet promoted the idea of the revolution as an environmental free-for-all. In an oft-cited passage Michelet describes how French citizens "began together the work of destroying our forests" during this period: "Trees were sacrificed for the most minor uses: one would cut down two pines to make a pair of clogs. At the same time, small livestock multiplied beyond counting and settled in the forest, harming trees, saplings, young shoots, and devouring hope."[58] Once the narrative of revolutionary-era destruction was accepted as fact, it became a gauge for environmental stress and a rallying cry in calls for reform. It was used to promote reforestation initiatives and stricter enforcement of forest legislation in Provence as well as French colonial Algeria. The French entrepreneur François Trottier drew on this image in his petition for a government-sponsored plantation of eucalyptus trees. His 1876 treatise on the role of the eucalyptus in Algeria claims that an amount of destruction equivalent to that described by Michelet occurred in Algeria "at the end of a single year."[59]

French protoconservationists were also influenced by key events to the north. Beginning in the mid-eighteenth century the rapid growth and spread of factories transformed England into the world's first industrial center. While France followed a more circuitous path to industrialization, it also experienced urbanization, mechanization, and the rise of factories during the nineteenth century.[60] These developments made a distinct impression on French thought. With its emphasis on machines and soot, the Industrial Revolution took people away from nature, which had previously been an intrinsic part of their lives. It generated new perspectives on and appreciation for the natural environment— the world outside smoke-stained cities.[61] Industrialization contributed significantly to the development of the conservation ideal. At the same time, it exacted major demands on natural resources and thus encouraged people to think practically about sustainability. These two ideas, conservation and sustainability, would prove critical to the conception of scientific forestry in France.

The nineteenth century heralded the spread of new perspectives on nature, and forests in particular, among the French intellectual elite. Trees began to be appreciated for more than just their commercial potential.[62] At the same time, developments in natural science promoted a deeper understanding of silviculture, while industrialization and urbanization drove up commercial demands on forests at a staggering rate. Although nineteenth-century forest legislation perpetuated the goal of sustainable exploitation that had been articulated in the Edict of 1669, it also reframed forest policy in self-consciously scientific terms, presenting itself as the product of precise, methodical scientific research. This rational approach to forest management gave forest engineers and administrative officials alike newfound confidence in their mission. It spurred them to enact legislation on a scale unforeseen in the history of forest management. In the nineteenth century French forestry strove to become not just scientific but also more systematic by expanding the reach of forest administration and implementing improvements in surveillance. In addition, sustainability, which had previously been superseded by commercial, military, and state interests, became a principal aim.[63] French foresters dreamed of creating vast environmental preserves throughout French lands. In the process they stepped up efforts against those who seemed to be impeding this initiative, with pastoralists heading the list.

The field of scientific forestry first developed in Prussia in the late eighteenth century, while France was mired in revolution.[64] In Prussia scientific forestry emerged as one branch of *Kameralwissenchaft*. Based on Enlightenment principles, these "cameral sciences"—named for the chamber (*Kammer*) reserved for royal council meetings—applied a scientific approach to economic, administrative, and social practices.[65] Indeed they might be considered the ancestors of today's applied sciences. By stressing the utility of science in administration, cameralism formed an important component of the governmentalized state. Its objectives were implemented and overseen by designated officials, in what Foucault has termed "cameralists' science of police."[66] By the last quarter of the eighteenth century the cameral sciences had become an integral part of the education system in Prussia and throughout German-speaking lands. In France cameralists inspired the genre of Enlightenment thinkers known as physiocrats, who argued for the centrality of natural knowledge to the agenda of political reform.[67] As in German-speaking lands, this explicit linkage of knowledge systems and government would prove critical in the development of state-sponsored scientific forestry.

As in France, German scientific forestry was motivated by growing fears of deforestation following environmentally devastating wars. The greatest impetus for systematic forest management was the Seven Years' War (1756–63), fought largely on German soil.[68] By the end of this conflict, the first forestry school had been established in the Harz Forest, in the center of what is now Germany, and *Allgemeines ökonomisches Forstmagazin*, the first journal devoted to forest science, had made its debut.[69] Johann Gottlieb Beckmann's *Beyträge zur Verbesserung der Forstwissenschaft* (Tract on the improvement of forest science) was published in 1763 as well, becoming the first book with "forest science" in its title, though it was not the first book on this subject.[70] Germans were well aware of their contribution to international forestry. A German encyclopedia of forestry, published in 1842, dared readers to "compare our literature and the number of our educated foresters to what there is abroad!" and declared boldly, "The beginnings of forestry science are entirely German."[71]

Nineteenth-century French forest scientists kept a close eye on developments among their "colleagues across the Rhine."[72] French foresters enthusiastically translated German texts into French and used them as the foundation of French forestry education. All students at the Royal Forest School at Nancy, moreover, were required to study German.[73] French forest engineers also adopted German methods of monoculture plantation, as well as the idea of sustained yield, which held that forests should be carefully managed to "always deliver the greatest possible constant volume of wood."[74]

At the same time, French scientific forestry followed a trajectory distinct from that of its German counterpart. The Forest Code of 1827, France's first self-consciously scientific piece of forest legislation, explicitly distinguished forest and hunting administrations, whereas German legislation continued to equate the two.[75] The French also lacked German foresters' dedication to order and mathematical computations. While France also instituted rigorous academic training for future forest agents, its program focused more on the natural sciences—such as the study of tree types and their environments—and their practical application. The 1870 forestry curriculum, for instance, included applied mathematics alongside elements of forest botany, forest economy, mountain reforestation, and forest law.[76] There were also certain technical differences in the two methods. For example, the French frequently converted old-growth forests into coppices (*taillis*) and coppices with standards (*taillis sous futaie*), while German foresters usually preferred to maintain high forest (*hochwald*) growths, though both methods were used in both cases.[77] Finally, in contrast to its northern European neighbors, France developed nineteenth-century forest legislation with Mediterranean landscapes and practices in mind. Thus, the French soon gained recognition as unrivaled experts on scientific forestry in the Mediterranean world.

The French Forest Code of 1827

In the early nineteenth century France emerged from the turmoil of the French Revolution and the Napoleonic era after having been battered politically, economically, emotionally, and environmentally. Capitalizing on this climate of weakness and uncertainty, the Bourbon ruler, King Charles X "the beloved" (r. 1824–30), passed a number of strict, sweeping administrative measures. On 31 July 1827 the French state promulgated a comprehensive new forest code. In some ways the Forest Code of 1827 was only the next in a long line of countless attempts at French forest regulation, but in another sense it was a watershed. Advocates considered its significance to be much broader than the mere preservation of forest resources. Jean-Baptiste Sylvère Gaye, a member of the Chamber of Deputies and a proponent of the new legislation, described forest conservation as "one of the primary concerns of societies, and, consequently, one of the primary tasks of governments. All of life's necessities are connected to this conservation; agriculture, architecture, nearly all industries seek in forests sustenance and resources that cannot be substituted."[78] A sweeping, ambitious document, the Forest Code of 1827 was the first French legislative document to adopt the principles of modern forest science. The code also sought to improve the application of the law through more methodical organization, enriched and intensified staff training, better surveillance, and local support. Finally, it placed communal, royal, state, and private forests under its jurisdiction, giving forest administrators control of nearly all French woodlands. This vast regime grew even greater through the code's novel interpretation of the term "forest." The Forest Code of 1827 gave the French central administration the power and the opportunity to make a momentous, measured impact on the environment.

The new code attempted to cover all aspects of forest use. Many of its clauses were recycled, sometimes word for word, from previous laws. The code specified similar regulations for clearing (*vidange*) and regular cuts (*coupes*) in the forests under its control and maintained forest agents' administrative role in both operations. Likewise, forest agents were still held personally responsible for violations, destruction, abuse, and grazing damage that occurred under their watch, but they also gained the right to accuse others of forest infractions without material proof. As before, miscreants were considered guilty until proven innocent, and many of the penalties for infractions were retained, including fines and confiscation of tools or livestock, depending on the offense.

The Forest Code of 1827 was in many ways an innovative and even revolutionary document. Instead of the poorly supervised, loose network of forest guards hired by private agents and the Crown, it envisioned a massive, complex, bureaucratic forest regime. In a gesture highlighting the host administration's commercial orientation, French forest administration was placed within the Ministry of Finance. The code also showcased the influence of forest science, stipulating that all forests be subject to regular maintenance (*aménagement*), a process described at length in subsequent documents. In addition, the code greatly expanded the purview of the forest regime. It henceforth governed royal, state, communal, and other public forests, as well as private ones, covering each type with a nearly identical set of restrictions.[79] The code placed nonwooded commons under the forest regime as well, including wastelands (*terres gastes, terres vagues et vaines*) and communal land classified as protected (*défens*), which the forest administration considered degraded and therefore aimed to improve.

Even as it enumerated usage rights (*droits d'usage*) for state and communal forests, the new code aimed to limit and ultimately eliminate these rights. Article 62 warned that no more concessions would be made in the future. The text also proposed the process of *cantonnement*, whereby forests were divested of customary rights at the discretion of forest agents.[80] Likewise, it sought to combat the traditional right of *affouage*, which guaranteed residents a portion of wood from their communal forests to be used as firewood. The new code strictly regulated the collection, sale, and distribution of firewood, and residents were no longer permitted to gather it themselves. The penalties for forest infractions, moreover, received greater attention and detail than in previous documents. A lengthy appendix posted a range of penalties depending on the type, age, size, location, and quantity of the trees or forest in question, as well as the nature of the offense.

The Forest Code of 1827 represented a major innovation in French forest administration in another significant way: it addressed the problem of enforcement. Indeed this aspect was probably its greatest long-term contribution. In contrast to previous forest legislation, the Forest Code of 1827 developed an extensive and systematic training program for foresters. No longer were they simply wounded or unwanted military men. Future forest *inspecteurs*, *sous-inspecteurs*, and *conservateurs*, the white-collar overseers of the forest regime, received training at France's royal forestry school at Nancy, which opened in 1824, and even lowly forest guards were expected to attend secondary vocational schools in forestry.[81]

Once they had completed their studies, graduates became part of a vast administrative network. They were sent to stations in every department of France, where the new forest regime established regional offices with a complex bureaucratic infrastructure, including a staff of forest guards of varying rank, an *inspecteur*, and a *sous-inspecteur*. A strict hierarchical organization required foresters to start with the lowliest tasks and gradually work their way up the ranks. To fight corruption, all forest guards and agents were subjected to scrutiny by higher-ranking members of the forest administration.[82] Provincial officeholders reported to the head of the forest administration, the national *conservateur* of forests, who was in turn overseen by the minister of finance. Within their jurisdiction, foresters strictly monitored exploitation, directed reforestation initiatives, and attended court cases involving violations of forest regulations. The number of personnel in the employ of the forest service increased significantly in the decade following the promulgation of the Forest Code of 1827, a trend that continued at least into the 1850s.[83] The forest administration thus not only cracked down on forest misuse on paper but also greatly increased surveillance and systematized punishment for offenders. For the first time, French forest legislation enjoyed a reasonable possibility of enforcement.

One of the most radical aspects of the new forest administration, at least for its Mediterranean constituents, was its treatment of pastoralism. Mediterranean pastoralism occupied a prime place among the forest practices targeted in the Forest Code of 1827. Indeed grazing sheep and goats in communal forests was *the* traditional customary right that those who crafted the forest code considered most detrimental to forest health. Through the new legislation, they sought to limit and eventually eliminate pastoralists from French forests. There was no ambiguity in the code's attitude toward this practice. Article 110 states, "In no case and under no pretext may the inhabitants of communities . . . bring into woodlands belonging to these communities . . . goats, ewes or sheep." This article applied to all communal and state forests, including those in which sheep pasturing had previously been permitted. Grazing was severely curtailed even in private forests, where the code restricted sheep and goats to sections that had been classified as *défensables*, meaning that they exhibited healthy growth and were not in need of protection.[84] Penalties for infractions included the seizure of any livestock found grazing illegally, as well as harsh fines or a stint in prison for those unable to pay.

The code did offer the possibility of granting exceptions to communities where the forest administration recognized access to communal pasture as "an absolute necessity," but it also reserved the right to remove or limit this and other privileges in any state, communal, or private forest through *cantonnement* or by converting pastureland into forest.[85] After a town had obtained a royal ordinance authorizing grazing, the next step was for forest agents to designate certain forest parcels as *défensables* and to determine how much livestock they would support. In typical fashion the Royal Ordinance of 7 June 1829, by which the town of Roquefort-la-Bédoule was granted the right to pasture sheep in its commons, contained various limitations. First, it restricted herds to cantons of communal woods "that will have been previously declared *défensables* by the agents of the administration of forests."[86] Second, it set a cap on the number of sheep to be admitted each year, to be determined annually by forest agents based on their evaluation of the forest. Finally, it reiterated the requirement that the commons be reserved exclusively to local inhabitants for their personal use. In the words of contemporary scholars, nineteenth-century forest agents did not sanction pastoralism; at best they "tolerated" it.[87]

In addition to restricting the locations where grazing was permitted, the forest code also placed strict limits on the type and number of beasts. Only sheep belonging to members of a given community were permitted to graze in its communal forests. Inhabitants were further restricted to grazing sheep exclusively for their personal use; communal land was not to be used for commercial pastoralism. Nor could one sell one's right to local communal land to someone living outside the community. The forest administration ordered additional grazing restrictions. It capped forest grazing at two sheep per hectare (about 2.5 acres), and it required that all livestock be branded and wear bells (*clochettes*).[88] Violations were punished with a fine levied per head of livestock and sometimes also with the confiscation of livestock and the imprisonment of the owner or shepherd. In cases involving protected lands or a repeat offense, the penalty was doubled. Moreover, anyone using the communal forest for pasture was required to pay an annual head tax, which supported the forest administration.

The new forest administration furthered its campaign against Mediterranean pastoralism by manipulating the very definition of the forest. In French the term "forest" (*forest, forêt*) has maintained the broad meaning of "a vast expanse of trees" since the medieval era.[89] Over the same period its legal designations have varied considerably. In the earliest French forest laws *forêt* referred specifically to royal game preserves, including deer parks and warrens (*garennes*).[90] This definition changed with the growing dominion of national forest administration, becoming both broader and more precise. As the author of an eighteenth-century French encyclopedia remarked, "The term 'forests' once signified waters as well as woods[;] currently it only means forests in the strict sense of woods, warrens, brush."[91] Accordingly, the Ordonnance des Eaux et Forêts of 1669 included woods and bushes as well as warrens ("*bois, buissons* & *garennes*") in its jurisdiction.[92] Throughout the history of forest legislation, moreover, the term "forest" implied a territory's productive potential, in other words, its exploitable trees. In the technical sense forests were distinguished from woods (*bois*) and copses (*taillis*) in French, as in English, by their extent as well as by the size and age of their trees. In practice, however, French accounts employed *forest*/*forêt* interchangeably with a variety of other terms, including *bois, taillis, futaie*, and *massif*, which in other cases had significantly different connotations. For instance, the eighteenth-century *Dictionnaire universel* states, "There is barely any difference between woods [*bois*] and a forest, unless a forest is a wood of larger extent."[93] The French definition of "forest" remains ambiguous even today. The French National Forest Office currently requires that only 10 percent of a region be covered in trees in order for it to be classified as forest, whereas the international standard is 20 percent.[94]

The authors of the Forest Code of 1827 saw the ambiguity of the French word *forêt* as an opportunity. Rather than adopting previous definitions—or delimitations—of forests, they undertook the formidable task of resurveying the French landscape.[95] Article 90 declared enigmatically that its jurisdiction would cover any "coppice [*taillis*] or high forest [*futaie*] . . . that will be recognized as susceptible to maintenance or regular exploitation by the administrative authority."[96] In the process, foresters and surveyors reinterpreted and generally broadened the qualifications for lands to be classified as "forest." This project was probably influenced by German forest science, which stressed the importance of clearly defining and identifying forestlands. Addressing this problem was indeed one of the first and foremost tasks of early German foresters.[97] Yet even before scientific foresters popularized monoculture plantations, German woodlands were as a rule much more homogenous than French ones. France's ecological diversity contributed to the vagueness and versatility of the word *forêt*. Its meaning varied considerably from the damp, temperate northern regions, to the high slopes of the Alps and the Pyrenees, to the sunny, dry coastal plains of the Midi. Although the Forest Code of 1827 established an interpretation of forests that included noneconomically viable tree species, the French forest administration's purpose remained to regulate and maintain as much territory as possible. By harnessing the definition of "forest" to state authority and power, French forest agents successfully recast this term as a tool of nineteenth-century governmentality.

By the mid-nineteenth century French forestry had begun to exercise a measurable impact on the Mediterranean landscape within and beyond the borders of France. French forest agents had successfully instituted regimes of scientific forest management in southern France, Algeria, the Ottoman Empire, and beyond. They traveled to such far-flung places as Greece, Switzerland, Denmark, Romania, Madagascar, Yemen, Arabia, Vietnam, Turkestan, and Indonesia in order to research and refine their forestry techniques.[98] They conducted extensive studies on forestry and forest legislation in diverse contexts, from Norway to Hungary to Japan.[99] French foresters expressed particular interest in the progress of forest administration in Mediterranean states that shared a mobile pastoral tradition, such as Spain and Italy, and in colonial contexts, such as India.[100] These initiatives generated considerable international prestige. Colonial powers and domestic governments alike regularly consulted French foresters in their implementation and development of forest management programs around the world. While French foresters increasingly gained international appointments, France's national forest school at Nancy became an international destination for forestry education. British forest agents destined for India completed their forestry education in France.[101] Gifford Pinchot, often considered the father of American forestry, studied at Nancy before returning to the United States to become the country's first professional forester.[102] The deeply rooted ties between French forest administration and Mediterranean mobile pastoralists helped to secure France's global reputation as the leader for forestry in Mediterranean and Mediterraneanoid environments.

Even as French forestry gained popularity among scientific, intellectual, and administrative circles abroad, however, it remained unpopular and problematic at home and throughout its Mediterranean spheres of influence. In this way the French forest regime exemplified both the strengths and the weaknesses of nineteenth-century French governmentality: its expansion of state power, sovereignty, and political knowledge on the one hand, as well as its crippling system of rationality, inflexibility, and cumbersome bureaucracy on the other. These problems intensified as the nineteenth century matured. In Provence, foresters continued to spar with local communities for the control and preservation of common lands. In French colonial Algeria, the forest regime, marred from the outset by competing interests and later commandeered by overzealous colonists, was quickly spiraling out of the central administration's control. In Anatolia, French foresters complained about a lack of support from the Ottoman state, while officials accused their French charges of lacking ambition and dedication to their task. Meanwhile, pastoralists around the Mediterranean continued to petition, negotiate, fight, and break the law to pursue their livelihood, reminding foresters and administrators that they did indeed count.

Part 2 | Growth and Transformation

CHAPTER 4

The Forest for the Trees

The Application of French Scientific Forestry around the Mediterranean

My friend, the most pleasing task in the mission of the forester is to create forests. For my part, I constantly devoted all of my efforts to populating the lands under my direction. Today, it gives me pleasure to think that I increased the wooded area of France by 4,000 hectares. Make it your goal to be able to say the same someday!
—Charles Vial, quoted in P. Carrière, "Prosper Demontzey"

In the decades following the promulgation of the French Forest Code of 1827, the influence of French scientific forestry spread across the Mediterranean. The implementation of the new code exposed both strengths and weaknesses in its formulation. Although the code proved much more successful than previous forest legislation, it was far from perfect, and it immediately generated hostile and lasting opposition. Local officials derided it for robbing them of revenue, while inhabitants bemoaned the loss of customary rights. Logging, mining, and agricultural enterprises protested the code's restrictions on their activities. At the same time, the code drew criticism from forestry experts, environmentalists, and other interest groups for not going far enough to safeguard French forests. Such concerns would drive the revision and transformation of French forest legislation for the rest of the century.

In 1838 the new code was applied wholesale to Algeria, playing a key role in French efforts to gain and consolidate control in France's nascent colony. Twenty years later French forest agents arrived in Istanbul, charged by the reformist Ottoman administration to establish a modern, scientific forest regime in the empire. Meanwhile, foresters stationed in Provence struggled to enforce the terms of the 1827 code. These three contexts shared common features, challenges, and players. In Provence, Algeria, and Anatolia local populations treated forest agents as outsiders and sparred with them over the meaning and value of forest preservation. For their part, French foresters in these three contexts quickly discovered the difference between policy and practice. Mobile pastoralism played a starring role in this conflict, as inhabitants contested foresters' efforts to limit or eliminate the industry. Such encounters ultimately fostered a process of negotiation in which the application of French scientific forestry around the Mediterranean inspired new perspectives on both forest administration and pastoralism.

The Forest Code of 1827 in Provence

The implementation of the new forest code was far from seamless even in the domestic context of Provence. Local administrators either failed to distinguish Mediterranean forests from other prevalent forms of vegetation, such as *maquis* and *garrigue* scrubland, or they attempted to exploit this ambiguity for their own ends and for the interests of their communities. In 1816, when the prefect of the Bouches-du-Rhône conducted a survey of the department's communal forests, many respondents claimed that their jurisdiction had few or no real woodlands left. "There are no communal woods or forests at all," declared the mayor of Allauch, "just wasteland used to graze herds of sheep."[1] Following the promulgation of the 1827 code, however, foresters and surveyors reclassified as forest many of the very zones that local officials had claimed just a few years earlier to be completely devoid of trees.

The nineteenth-century forest regime justified its expanded jurisdiction by reconceptualizing Provence's environment. In his *Statistique des Bouches-du-Rhône*, published shortly after the promulgation of the Forest Code of 1827, the Comte de Villeneuve had presented pine, downy and holm oak, and other light- and soft-wooded trees (*bois blanc*) as hallmarks of the Provençal forest.[2] The new code, taking its cue from the Bouches-du-Rhône's strict Napoleonic-era prefect, Antoine Claire Thibaudeau, added to this description the ubiquitous kermes oak.[3] The forest regime also gave this classification much broader significance, applying it to all state, royal, communal, public, and private forests, tethering it to additional regulations, and, through increased surveillance, making it possible to enforce. By including the kermes oak in the description of forest vegetation, forest agents effectively expanded the definition of the forest to include regions previously considered wasteland and used to graze sheep and goats. The new designation placed these zones under the same rules as forests, thereby either limiting their use by pastoralists or excluding them altogether. This reclassification was particularly detrimental for the owners of goats, who were generally banned from forests and thus depended on wastelands. In redefining the forest, the forest regime removed the goatherd's last recourse.[4]

The Forest Code of 1827 somewhat tempered its ban on forest grazing by allowing it "in certain locations, by a special royal ordinance." Many municipalities seized on this clause, clamoring for exemptions and stressing pastoralism's value to their society. As the mayor of one village explained, "the denial of this authorization would deprive the town of revenue . . . and the owners of herds would not know where to lead them to pasture."[5] Many Provençal communities did gain special permission to graze sheep on communal lands in the years following the promulgation of the code, but forest agents ensured that this practice would become much less profitable for communities and pastoralists.[6] Under the new forest code, profits from grazing violations and the head tax went to forest agents as partial compensation for their services.[7] The forest regime also added to pastoralists' financial burdens by stepping up fines, taxes, and enforcement. At the same time, foresters worked actively to extend forest cover throughout Provence, attempting afforestation even in the rockiest, most degraded soils, irrespective of the local economy.[8] Former pasture became future forest. Hit with the loss of pastoral revenues, many communities turned to logging and agriculture, which the forest regime generally preferred over pastoralism.[9]

The forest administration also limited access to pasture in Provence by redefining or rezoning land. This practice stemmed largely from the ambiguity of the forest code, which obscurely claimed for the forest administration all lands "susceptible to maintenance."[10] Many local inhabitants and officials protested that their "communal forests" did not fit this description, since they were not regularly maintained. In some cases the transformation was as simple and blatant as a new vocabulary. In 1830, for example, the municipal council of Roquefort-la-Bédoule abruptly began referring to grazing rights in local "communal woods" (*bois communal*) that previously had been called "wastelands" (*terres gastes*).[11] Moreover, the forest regime often categorized new acquisitions as degraded and placed them under special protection (*en défens*), which further restricted their use.[12] Forest agents conducted regular inspections of public forests in their jurisdiction. In forests they determined to be endangered, they restricted customary rights or removed them altogether.[13] The forest regime also protected forests from customary rights and exploitation for cultural, religious, and recreational purposes. In this way a number of communities temporarily or permanently lost permission to graze animals in their communal forests.

The code's grazing restrictions generated persistent local resistance in Provence. The abundance of court cases adjudicating grazing violations in the mid-nineteenth century suggests that forest agents were attempting to enforce the legislation and that inhabitants were continuing to disregard it. Those accustomed to violating such regulations with impunity were all the more outraged when they were caught. The most commonly reported offense was unauthorized grazing in forests and other common lands.[14] Even in communities that had been granted an exception to the code's restrictions on grazing, the process of authorizing and regulating pastoralism proved problematic.[15] The code's stipulation that commons be reserved exclusively for local use presented a special challenge for foresters, who frequently complained that pastoralists were exploiting their common lands for profit. For example, the inhabitants of Roquefort-la-Bédoule were accustomed to inviting pastoralists from nearby La Ciotat, a coastal town limited in communal pasture, onto their land, which provided Roquefort with a considerable part of its annual revenue.[16] Apparently such practices were so widespread that the forest regime threatened to discontinue granting concessions altogether. Well aware of what this would mean for his constituency, the Bouches-du-Rhône prefect sternly reminded the mayor of Roquefort that such exploitation was "completely contrary to the spirit of the exception."[17]

Local inhabitants fought and foiled the forest regime in other ways as well. Pastoralists often exceeded the maximum number of sheep authorized on communal land or grazed their herd beyond accepted boundaries, hoping that forest agents would not notice. Local officials often turned a blind eye to such activities and generally sided with their constituency in cases of forest disputes, commonly underreporting annual grazing figures and going above the head of the regional forest administration to petition the prefect.[18] This local political support for pastoralism should come as no surprise given that many of Provence's mayors and municipal council members possessed flocks of their own. Likewise, some of the prefects who presided over the Bouches-du-Rhône Department expressed more sympathy for local interests than they did for the forest code that they had sworn to uphold.[19] From the perspective of this largely pastoral society, reforestation was an unnecessary burden that would bring, in the words of one community, "great hardship to [its] inhabitants."[20] Others complained that forest agents were too strict. In this vein the mayor of Gémenos proclaimed in 1856, "The iron hand of the employees of the forest administration has weighed on us for too long."[21]

Communities unable to escape the forest regime sought ways around it. While local authorities denied mismanagement of their communal forest, they also understated its economic potential for logging in the hope of preserving access to pasture.[22] In addition, they regularly underreported forest exploitation. Forest agents also faced staunch resistance to reforestation initiatives. In a typical case the regional forest inspector of the Bouches-du-Rhône described an act of vandalism in the community of Orgon, where a new plantation of healthy saplings had been destroyed in protest by "parties unknown."[23] The forester complained, "This unfortunate incident may well be due to the opposition of the municipal authority, which fears a reduction in pastureland. It must be admitted that we will encounter there a most unwelcome resistance."[24]

Forest agents initially had few means of defense against the tide of local opposition. Despite improvements, forest surveillance remained inadequate. Although agents were better distributed following the implementation of the 1827 code, forestry remained a dangerous job. During the period from 1841 to 1890 at least eleven French forest guards died in service every year.[25] Moreover, the forest regime effectively reduced incentives for local inhabitants and communities to police their neighbors when it accepted the responsibility for and profits from forest protection.

At the same time, the forest administration was sensitive to its reception. In 1849 it launched a critical survey of all forests placed under the forest regime. To promote impartiality, the review was conducted by a mixed panel of foresters and nonspecialists.[26] Although the reviewers' findings had little legislative impact, the initiative was indicative of a changing attitude toward the application of the law. Over the following years the forest administration increasingly tempered its goals and granted exceptions to the rule. In a key example it allowed communities to remove certain communal lands from the oversight of the forest regime when their exploitation was particularly vital to the local economy.[27] In 1872 the administration approved this exception for about 7 percent of the communal land belonging to the town of Egalyiers.[28] This concession followed years of tenacious requests by the mayor, who claimed that it was necessary to "come to the aid of poor farmers."[29] Such *distraction* from the forest regime, as it was termed in French, caused problems of its own. In 1854 the forest administration freed 740 acres of communal land in the coastal village of Ceyreste, but when pastoralists tried to lead their livestock to this new pasture, they found it surrounded by protected areas, making it impossible to access legally.[30] For many forest advocates, distraction was the antithesis of sound forest management. Commenting on this practice, Louis Tassy asked rhetorically how local communities could be trusted to protect their own forests against personal interests and infractions when even the central administration, "with all the authority invested in it," still did not always succeed.[31] In any case, requests for removal were granted only in exceptional cases. More typical was the finance minister's callous response to an appeal by the sub-prefect of Arles to liberate commons in the village of Mouriès. The minister denied the request in no uncertain terms, warning that for any land freed from the purview of the forest regime, "complete destruction" would result.[32]

In addition to granting exceptions in public forests, the forest administration quickly backed away from its original claims over private forests. The Forest Code of 1827 had given forest agents many of the same powers in privately owned forests as in state and communal forests. This provision did not sit well with landowners previously authorized by the Code Napoléon to exercise free rein throughout their property. One clause proved particularly irksome: the prohibition of sheep and goats in private forests, stipulated in Article 78.[33] Many landowners drew considerable profit from renting pasture, especially in the wake of the shrinking commons. Like public communities, owners had the right to petition for concessions from the forest regime, which would then determine which parts, if any, of their property were suitable for grazing. In practice this was a process that neither proprietors nor foresters enjoyed, and it seems that both sides tacitly chose to ignore it. Finally, in 1866 the general director of forests published an act, followed by a memorandum in 1868, formally sanctioning this attitude of disregard. It effectively stated that Article 78 would remain valid but would not be enforced.[34]

Even more representative of the forest regime's evolution over the course of the nineteenth century was a law passed in 1859 that included several amendments to the forest code and authorized forest agents to ignore or abandon minor cases that they considered not worth pursuing.[35] It also enabled them to compromise with transgressors by lowering the fine in cases where the accused proved unable to pay.

Previously, insolvent offenders had been thrown into a debtors' prison. Under the new system, shepherds caught grazing their herds illegally but unable to pay the standard fine might be given the option to pay less or to convert part of the sum into an obligation of labor.[36] There were practical motives behind these measures. Far from adding to the revenue of the forest regime, the former system had cost it money. In formulating the 1859 law, the forest administration reasoned that some profit was better than none.

The 1827 law's status in the mid-nineteenth century exposes critical distinctions between the formulation of the law and its application in Provence. The implementation of the code involved compromise and accommodation among foresters, pastoralists, officials, and other stakeholders. In turn, this process of negotiation served to reshape French forestry.

French Forestry in Algeria

In the conclusion to the Forest Code of 1827, King Charles X commanded French citizens to observe the law "in our entire realm, lands and countries of our control [*obeissance*]."[37] This territory was about to grow significantly. Just as the French government was putting the final touches on its new forest code, the *dey* (governor) of the Ottoman regency of Algiers insulted a French envoy by slapping him with a flyswatter.[38] This seemingly minor incident led to a breakdown in diplomacy between the two formerly friendly empires. By the end of the year communications and trade had reached a standstill. The "fly whisk" episode, as it is popularly known, offered an international corollary to a host of domestic problems haunting Charles X's reign. Not least among these was his growing unpopularity, and the recent passage of strict forest legislation did not help. In this context his minister of war suggested that an Algerian campaign might provide "a useful distraction from political trouble at home."[39] Thus, on 14 June 1830 a contingent of thirty-seven thousand French troops reached the Algerian coast, and by 5 July they had forced the Ottoman governor to surrender the key port city of Algiers. Almost simultaneously, on

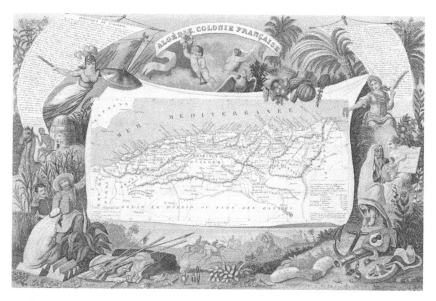

4. Illustrated map of French colonial Algeria, 1856. Archives Nationales d'Outre-Mer (ANOM), 16_8FI_536_V042N067: A. M. Perrot et Raimond Bonheur, "Algérie, colonie française," région du sud N 87, map, 1856.

26 July 1830 a popular revolution broke out in France, forcing King Charles X to abdicate the throne in favor of his distant cousin, Louis-Philippe. Hampered by political unrest in the metropole as well as staunch resistance from the indigenous population, the conquest of Algeria proceeded slowly but steadily. Beginning with cities, coastal regions, and the Tell, French troops gradually pacified Algeria and brought it under their control. Many remote and mountainous regions did not fall until after midcentury, and the French never gained full authority in the desert zones south of the Atlas.

Control of Algerian woodlands was a major priority for the early French colonial mission. Ongoing campaigns, which expanded French control slowly in the middle decades of the nineteenth century, required a steady supply of timber. For this reason Algerian forests were initially administered by the military and oriented to meet its needs.[40] Timber was also necessary to construct and solidify the French colonial presence, as well as to rebuild the edifices that French soldiers had razed in the heat of conquest. The French state, moreover, was eager to reap benefits from its new possession's natural resources, especially in the wake of alarm over deforestation at home. Meanwhile, preliminary surveys of the Algerian landscape helped to convince French officials of the urgency of systematic, "modern" forest management. The Algerian Forest Service was thus established in 1838, on the heels of conquest.

Concern for Algeria's environment intensified during the early years of colonial rule. Following the mixed results of the official "Exploration scientifique" from 1840 to 1842, France ordered subsequent commissions, statistics, and censuses of Algeria's forests. Such studies varied in their presentation of the environmental state of Algeria.[41] In 1848, for example, one French geographer called the massifs of the Kabylia region "an almost inexhaustible mine of wood."[42] Another praised the forests of the Tell as an "immense resource."[43] At the same time, however, some European observers were already characterizing Algeria as "devoid of forests."[44] Conflicting representations of the Algerian environment resulted partly from limited knowledge. Only a tiny fraction of the colony's woodlands had been surveyed by midcentury, and surveying was still under way in the final decades of the nineteenth century. But early colonial reports had one important feature in common. They consistently portrayed forest management on the eve of the French conquest as nonexistent, and they urged the French administration to take immediate action to save what forest cover Algeria had left. One reconnaissance report warned that, if left unchecked, traditional practices would "destroy precious resources for the future," while another prophesied that "this green, wooded country would soon be converted into a sterile and burning desert."[45]

Colonial agents' interest in Algeria's forests was more than just commercial; they explicitly equated forests, civilization, and empire. As early as the 1840s the French minister of war stressed the "necessity of conserving trees and brush" in France's new possession and of developing forest plantations vital to "the success of all colonization."[46] Likewise, in an 1867 report to the governor-general, the agronomist F. Robiou de la Trehonnais asserted, "Societies cannot exist without wood," pressing this point throughout his opening chapter, titled "Forestation and Colonization."[47] Convinced that Algeria's forests constituted one of its greatest resources and that these woodlands were desperately ill, the French state gave the forest regime a starring role in the colonial enterprise.

In 1838 the French administration applied the Forest Code of 1827 to the whole of Algeria.[48] The code's implementation offered a quick fix to a number of administrative problems. First, it sanctioned the immediate control and exploitation of forest resources by the state. This feature facilitated the process of conquest and consolidation by placing Algeria's environment in the service of the military, which made far greater demands on Algerian forests than any other patron in the early years of the French occupation.[49] The colonial forest administration also systematized and streamlined many of the complexities of settlement, including property distribution, customary rights, and demands for concessions. Finally, it provided a fast and firm response to the supposedly destructive habits of indigenous inhabitants, as well as a legitimate way to unseat them from prime real estate. Colonists began pouring in long before the dust of conquest had settled. With the implementation of the forest regime, French foresters also began to arrive, ready to take up the defense of trees. By contributing to state bureaucracy, centralization, surveillance, and repression, the colonial forest administration represented a key component of French governmentality.

As in Provence, the French Forest Code of 1827 faced significant stumbling blocks in France's new colony. French foresters abroad, in contrast to their colleagues in the metropole, encountered substantial local endorsement of their initiatives; throughout the colonial era most European settlers, together with a handful of indigenous agriculturalists and entrepreneurs, supported the French forest regime. Yet such parties perpetually sought to curb restrictions on their own exploitation of Algerian land, even as they pressed for stricter policies and penalties for others. As a result, the burden of forest restrictions in Algeria was absorbed almost entirely by its indigenous population. Again as in Provence, French imperial agents drew on environmental anxieties and the cause of reforestation, citing the same arguments that scientists, foresters, and officials were using to justify the marginalization of mobile pastoralism in Mediterranean France. Such policies posed new challenges for so-called Kabyles, or Berber populations whose primary industry was agriculture, but for nomadic tribes who traditionally made their living from sheep, the consequences were generally much worse.

The process of implementing the French forest code in Algeria was slow and fraught with challenges. One major hindrance was the code's incongruity with Algeria's environment. The 1827 code was designed primarily to address the concerns of metropolitan France, where reforestation efforts focused on preventing erosion, flooding, and other problems of excess water.[50] In Algeria ecological issues centered on the lack of water—problems such as desiccation and drought. For the same reason, the environmental issues attendant to mountain deforestation and afforestation did not translate to Algeria, where an equal if not greater concern was deforestation in valleys and on plains. In addition, Algeria's distinct flora, fauna, and soil composition hampered efforts at reforestation as well as cultivation. Yet French colonists, foresters, and officials often overlooked such differences, viewing the Algerian environment as degraded rather than different. They attempted to improve the landscape by experimenting with exotic plant species and by cultivating orchards and vineyards, with little success.[51] France's ignorance of Algerian forests also set the colonial forest regime apart from its metropolitan counterpart. As in continental France, the forest regime for Algeria governed all lands designated as forests. In the mid-nineteenth century, however, very little was known about the extent or nature of this territory. Surveying forests kept colonial agents busy.[52] Even in the final decades of the nineteenth century foresters continued to respond noncommittally to requests for statistics on forest cover in their jurisdiction.[53]

Other problems with the French code arose from ignorance of Algerian society. Many if not most Algerian woodlands were inhabited, but they remained subject to the same usage restrictions intended for uninhabited forests in metropolitan France. Residents were banned from traditional and essential practices such as building structures or lighting campfires in forestlands and from cutting or gathering wood and other forest products. Some clauses seemed a direct affront to indigenous groups. For example, the code permitted pigs to be pastured in forests, which was of little value to Algeria's predominantly Muslim and Jewish population, but it prohibited sheep and goats.[54] As in the metropole, the forest regime amplified the effect of its rules by defining woodlands broadly. French entrepreneurs complained about the inclusion of vast tracts of "brushland" in state forestlands, "even though one never sees any groves of trees there."[55] Even forest agents acknowledged the dubious nature of "forests" in their jurisdiction. For example, an 1863 survey described the countryside of Constantine as "not always wooded, but to be considered as forest."[56] This liberal designation of woodlands served as a particularly great source of frustration for mobile pastoral groups, who bore the blame for the majority of forest violations. In general the code's environmental reasoning resonated poorly with indigenous Algerians, who understandably tended to view it as a tool of imperial domination.

The composition and characteristics of forestry personnel also distinguished the Algerian forest service from its metropolitan neighbor. In continental France the forest regime lacked sufficient staff to implement its ambitious goals with complete success. The problem was greatly exacerbated in Algeria, where in the first few decades of conquest even basic administration was severely limited. How could foresters enforce unpopular French laws when even the military had failed? In an 1844 letter to the governor-general of Algeria, the French finance minister criticized the colonial government as "quite weak and powerless relative to the conservation of wood so useful in Africa," explaining that "devastations" by indigenous groups "remain unpunished, most often due to the manifest lack of personnel."[57]

To lure French forest school graduates to Algeria, recruiters offered significant incentives, which were effectively combat pay. Those who accepted posts in Algeria received an initial bonus as well as a higher salary than guards in the metropole.[58] In Algeria the colonial forest service also supplemented French guards with indigenous guards, who were required to accompany their French associates, particularly in remote or dangerous regions.[59] Indigenous guards acted as intermediaries between the local population and the forest service, which benefited from their knowledge of local geography, language, and customs. At the same time, their privileged position with the French colonial administration placed them under significant pressure. It theoretically offered advantages over the poor situation of other Algerians, but they were often seen as traitors and ostracized by their own communities, while also failing to gain acceptance by the French colonial population. In addition to their local escort, forest guards traveled armed with well-maintained pistols and rifles supplied by the French military. These weapons were in part a symbol of the guards' additional function as auxiliary forces, but they were also a necessity.

Despite significant incentives for employment, the Algerian forest administration chronically failed to attract a sufficient number of foresters.[60] The already sparse ranks of French citizens who joined the Algerian forest service were spread thinly across a vast range of tasks as well as territories, and their limited numbers constantly forced their superiors to play triage. In addition to their myriad duties as foresters, they were expected to guard the frontier and offer military service as needed.[61] Such factors continued to weaken colonial forest administration for the rest of the nineteenth century. In 1889 the Algerian forest conservator wrote to the governor-general to complain that forest guards assigned to supervise tribal activities had been requisitioned for other tasks "in one of their vast domains, and the Arabs know how to profit from it," adding, "the destruction grows and the forests suffer."[62] According to one calculation, the ratio of guards to forests theoretically placed each person in charge of nearly forty square miles of land.[63] The reality was often worse; because guards were required to work in pairs for security reasons, many areas were neglected out of necessity. French foresters, moreover, arrived in Algeria woefully underprepared and ill equipped to confront the tensions underlying relations between European and indigenous inhabitants. Anticipating the dangers of the job, many simply chose not to enforce the law or to enforce it selectively. In this way they formed a tacit compromise with local inhabitants that served to temper the impact of the colonial administration.

Even in areas with adequate surveillance, foresters found many clauses of the Forest Code of 1827 difficult to enforce. Some complained that the practice of imposing fines for forest infractions was ineffective in cases involving indigenous Algerians because they either could not or would not pay. Others decried the alternative punishment of forced labor, citing Arab laziness or recalcitrance. In a typical case the forest conservator complained, "The natives, who are by nature very disobedient, refuse to complete the assigned tasks."[64] Sequestering livestock, a third option in cases of pastoral violations, generated comparable criticism. Indeed it seems that this penalty was no more expedient in Algeria than in France. Animals frequently disappeared after being confiscated, causing forest agents great embarrassment and forcing them to drop charges against the owner.[65] More broadly, colonial foresters struggled to balance sanctions against indigenous pastoralists with encouragement for European ventures in animal husbandry.[66] While the distinction between these two activities made sense to colonial officials, they were not easily separated in ostensibly race-blind legal documents.

For these reasons the application of the French forest code in Algeria quickly became a perpetual source of debate among colonial agents. Military leaders, who witnessed firsthand the dangers of attempting to impose foreign legislation on hostile and poorly controlled indigenous groups, sought concessions and exceptions to the code. Even foresters protested. In 1845, while the conquest was still under way, the sub-inspector of Bône, in Constantine, challenged the central forest administration's reforestation policy, claiming that it was not an effective way to gain control of Algeria's indigenous population. He surprisingly recommended that the forest regime recognize and respect the local practice of pastoralism, in spite of environmental costs. "If a section is devastated by fire or grazing," he reasoned, "we might hope that the damage is not irreparable."[67] Other members of the colonial forest service went further, proposing extensive amendments to the French code or the promulgation of an entirely new piece of legislation unique to Algeria. Early drafts of an Algeria-specific forest code attempted to address the challenges of administering tribal confederations of nomadic pastoralists and their herds and to set the terms for regulating the use of forests for pasture, rather than shutting pastoralists out completely.[68] In some cases such initiatives won sympathy from local or regional administrators, but efforts to modify the French code were resisted all the way by the central forest administration as well as by a growing lobby of European settlers. Although formal discussion of an Algerian forest code began in the early 1840s, just five years after the implementation of the French Forest Code of 1827, Algeria would not receive an independent code until 1903.

The Development of Scientific Forestry in Anatolia

While French colonial foresters wrestled with the task of implementing scientific forestry in Algeria, their colleagues faced an even greater challenge in the eastern Mediterranean context of Ottoman Anatolia. In the fall of 1857 French foresters Louis François Victorin Tassy and Alexandre Stheme arrived in Istanbul in the employ of the Ottoman state.[69] During the years that followed, they directed an effort to survey Ottoman forests and develop a system of modern, scientific forest administration for the empire.[70] In many ways these agents had their work cut out for them. The extent of forest cover was not nearly as well documented in the Ottoman areas as in France, but the accounts of early modern witnesses together with Ottoman state-building initiatives suggest a general picture of plentiful forests gradually extinguished. As the Ottoman Empire expanded in the fourteenth through seventeenth centuries, it gained control of a wide range of natural resources, which contributed to its military prowess, power, wealth, and longevity. At its peak the Ottoman state was almost completely self-sufficient, and its timber resources rivaled or surpassed those of other Mediterranean empires.[71] In a perfect representation of the Ottoman Empire's cyclical system of resource management, the bulk of Egypt's timber supply came from Anatolia, which was also the primary recipient of Egypt's foodstuffs.[72] In the sixteenth and seventeenth centuries, when France was already beginning to fret over deforestation, the Ottomans still enjoyed ample forest resources in the Balkans, along the northern Black Sea coast, and along the Mediterranean coast in the west and south.[73] Anatolia's extensive forests were a source of national pride. The seventeenth-century Turkish traveler Evliya Çelebi described the region just south and east of Istanbul (Kocaeli Province) as a "sea of trees" (*ağaç denizi*), an epithet used through the late nineteenth century.[74] In the mid-nineteenth century the Ottoman Empire was still harvesting wood from these and other Anatolian forests for export to France and other international destinations.

Ottoman restrictions on forest exploitation in the early modern era were limited, especially compared to those of France. As in Europe, mining, heating, construction, timber extraction for military and commercial shipbuilding, clearing for agriculture and pasture, fires, wars, and urbanization all impacted Ottoman forests.[75] The administration regulated some activities, such as hunting and fishing, and closed certain forests to public use, but such moves were motivated by property concerns rather than environmental ones. Moreover, inhabitants had for centuries enjoyed liberal access to forests not under private ownership (*cibâl-i mübâha*).[76] Prior to the mid-nineteenth century the Imperial Naval Arsenal oversaw forest management. As a rule, it saw fit to protect only those forestlands designated as reserves for naval construction, hunting, or other use by the sultan.[77] Many of Anatolia's woodlands lacked commercial potential because there were no nearby, convenient waterways to facilitate the transportation of timber.[78] Thus, forests in the Taurus Mountains of southwestern Anatolia remained relatively well preserved into the twentieth century. Others were transformed through clear-cutting, slash-and-burn agriculture, or pastoralism.[79] Those in the empire's European provinces and along Anatolia's northern and western coasts were particularly susceptible, but burning for agriculture occurred throughout the empire with the tacit acceptance of the central administration. In the eyes of the early modern Ottoman elite, forestlands were more profitable once they were cleared.

Scientific forestry arrived in the Ottoman Empire through the channel of the Tanzimat, a period characterized by Western-leaning and centralizing reforms in Ottoman administration. It formally began in November 1839, when Sultan Abdülmecid I issued the Hatt-ı Şerif of Gülhane (Rose Chamber Edict), a decree outlining basic rights for Ottoman citizens and promises of reform. The Rose Chamber Edict was aimed as much at appeasing and impressing the Great Powers of Europe as it was at the Ottoman population. The sultan's foreign minister, Mustafa Reşid Pasha, who helped institute the reforms, read this document before an assemblage of Ottoman elites as well as foreign diplomats and dignitaries.[80] Although its value was largely symbolic and it was not legally binding, this decree had significant political consequences. The Tanzimat era that followed witnessed a proliferation of legislative measures designed to increase efficiency and productivity in the transportation of imperial resources and the delivery of profits to the imperial treasury. It was this context that gave rise to the concept of the Sublime Porte as a central administrative bureaucracy directed by the sultan's top ministers.[81] Over the thirty years that followed, these figures drove reform by developing transportation and communication systems, encouraging agricultural expansion, restructuring the commercial exploitation of natural resources to guarantee a profit for the state, and investing in modern resource management, including scientific forestry.

One year after the Gülhane proclamation, Sultan Abdülmecid I established the empire's first forest directorate, under the Ministry of Trade.[82] This first institution proved ineffective, and he annulled it less than a year later. Abdülmecid made a second attempt to reform forest management in 1857, just after the conclusion of the Crimean War, a conflict that had exposed severe inefficiencies in Ottoman administration. He established of the Council of Public Works, which oversaw the protection and regulation of the empire's natural resources, including forests. In the same year, the state began appointing foreign experts to help develop and modernize various aspects of Ottoman economy and industry.[83] Forestry was one of the targeted areas. From the council's inception until 1878, nearly all of its forest specialists were French.[84]

The selection of French experts was in many ways a natural choice. Throughout the modern period France had worked with the Ottoman state in diplomatic, economic, and military alliances. From the mid-sixteenth through the eighteenth century merchants from Marseille played a prominent role in both intra-Ottoman trade and Ottoman trade with France.[85] In the seventeenth century the phenomenon of *turquerie* exploded onto the French cultural scene. Turkish influence became fashionable in clothing, accessories, food, literature, art, interior design, and music. Even King Louis XIV caught the bug. Following an Ottoman diplomatic mission to France in 1669, he commissioned Molière to include a Turkish scene in his latest play, *Le bourgeois gentilhomme* (1670).[86] During the same period, French society and culture exercised an increasing influence on the Ottoman elite. By the late eighteenth century French had become the language of diplomacy as well as that of culture and learning. The Ottoman language began to borrow extensively from it, incorporating terms such as "civil" (*sivil*), "police" (*polis*), and "économie" (*ekonomi*), even when autochthonous synonyms existed.[87] When Selim III undertook military reforms in the late nineteenth century, he based them on the French model, importing instructors as well as an entire library from France.[88]

These friendly relations endured into the nineteenth century, though they began to falter following French military campaigns in the Middle East, such as the invasions of Egypt and Algeria, as well as French support for the Greek War of Independence.[89] In 1850, moreover, President Louis-Napoléon claimed French custody over Christian holy sites in the Near East, a move that the Ottoman state was loath to support.[90] As the Ottoman administration proved less and less capable of competing militarily and financially with European states, it grew increasingly apprehensive about European ambitions in its territories.

The Crimean War of 1854–56, in which France and Britain helped to defend the Ottoman Empire from Russian aggression, heralded a new age of friendship between the Sublime Porte and France. Following the conclusion of this conflict, the French and Ottoman administrations began negotiating the resumption of diplomatic relations. In a letter dated August 1857 the Ottoman minister of foreign affairs assured the French ambassador "of the desire of the Sublime Porte to consolidate more and more the ties of the cordial friendship that so happily unites the two empires."[91] The engagement of French forest engineers was an integral part of this effort. The selection of French agents was also due to the growing international reputation of French forestry. Since the promulgation of the Forest Code of 1827, France had gained international renown as a pioneer of modern forest science. French foresters, moreover, represented a Mediterranean connection, in contrast to their German rivals. In the middle decades of the nineteenth century these characteristics contributed to increasing interest in and respect for French methods within Ottoman intellectual circles.[92] Thus, while the Ottoman administration chose delegates from various European states to head other newly created offices in its economic and industrial bureaucracy, it wanted imperial forest specialists to be French.

The French foresters assigned to the Ottoman Empire were under no illusions regarding the difficulty of their undertaking. Even before the establishment of their mission in Istanbul, the French journal *Annales Forestières*, the predecessor to the *Revue des Eaux et Forêts*, had described Turkish forests as an elusive but worthy prize. "Considerable difficulties and obstacles must be overcome in order to take advantage of [Turkish] forests," warned one article on the subject. Nevertheless, its author urged that "improvements can and must be attempted," and he counseled his compatriots to proceed cautiously, noting that "not without excessive prudence will we succeed in planting in Turkish forests the seeds of European forest science."[93] Those who traveled to Istanbul a few years later seemed to take these words to heart.

The French forest mission brought technological expertise as well as social and environmental perspectives, including narratives of Mediterranean environmental decline. Much like their counterparts in Algeria, French forest agents in Anatolia described the landscape as both full of potential and severely degraded. On the one hand, they viewed the region as a haven of "great, beautiful and abundant woodlands of all varieties."[94] On the other, they regarded the history of Ottoman forest management in unequivocally negative terms. "It was the regime of absolute freedom," grumbled a member of the French mission, explaining, "that is to say, the absence of any control, due to a frightening disorganization, unchecked abuse with the prospect of complete ruin in the near future."[95] Likewise, the editors of *Annales Forestières* charged that Turkish forests had been "all but abandoned to the forces of nature and the discretion of their inhabitants."[96] A late nineteenth-century report in the *Revue des Eaux et Forêts* agreed, recalling, "From the point of view of forestry, the disorder was complete. [The administration] had only the vaguest notions about the location and consistency of wooded areas."[97] For these French observers the issue at stake was not so much environmental degradation—they admitted that the Ottoman Empire maintained robust forest resources—as it was poor administration. From the highly bureaucratized, governmentalized perspective of France's Second Empire, the Ottoman state apparatus appeared anachronistic, despotic, chaotic, and weak. Therefore, French efforts to institute scientific forestry in the Ottoman Empire formed part of a broader initiative to develop new conduits of power and knowledge for the Ottoman state and for France.

The French forest mission to the Ottoman Empire included some of the same figures who influenced forestry in other parts of the Mediterranean world. Louis Tassy, one of the two initial forest engineers assigned to the empire, was a native of Aix-en-Provence. Although his work in the Ottoman Empire introduced him to a different language, society, and culture, the environment of southwestern Anatolia would have seemed comfortingly familiar. After his initial stint as head of the Ottoman forest administration from 1857 to 1862, he was assigned to Corsica, another semiarid Mediterranean land, where, according to one of his biographers, he "rediscovered the beautiful Mediterranean climate."[98] He later returned to Istanbul in the capacity of vice president of the Council of Public Works.[99] Tassy's legacy in Ottoman forestry was significant. He initiated the first systematic survey of Ottoman forests, and his 1858 book, *Études sur l'aménagement des forêts*, became the standard textbook for budding Ottoman forest agents. In 1865 the

Revue des Eaux et Forêts boasted about his reappointment, remarking that the Ottoman government had "never ceased to regret the loss of his excellent service."[100] By the time he settled back in France in 1868, Tassy had gained prominence within the French forest administration. His reputation as an expert on Mediterranean forestry won him the task of conducting an extensive investigation of Algerian forestry in 1872. His report drew on his professional experience in Anatolia and Provence. Thus, Tassy represented a direct link among these three Mediterranean contexts and one way in which they influenced the core of French forest administration.

Tassy's successor as director of the Ottoman forest commission also had a Mediterranean connection. A fellow graduate of the French forestry school at Nancy, Louis Adolphe Bricogne originally hailed from Montpellier.[101] During his tenure he conducted an extensive survey of Anatolian forests, reporting his findings to both a Turkish audience and to foresters back in France. In keeping with dominant narratives about the Mediterranean environment, Bricogne provided a grim assessment of Anatolian forests. He described most of the province of Konya as "completely deforested."[102] To the south, the northern slopes of the Taurus Mountains that divide Konya from the province of Antalya were in his estimation "ruined," as was the hinterland of Antalya Province. Bricogne characterized the landscape of Alanya, a coastal town east of Antalya, as "the poorest, rudest of all the coast of Karamania [southwestern Anatolia], its forests reduced to nearly nothing."[103] In explaining Anatolia's supposed environmental decline, Bricogne and his colleagues drew on contemporary perspectives about the Mediterranean region's natural environment, society, and progress. While he acknowledged that certain forests had long been exploited for shipbuilding and commercial logging, Bricogne blamed nomadic pastoralists for much of the damage. His reports make frequent reference to Anatolian nomads' "ravages" and their "detrimental effect on forests."[104] Among their destructive practices, Bricogne listed the mortal bite of their sheep and goats, their custom of haphazardly cutting down old-growth trees "to make a minimal profit," and "burning the underbrush of forests" to create and maintain pastures. As a result, whole stands were "burned, overgrazed, trampled by livestock [and] can no longer regenerate naturally."[105]

The members of the French forest mission offered various recommendations for the problems they identified within Ottoman forest administration. An 1865 report published by the commission targeted both weak legislation as well as limited surveillance and enforcement. Offering a remedy, it called for the immediate adoption of extensive reforms, greater restrictions on forest use, and a significant increase in staff.[106] This proposal included a detailed plan to increase revenue for the state.[107] In his own reports Bricogne emphasized the need for stronger legislation and enforcement, but he also recognized that regions populated by nomadic tribes would be difficult to regulate because nomads had long enjoyed free access to forests through customary rights.[108] In a telling allusion to the similar problems facing Provence, his account refers to these rights as *droits d'usage*, the terminology used in France. In the "deforested and ruined" environs of Alanya he judged the forest service to be "more powerless than anywhere else," citing the "violence," "savage customs," and "well-known" intractability of its tribal population.[109]

In 1870 Bricogne reported on the success of the Ottoman forestry mission in optimistic terms. "I am pleased to observe," he wrote, "that the work begun twelve years ago, by our excellent director Monsieur Tassy, is on the path of prosperity."[110] He and his colleagues had indeed made significant steps toward implementing French scientific forestry in the Ottoman Empire. The first Ottoman forest institute opened its doors in 1857 to a handful of students plucked from the ranks of Ottoman administration.[111] Three years later the school released its first graduates, and the Ottoman press crowed that the empire now possessed nine capable agents who, "when a contingent of guards is placed under their command, will be able to manage 100,000 hectares of forests to good effect."[112] Bricogne reported that the school was continuing to prosper in 1870, and by 1878 it had granted degrees in scientific forest management to fifty-eight Turkish graduates.[113] Meanwhile, Louis

Tassy's 1858 textbook, *Études sur l'aménagement des forêts*, was translated into Ottoman Turkish in 1861. This book introduced its Turkish audience to precepts inherited from German forestry. It promoted the idea of sustained yield; stressed statistics, surveys, and classification; and introduced a range of management techniques, from monoculture plantations to coppicing (*baltalık*) and clear-cutting. It also presented French environmental perspectives, including Mediterranean environmental declensionism and the relation of forests and civilization.[114] The French mission also helped to craft the Ottoman Empire's first independent forest bill, passed in 1861. In many ways the content of this bill reflected its French influence. It reminded Ottoman subjects that trees could legally be cut or removed from state forests only with a proper license.[115] It prohibited grazing in certain forest areas, and it outlawed lighting fires.[116] For these and other infractions, it established penalties comparable to those outlined in the French Forest Code of 1827, including fines and incarceration. Like other legislation of the Tanzimat era, the 1861 bill was designed with the profit of the state in mind, but it also set the terms for a hierarchical organization of forestry personnel similar to the forest regime in France.[117] In addition, the bill paved the way for the passage of more comprehensive Ottoman forest legislation.[118]

Largely through the efforts of the French commission, the survey of Ottoman forests was well under way by the late 1860s. Surveys completed in 1866–77 suggested that the empire still contained extensive and varied forest reserves.[119] The forest administration was most interested in studying the timber resources of the empire's European provinces because of the relative marketability of that timber, but it conducted surveys of forests throughout Anatolia as well. In the late 1860s Anatolia's total forest cover was estimated at nearly twelve thousand square miles, or twice that of Ottoman Europe.[120] Anatolia's population density, which was on average much lower than that of the empire's European provinces, further offset this figure. Such encouraging statistics, however, did not make French foresters less anxious about the fate of Ottoman forests. On the contrary, French agents grew even more convinced of the importance of protecting this valuable resource for the future of the Ottoman state and, as France remained a buyer of Ottoman timber, their own national future as well.[121]

Beyond their limited achievements in Ottoman forestry, however, the members of the French mission encountered a host of frustrations. Insufficient funding prevented the Ottoman state from securing positions for most of the Turkish graduates of the new forest school, even as the French forestry directors begged for more staff.[122] As a result, they struggled to recruit students for the institute, and those they found were often less than ideal. According to Bricogne, few of the students entering the school spoke French, and the majority possessed no more than a basic understanding of arithmetic and lacked any type of scientific background.[123] These ragtag agents did little to improve the state of Ottoman forest surveillance. In addition, the 1861 bill was never formally applied. It ultimately represented no more than an elaborate wish list, a potential guideline for future legislation. For French forest delegates, however, the greatest sore point may have been their relegation to commercial matters. The Ottoman state's interest in forestry was aimed primarily—if not exclusively—at maximizing forest profitability. For the French foresters in its employ, this meant that their lofty principles of environmental conservation and reforestation remained off the table, at least for the time being.

In many ways the development of forest administration in nineteenth-century Ottoman Anatolia echoed the transformation of the French forest regime and the challenges it faced in Provence and Algeria. Like the French, the Ottomans attempted to deal with forest misuse, violations, and overexploitation through bureaucratization and legislation. In Anatolia as in Provence and Algeria, mobile pastoralists became prime targets of modern forest legislation. The French foresters who cultivated Ottoman scientific forestry helped to marginalize the practice of mobile pastoralism in Anatolia by importing homegrown social and environmental narratives and by incorporating antipastoral bias into their reports and recommendations. Because its principles were so poorly enforced, Ottoman forest administration ultimately had little direct impact on the sedentarization of nomadic tribes. As subsequent chapters show, however, a combination of other factors would perform that task swiftly and effectively. Yet, beyond its threat of punishment for infractions, Ottoman forest administration served to legitimize the sedentarization process, particularly in the eyes of powerful Western critics. Indeed shortly after the arrival of the French forest mission in 1857, the administration began to frame accusations against nomads in new, environmental terms. In 1858 it ordered the resettlement of tribes that had damaged public forests to "more suitable areas."[124] Thus, even if it failed to intimidate Anatolian nomads, Ottoman environmental legislation succeeded in winning the Sublime Porte the support necessary for what would become a ruthless sedentarization campaign.

The cases of Provence, Algeria, and Anatolia illustrate the extensive reach of French scientific forestry in the mid-nineteenth century, a reach that affected politics, populations, economies, and environments across the Mediterranean world. They also show how this institution evolved through its various Mediterranean applications and how these contexts were connected. In all three cases French foresters managed people as well as landscapes. Guided by preconceptions about Mediterranean environments and pastoralism, they aimed to exclude shepherds and their livestock from woodland areas, to reserve the forest for the trees. In the process, however, foresters and other officials lost sight of both their human charges and the environment. The application of French scientific forestry around the Mediterranean Sea contributed to the marginalization of mobile pastoralism, but Mediterranean pastoralists also influenced the development of French scientific forestry through passive and active resistance. At the same time, the remaining practitioners of this tradition were becoming increasingly improbable scapegoats for Mediterranean deforestation, which French scientific forestry had so far failed to check. As the next two chapters show, the application and evolution of French forestry in the mid-nineteenth-century Mediterranean world occurred in a broader context of social, political, and environmental change that placed additional pressure on Mediterranean pastoralists and drove them into retreat.

CHAPTER 5

Against the Grain

The Transformation of Land Use and Property

> Nothing is more dangerous in a new country than the frequent use of
> forced expropriation.
>
> —Alexis de Tocqueville, *Writings on Empire and Slavery*

French forestry did not mature in a vacuum. In nineteenth-century
Provence, Algeria, and Anatolia land use and property administration
evolved alongside the application of scientific forestry and formed a
complementary element of governmentality. The mobility of Mediterranean pastoralists made them acutely susceptible to this trend. In
Provence private property eclipsed the commons on which small-scale
pastoralists depended, ultimately excluding them and transforming the
industry. In Algeria new property legislation, together with changing
interpretations of landed property and ownership, served to dispossess
indigenous pastoralists in favor of European colonists. Like northern Algeria, southwestern Anatolia had to accommodate an influx of
settlers in the mid- to late nineteenth century, in this case refugees
from former Ottoman provinces lost in war. This population pressure
placed additional constraints on nomadic groups, especially as their
new neighbors began converting pasture to agricultural fields. At the
same time, property legislation formally excluded nomadic pastoralists from lands on which they had long depended. In all three cases
nineteenth-century developments in land laws and land use served

to marginalize Mediterranean mobile pastoralists, hitting hardest the small-scale producers whose voices carried little political weight. It is no coincidence that the nineteenth century witnessed the waning of both "no-man's-land" and nomads' land throughout the Mediterranean world.

Revolution, Property, and Pastoralism in France

French sheep traditionally exploited a variety of landed property types. Many pastures were owned by private landowners, the state, or, prior to the French Revolution, the Catholic Church and rented to outsiders and commercial pastoralists for profit. In many communities of Haute-Provence the rental of summer pastures to transhumant shepherds from the Bouches-du-Rhône and the region surrounding Arles constituted a principal form of income.[1] Even the massive flocks of large-scale commercial agriculturalists, which typically enjoyed access to their owners' pastures for part of the year, had to be provided with rented pastures during their seasonal commute. Many French pastoralists benefited from the right to *vaine pâture*, which allowed them to graze their herds, most often during the fall or winter, on local fields that farmers left fallow. This practice provided agriculturalists with additional income as well as a good source of fertilizer.[2]

Prior to the nineteenth century, a small but significant portion of the territory dedicated to Provence's pastoral industry was communal land.[3] Residents' use of lands held in common by their community was regulated by customary rights (*droits d'usage*).[4] In Provence communal lands typically included designated wastelands (*terres gastes*), unsown agricultural fields or territory deemed unsuitable for agriculture (*terres vaines* and *terres vagues*), communal forest—variously defined—and other vacant lands. While national forest legislation regulated and limited the use of communal forests long before the turn of the nineteenth century, it also granted local inhabitants access to these spaces as well as basic rights of exploitation. Customary rights generally allowed inhabitants to collect deadwood, herbs, and other forest products and to pasture their livestock in common lands. They specified the number and type of animals that could be admitted into the communal forest, where they were allowed, and regulations governing their activities.[5] Customary rights on communal lands were reserved solely for members of a community for their personal use. Even those living in the surrounding countryside were excluded, and residents were forbidden from selling or transferring their usage rights to anyone else. In Arles pastoralism constituted such an important part of the city's local economy that special laws dating at least to the early medieval period regulated access to communal pastures. Throughout the early modern era a traditional customary right called the *droit d'esplèche,* granted exclusively to the local population, allowed them to graze their flocks free of charge on the nearby Crau. This privilege disappeared in the modern era, however, as grazing access and the pastoral industry in general became more oriented to profit.[6] The extent of communal pastures also declined steadily in the eighteenth and nineteenth centuries.

Communities gained revenue from local pastoralism in three principal ways. The most significant and consistent profit came from the head tax. Communities collected this tax annually from local pastoralists who grazed their livestock on communal land. The tax was determined per head, based on the type and number of animals. It provided a major source of revenue for the numerous villages in Provence with a large population of sheep and goats. The rental of communal pastures provided another source of profit for the communities of Provence. Seasonal access to pastures was regularly rented to owners of transhumant herds both in the Alps and in Basse-Provence, where Arles and a handful of other communities with abundant pastures offered prime destinations for wintering sheep.[7] In addition to the steady revenue that the head tax and the rental of pastures provided, grazing violations on communal land provided a supplemental source of profit. Municipalities fined owners of livestock found grazing illegally on communal land, and the fines varied according to the number, location, and type of beast, as well as the nature of the violation. These animals, moreover, could be confiscated by the community and held until the fine was paid. Under certain conditions, or if the owner defaulted on the payment, the wayward livestock became the property of the municipality and could be sold to generate additional profit. The prevalence of court cases from the eighteenth and early nineteenth centuries regarding communal grazing violations shows that this system encouraged local vigilance and was used frequently.[8]

Both customary rights and the extent of common lands expanded significantly during the French Revolution. Prior to 1789 the Catholic Church had been the largest landowner in France, possessing 10 percent of all French property. In the wake of revolution feudal rights were abolished, a new national church was established, and nobles and clergy fled France. The National Convention redistributed church and feudal holdings to local communities and the state. While much of that property was auctioned to private owners, other parcels became publicly owned. The law of *partage* (distribution), passed 10–11 June 1793, declared, "All communal lands, known throughout the Republic under various names . . . belong by their nature to the body of inhabitants or members of communities on the territory of which these lands are located."[9] The revolutionary era thus significantly expanded French citizens' access to and exploitation of communal lands.[10] Landed property shifted again following the Restoration of 1814–15, as many former owners returned to France and managed to reclaim their land. Such changes also introduced new patterns and systems of regulation for grazing, thus transforming the pastoral industry. Ultimately they gave the French forest regime much greater control over the practice of pastoralism.

The revolutionary government also reassessed customary rights. In 1791 the passage of a new rural law code reaffirmed the age-old tradition of *vaine pâture* (fallow pasture), authorizing inhabitants to pasture their livestock on the unenclosed fallow fields of their neighbors, provided that they admit a comparable number of beasts to their own unseeded property.[11] These stipulations technically limited this privilege to landowners and entitled those who possessed the most land to make the greatest use of others' vacant lands. Landowners had the choice of opting out of this system by fencing off their property, but doing so divested them of the opportunity to exploit the fallow fields of others. The 1791 law ultimately expedited the decline of this practice. It remained in force until the late nineteenth century, when *vaine pâture* was formally abolished.[12]

At the same time, the system of communal property began to receive greater scrutiny. In the mid-eighteenth century the Comte de Buffon had argued that forests would be better safeguarded if owned privately rather than communally.[13] Others took up his call, and in the late eighteenth century a few Provençal communities began to experiment tentatively with privatization by divvying up communal lands among inhabitants.[14] This perspective gained traction following the French Revolution, when an abundance of literature appeared claiming that the abolition of feudalism had triggered an irresponsible pillaging of the land. Ironically, the end of feudalism also fueled the trend toward privatization by opening up space for private property. In broader terms, nineteenth-century critics of commons maintained that private ownership would encourage better environmental stewardship. According to one observer, "Every undivided property is doomed to indifference" because "the collective being, in its relations with the land, envisages no improvement, responsibility, or future." For these reasons, he concluded, "communal pasture imposes an inevitable sterility on the soil."[15] By the early nineteenth century, moreover, France's rival—England—had methodically eliminated its commons through parliamentary acts of enclosure, and other Western nations had begun to follow suit. Treatises in the burgeoning field of economics offered scientific evidence for the benefits of privatization, both for individuals and the state. Consequently, French intellectuals and officials came to associate privatization with progress, prosperity, and modernity.

Private property, however, did not necessarily fare better than communal land during and following the revolution. As one official warned in 1798, "it is an evil not to have subjected to the police the forests that were sold by the Republic. The purchasers have razed them entirely, and timber, which is becoming rare, will one day be lacking."[16] One hundred years later many still blamed revolutionary-era environmental destruction on private as well as communal overexploitation. Describing private property's independence from the forest regime following the law of September 1791, one early twentieth-century scholar claimed, "The abandonment of rules of use and management [and] the authorization of unlimited cutting were the immediate and unfortunate consequences of this abrupt liberation."[17] In the eyes of some observers the ideal solution was private ownership under the supervision of the forest regime, the very system that the Forest Code of 1827 sought to establish.

As the privatization movement gained speed in the early nineteenth century, it also drew legislative support. In 1802 Napoleon initiated an extensive reform of the traditional Rural Code (*code rural*) that aimed to reduce customary rights, targeting communal pastoralism in particular.[18] Although it proved unenforceable at the time, the measure represented an early attempt to ban the practices of *parcours* and *vaine pâture*. Two years later the emperor passed the Civil Code, which accorded private landowners an unprecedented amount of independence by granting them the right to dispose of their land as they saw fit.[19] While the law scratched many initiatives of the French Revolution, including the requirement that inheritance be divided equally among one's children, it continued to encourage the division of land parcels by giving landowners such license and by upholding the revolutionary ban on primogeniture. Finally, Napoleon encouraged privatization and the fragmentation of French estates through his creation of a new nobility, which involved handing out titles such as the newly created Legion of Honor and awarding generous land grants to his loyal followers. At the same time, aristocratic émigrés, reassured by the emperor's overtures and social conservatism, began to return to France to reclaim their former possessions.[20] These disparate members of the new French elite worked together to reverse the tide of popular exploitation that had taken hold of French territories in the previous decades, and they uniformly condemned that public excess as a destructive free-for-all. The privatization of French forests proceeded even more swiftly during the Restoration, when the effort was recognized as a valuable source of profit for starving state coffers. Indeed from the time of the French monarchy's return in 1814 to the establishment of the Third Republic in 1870, France surrendered more than a fourth of its woodlands, an area the size of Corsica, to private buyers.[21]

By the time Charles X promulgated the Forest Code of 1827, the nature of French property already had irrevocably shifted. Pastoralists who had long depended on communal lands had watched those lands shrink steadily over the previous decades. Some gave up on the commons and began to rent pasture from private landowners. The new forest code encouraged this shift by making the exploitation of common lands less appealing; it saddled users with additional fees, stipulations, and bureaucracy. In many cases, moreover, they closed communal forests to grazing altogether, forcing pastoralists to seek pasture elsewhere. The declining use of commons sent communities into a vicious cycle. Each year brought fewer sheep, which meant less revenue from the annual head tax. In order to meet expenses, many municipalities felt compelled to sell off common lands, which in turn contributed to the trend of privatization.[22] Regardless of how they responded to these pressures, under the Forest Code of 1827 pastoralists found themselves subject to the same restrictions and limitations whether they grazed their flocks in public or private woodlands.

Far from healing old wounds between pastoralists and farmers, the evolution of property tended to create more conflict. Provence emerged from the revolutionary era's jumble of changes in property law with an uncommonly large number of private forests and less than its share of communal land.[23] Peasants, pastoralists, and other parties all clung jealously to what they had left. Local officials were generally sympathetic to pastoralism since it formed a vital part of their economy, but sometimes even they thought the exploitation of common lands went too far. In 1830, for example, the mayor of Roquefort-la-Bédoule addressed the municipal council with a bitter complaint against local pastoralists' domination of the town's commons. The previous year, in 1829, pastoralists had been granted permission to graze their livestock in communal forests by a royal ordinance, following the stipulations of the 1827 code. According to the mayor, these sheep had since taken over all exploitable lands "without distinction," which, he protested, "deprived the community of an annual revenue of 400–500 francs, but which gave a major advantage to the owners of herds."[24]

In nineteenth-century Provence infractions on common lands most often concerned one of a handful of crimes: shepherds herding more sheep than they were allowed, pasturing sheep without having previously declared them on the municipal register, herding goats, or allowing animals to graze in protected areas.[25] These issues, however, rarely arose in suits involving private lands. Instead landowners' main concern was pastoralists' presence on their property. In rural Provence poaching pasture numbered among the most common sources of litigation. The problem could run the other way as well. In 1818, for example, pastoralists in the community of Allauch lodged a complaint that private landowners were illegally using communal land for cultivation.[26] Most grazing offenses brought to court in the early nineteenth century involved property disputes, rather than violations of forest legislation. Incidents of sheep and goats grazing illegally on private land were particularly prevalent following the restoration of the monarchy in 1815, when the return of émigrés limited the amount of land available for grazing.[27] Violators were usually discovered by the property owners themselves or by their agents rather than by public forest guards, who were conspicuously absent from most of these cases. Grazing sheep illegally on private lands was not a new phenomenon, but as the balance shifted from public to private lands and fewer landowners opened up their fallow fields to grazing, conflicts between shepherds and farmers grew more frequent.

The regularity of property violations may have been exacerbated by the obscurity of property divisions. In the case from Allauch referenced above, the mayor claimed he was powerless to act because the town lacked sufficient revenue to effectively survey the boundaries of the commons. The lack of clear borders between public and private property, and between exploitable and protected communal lands, remained a widespread problem long after the passage of the Forest Code of 1827. French foresters were responsible for surveying and delimiting the commons, as well as for marking territory to be protected or kept in reserve.[28] In addition to the challenge of identifying the Provençal forest, this task involved navigating disputes between private landowners, local administrations, and other interested parties. It also required trekking through brambly Mediterranean woods and shrublands while transporting a sufficient supply of boundary stones.[29] Meanwhile, the borders of public forests were constantly shifting—and shrinking—as communities and the state sold them off for fast cash. For these reasons, surveying kept foresters busy for much of the nineteenth century. In its final decades they were still struggling to delineate communal forests.[30]

Some critics of pastoralism protested the free rein that livestock apparently enjoyed in lands not covered by the forest regime. These were the lands ambiguously abandoned by Article 90 of the Forest Code of 1827, which specified that the forest administration would oversee only those woodlands "recognized as susceptible to maintenance or regular exploitation by the administrative authority."[31] This clause allowed foresters to identify dubious territory as forest, but it also freed other communal lands from the protection of the forest regime. Worse, such unregulated lands were often those in the poorest condition, judged to be beyond the help of reforestation efforts. In some instances Article 90 led to the neglect of small forest plots, which were particularly prevalent in Provence, because they were deemed too insignificant to merit the attention of the forest service. Either way, as one scholar observed at the turn of the twentieth century, it was "just those [forests] that demanded the most serious, urgent, and necessary protection" that were abandoned to public use.[32] In general the passage of the Forest Code of 1827 relegated communal grazing either to these discarded lands, which were least able to sustain it, or to the healthiest, most robust woodlands, classified as *défensables* (the opposite of those *en défens*). In both cases pastoralism ultimately triggered a more negative environmental impact than it had prior to scientific forest regulation, and it consequently became even more unpopular among outsiders.

Pastoralism's environmental impact was somewhat counterbalanced by the number of sheep exploiting common lands; their numbers fell steadily over the course of the nineteenth century. In the village of

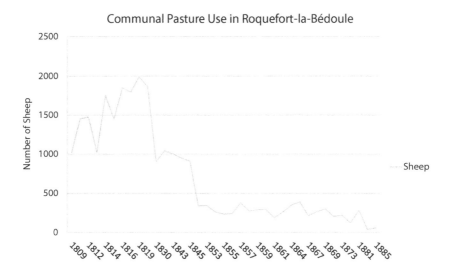

Communal Pasture Use in Roquefort-la-Bédoule

5. Communal pasture use in Roquefort-la-Bédoule, 1809–85. Archives départementales des Bouches-du-Rhône (BDR), 150 E 1N 4: "Roquefort la Bédoule. Pacage. États des propriétaires dont les troupeaux paissent sur les terres gastes et les bois communaux, 1809–1884."

Roquefort-la-Bédoule the total number of sheep grazing in the communal forest and *terres gastes* dwindled from a peak of close to two thousand animals just before the implementation of the Forest Code of 1827 to a negligible amount in the final decades of the century. By the mid-1880s the village commons had a single patron with a paltry flock of a few dozen sheep.[33] In many communities the forest regime reconciled itself to the presence of the few shepherds who still used communal forests, and it relaxed restrictions. In Les Baux-de-Provence, for example, communal grazing occurred alongside reforestation projects throughout the 1880s and 1890s.[34] As commons faded from the landscape, so did the practice of forest grazing. Small-scale pastoralists also disappeared, since the use of these lands was vital to their existence. By the end of the nineteenth century there was hardly anyone left to fight for the preservation of common lands. A law passed on 22 June 1890 outlawed the customary right of communal grazing (*vaine pâture*) throughout France, except in communities that petitioned successfully to preserve it.[35] The law generated no public outcry, and only a handful of communities elected to preserve their traditional grazing rights. In this way the long-cherished privilege of communal pasture access all but disappeared from Provence and, with it, the common ground of foresters, farmers, and pastoralists.[36]

The nineteenth-century forest regime also sowed conflict between pastoralists and agriculturalists by supporting the conversion of pasturelands to fields for cultivation. This program was one result of the 1827 code's reference to "the improvement of pasture lands."[37] In the context of French scientific forestry, "improvement" (*aménagement*) referred to the process of organizing and facilitating regular cuts. In contrast to the term *ménagement*, it implied commercial exploitation rather than conservation.[38] The expansion of agriculture in Provence was particularly pronounced throughout the Crau and the Camargue. These territories provided winter pasture for about half of the total number of sheep in the Bouches-du-Rhône Department.[39] They singlehandedly had made the nearby city of Arles the center of Provence's pastoral industry in the early modern era. While Arles continued to dominate pastoralism in Provence through the nineteenth century, other industries steadily trickled into the region. The forest and agricultural administrations, both governed by the Ministry of Finance, promoted cultivation in these areas through the expansion of irrigation and drainage systems. As a result, the size of the Crau's steppe diminished considerably, as did the marshes of the Camargue, and both regions fell subject to the plow. By the late nineteenth century much of their traditional grazing ground had been converted to wheat fields, olive and almond groves, orchards, vineyards, and rice paddies. Despite the decline in area available for pasture, however, the number of beasts increased steadily. In the late eighteenth century the lands in these regions were carrying approximately 150,000 sheep.[40] By the mid-nineteenth century this figure had risen to more than 300,000 in the Crau and Camargue alone.[41]

The rise of privatization had a dramatic impact on Provence's pastoral

6. The Crau in the 1930s. Archives départementales des Bouches-du-Rhône (BDR), 6 Fi 12205: "Plaine de la Crau," photograph, c. 1930.

industry and its environment. As the nineteenth century progressed, pastoralists increasingly competed for space with farms, groves and orchards, and other industries. They also faced new obstacles to seasonal migration, such as the challenge of navigating across a host of property lines. Together with the disappearance of commons, this shift encouraged transhumant shepherds to patronize private pasturelands. Landowners were generally less strict than the government in regulating grazing, but renting their property was expensive, and increased use placed greater strain on these lands. In addition, this trend effectively moved Provence's pastoral industry outside the realm of the forest regime and beyond the concern of forest agents. Other pastoralists chose to adopt sedentary sheepherding, stabling their sheep rather than facing the new challenges and risks of mobility. Consequently, the lush *coussouls* of the Crau and the verdant pastures of Haute-Provence, both of which had formerly spent most of the year in unencumbered regeneration, were now used extensively year round. Such intensified exploitation wrought a much greater environmental footprint. In this way the conviction that Mediterranean pastoralism was environmentally unsustainable became a self-fulfilling prophecy.

Land, Production, and Displacement in French Colonial Algeria

French debates about land use and ownership quickly spread across the sea to French colonial Algeria, where they also gained new significance. Both the territory and its inhabitants presented major administrative challenges, and, as colonial agents quickly discovered, land and property could be key catalysts in the establishment and consolidation of French rule. The first step toward controlling Algerian land was understanding it. The French administration pursued this goal by commissioning investigations of Algerian history, society, and ecology and then using this information to pave the way for European settlement and exploitation. Almost from the date of conquest, French concerns about Algerian forests were linked to the issue of property. The *Statistique générale* of 1844, France's earliest comprehensive report on Algeria's environment, branded local inhabitants as poor stewards of their forests and recommended that forested lands be removed from their control. In this way French discourse on environmental practices helped to pave the way for the French colonial agenda of indigenous dispossession.[42]

The French conquest proceeded steadily but slowly through the middle decades of the nineteenth century, and Algeria remained under partial military rule until 1870.[43] Although the government's initial response to concerns about Algeria's environment was to apply the French Forest Code of 1827 without alteration, its impact proved even more limited than the modest extent of French control. Even in the 1850s, when most of northern Algeria had been subdued, the Forest Code carried very little weight. While colonial officials viewed forest administration as an excellent goal, they acknowledged that, at least in the short term, stronger stuff was necessary. Control of the countryside was ultimately achieved through forest administration, property legislation, and the plow. Forests occupied center stage in the colonial discourse on property. These were the territories that the French central administration considered most endangered and in need of protection. While forests provided resources, pasture, and shelter for indigenous pastoralists, for colonists they represented prime real estate and opportunities for exploitation. As in France, the French forest regime in Algeria was tasked with identifying and classifying woodland areas, distinguishing them from other types of land. Thus, throughout the nineteenth century French forest agents strove to apply property divisions that would best protect and preserve Algerian forests, while settlers fought for legislation that would increase their access to forests while limiting indigenous usage rights.

At the time of the French invasion the inhabitants of Algeria recognized several categories of land ownership, including private property (*mülk*), collective or communal property (*arch*), state property (*miri*), and unclaimed land. These divisions and their associated usage rights had origins in local customs, Ottoman law, and Islamic tradition. French colonial officials made a sincere effort to comprehend Algeria's traditional forms of property ownership and its methods of land management and exploitation.[44] In some places they even preserved precolonial structures, especially where a lack of access or resources made colonization impracticable or undesirable. Vast swaths of Kabylia in the province of Constantine, for example, remained largely untouched by colonial rule through the 1850s.[45] The French administration designated these and other remote territories as *communes mixtes*, a label that implied integration and coexistence but more often referred to indigenous communities.

Despite efforts to understand and adopt local property traditions, however, most colonists remained ill informed about indigenous practices, and their attitudes toward Islamic law ranged from ignorance to contempt.[46] Meanwhile, the colonial administration increasingly promoted a narrative that "property exists only exceptionally among the Arabs."[47] Such an argument encouraged settlers to move in and claim land. Colonial perceptions of Algeria were also influenced by contemporary European property debates. For many observers the

Algerian environment was a classic consequence of what Garrett Hardin would later term the "tragedy of the commons"; it reflected centuries of communal use and abuse.[48] Advocates of this perspective argued that Algerian land could be saved only by converting it to private property, with European caretakers. Reports catalyzed interest in settlement and colonial agriculture by promoting the "fecundity" of Algerian soil and providing promising production estimates for a multitude of crops, including grain, cotton, and tobacco.[49] Such factors placed the early colonial administration under significant pressure to free up land for European settlement.

Privatization and indigenous displacement in Algeria gained momentum in the wake of political tumult in mid-nineteenth-century metropolitan France. In 1848 another revolution swept King Louis-Philippe from the throne and extinguished the monarchy. Louis-Napoléon Bonaparte, the nephew of Napoleon Bonaparte, was elected president of the newly created French Republic in a landslide victory. On 16 June 1851 the new leader, who would soon become Emperor Napoleon III, passed a law that greatly expanded the territory available for French colonial exploitation in Algeria, inviting French settlers—and in theory indigenous Algerians—to claim land through the process of *cantonnement,* or division into private holdings.[50] Proponents of the idea insisted that adopting European land-use practices would civilize the indigenous population. "Without the integration of races [there can be] no progress, and without cantonnement, no integration of races," reasoned the editor-in-chief of the Algerian journal *La Colonisation* in 1858.[51] In practice the law aimed to use new administrative tools to expand government control over Algeria's inhabitants and environment. It designated three types of geographical features as state property exempt from expropriation: forests, mines, and key waterways. Of these three categories, forests were by far the most extensive, as well as the most disruptive to private and indigenous property claims. As in France, many of the so-called "forests" protected by this clause did not contain a single tree.

Louis-Napoléon hoped that the law would win him popular support in France's prized colony and restore stability there after years of agricultural hardship and unrest, but it seemed to please no one. While the law's redistribution of land offered European settlers a boon, it also substantially curtailed their access to Algerian land. Colonists complained that most of the land grants were snatched up by powerful speculators and financial institutions, depriving them of any potential benefit.[52] Many French entrepreneurs, for their part, felt that the law placed oppressive limits on their ability to exploit the Algerian environment, and they continued to press for concessions. French colonial administrators criticized the law as unrealistic and impractical. In the 1850s very little Algerian territory had been surveyed, so it was often impossible to say what was forest and what was available for colonization. As late as the 1880s the governor of Algeria complained that the obscure boundaries set by the forest regime continued to hamper the implementation of cantonnement.[53]

The law of 1851 dealt a heavy blow to indigenous tribes. It assumed that they had an abundance of land and offered to formally recognize part of this territory as theirs if they voluntarily gave up their extra land to the French state. As an obvious consequence of this measure, such groups were pressured into sacrificing much of their territory to European settlers. In the province of Constantine, cantonnement claimed 50 to 85 percent of indigenous tribes' ancestral lands, leaving them with little other than remote mountain regions.[54] Many pastoralists lost their traditional livelihood because it required access to extensive territory. Ironically the law drove some Algerian peasants to nomadism by robbing them of access to fields. The system of cantonnement also enabled the French administration to pick its parcels in the exchange, so that indigenous inhabitants often were forced to cede choice forests and agricultural plots. As a result, the "most beautiful of tribes" were, in the words of one observer, "diminished by half and ruined completely" over the course of a decade.[55] For all its costs, the law did little to improve Algerian woodlands. While forest agents struggled to survey the vast expanses of Algerian territory that had come under French control, the use of unclassified forests continued unregulated and unabated. Even the process of identifying and protecting forestlands fueled destruction because it meant banishing pastoralists to territory that was more crowded and often less able to sustain their impact.

After a brief tour of Algeria in 1860, Napoleon III abruptly changed his policy on colonial administration. He proclaimed the French colony to be an "Arab kingdom," and he argued that "the natives, like the settlers, have a legal right to my protection, and I am just as much the emperor of the Arabs as the emperor of the French."[56] In place of the civil administration he had helped to institute less than a decade earlier, Louis-Napoléon restored the full authority of the military through the office of governor-general and the Bureaux Arabes, considered more sympathetic to indigenous perspectives.[57] In the decade that followed, Napoleon III pursued policies proposing to uplift indigenous Algerians and acclimate them to French colonial administration. He now condemned cantonnement for its negative impact on these groups, and on 22 June 1863 his administration passed a law designed to stop colonists from claiming land belonging to those local groups.[58] For the first time since the French conquest, indigenous tribes were formally recognized as "the owners of the territories in which they exercise permanent and traditional use."[59] The law put a temporary halt to the process of cantonnement, and it theoretically offered tribes the opportunity to gain French legal status for their land through property concessions.[60] Finally, as if to validate the emperor's newfound faith in the civilizing mission, as well as France's predilection for private property, it promoted the division of tribal lands among individual members.

Such measures earned Napoleon III even greater scorn among settlers. Most Europeans who had immigrated to Algeria had gone in the hope of personal aggrandizement, and they had little inclination to subordinate their own success, let alone survival, to the altruistic betterment of the indigene. The French-Algerian journal *La Seybouse* expressed this sentiment clearly. "In our opinion," wrote the editors, "there is in Africa only one interest worth respecting, and that is the colonist's—ours; only one important law—ours."[61] The law of 22 June 1863 contributed to the factionalization of the colonial administrative apparatus.[62] The Bureaux Arabes, which had welcomed cantonnement just a decade before, now considered it a duty to protect indigenous groups against the colonial onslaught.[63]

In the long run the law also negatively affected indigenous Algerians, as colonists used their greater familiarity with French law to maneuver around measures designed to safeguard local land claims. The implementation of the 1863 law proved particularly harmful for mobile pastoralists. Because their use of land was not "permanent," settlers and officials often disregarded the locals' property claims. Even those pastoral tribes lucky enough to gain formal ownership of part of their territory were now confined to it year round. They suddenly found their annual migration cycle restricted, if not closed off completely. The range and options of these tribes dwindled even more as once-communal territory was systematically divided up among individuals, many of whom, out of confusion or desperation, sold off their parcels to speculators. This process of dispossession helps to account for a rapid rise in the number of settled Arabs during the mid-nineteenth century. According to French census reports, the population of tribal Arabs (*Arabes des tribus*) grew by less than 9 percent from 1856 to 1861, while the population of settled Arabs (*Arabes des villes*) nearly tripled.[64]

In 1870 and 1871 French dissatisfaction with Napoleon III, combined with loss and humiliation in the Franco-Prussian War, triggered another revolution that brought an end to the Second Empire and the reign of Napoleon III. Political tumult and the return of republicanism in the metropole sent shockwaves through Algeria, where the establishment of the French Third Republic signaled the end of Louis-Napoléon's "Arab kingdom" era and the reinstatement of civil administration. The new regime supported settlers' rights, deepened the rift with the indigenous population, and precipitated the greatest insurrection that the French administration had faced since the conquest of Algeria. The revolt, which erupted in early 1871, was prompted by indigenous Algerians' growing desperation and hostility toward the colonial regime, as well as by three years of particularly dire environmental conditions, including drought, famine, and locust invasions.[65] It began as an uprising of powerful tribal leaders and quickly spread to the masses. Yet the rebel fighters lacked coordination and proper weapons, and by September 1871 the majority had surrendered.[66]

The rebellion ultimately proved detrimental to Algeria's indigenous inhabitants. Thousands of rebels died fighting, and these losses, combined with the cost of disease and famine in the preceding years, decimated the indigenous population. In 1871–72 the total number of indigenous inhabitants fell to roughly 2.1 million, a loss of more than half a million people in just ten years.[67] Those who survived were significantly worse off after the revolt. The French administration saddled guilty tribes with indemnities that they could never afford to pay, forcing them to sell off much of their remaining territory. As a result, European settlers gained 446,000 hectares of confiscated land, an area roughly the size of the Bouches-du-Rhône Department.[68] In addition, the rebellion served to justify a regime of repression characterized by the views and initiatives of settler-rights champion Auguste-Hubert Warnier.

Warnier had arrived in Algeria in 1834 as a young military surgeon and a member of the scientific commission. He initially shared the idealistic Saint-Simonian convictions of many of his peers, but half a century of colonial life changed his perspective. In various mid-nineteenth-century publications he helped to codify the "Kabyle myth" by contrasting Berbers and Arabs. While he characterized Berber society as relatively orderly and civilized, he condemned "the Arab" as lazy, corrupt, and degraded, a "devastating torrent," and a brutal, unwelcome invader.[69] By the 1860s he had become a prominent advocate for settlers' rights, a stance that won him election as an Algerian delegate to the French senate in 1871. On 26 July 1873 he succeeded in passing a law that removed the indigenous property concessions granted by Napoleon III, thus allowing colonial appropriation to resume uninhibited.[70]

The law of 26 July 1873, often called the Warnier Law, was much more than a backlash against the centralizing, bureaucratizing, and *arabophile* politics of Napoleon III. It symbolized the arrival of a new era in Algeria, when settlers' interests definitively triumphed.[71] Indeed an Algerian official would later unabashedly call it the "settlers' law."[72]

Even more than its predecessors, the Warnier Law championed private property and the fractionalization of communal land. Jettisoning any semblance of respect for local precolonial customs, it subjected all Algerian territory to French civil law. Tribal territory that remained undivided was opened up for bidding, including *arch* as well as shared *mülk* lands.[73]

Aftershocks of the 1871 rebellion reverberated through the late nine-teenth century in various minor uprisings, but they were quickly and effectively suppressed. In the wake of 1871 few Europeans dared advocate for the indigenous population. The Bureaux Arabes, Algeria's traditional interlocutors between colonials and colonized, disappeared with the reinstatement of civil administration. While some colonial apologists continued to use the vocabulary of *la mission civilisatrice*, it now sounded hackneyed and perfunctory. For the rest of the nineteenth century European settlers and a new generation of Algerian-born Europeans (later known familiarly as *pieds-noirs*) dominated the political scene. Through the Warnier Law, its subsequent amendment in 1887, and other legislation, they promoted the supremacy of the colonial population. In this context colonial enterprises flourished, often at the expense of the indigenous population and the environment.

Landholding, Resettlement, and Sedentarization in Anatolia

Across the Mediterranean in southwestern Anatolia, shifts in property designation, taxation, and legislation were also dramatically reshaping mobile pastoralists' lives and assisting sedentarization efforts. Beyond the bustling metropolis of Istanbul, early modern Anatolia presented an almost entirely rural landscape. Most of this territory was *miri*, or state land, though private ownership and other property designations did exist.[74] Although *miri* land theoretically belonged to the sultan, in practice the administration of rural territory was more complex, and it varied greatly from place to place. Even Anatolia, the heartland of the Ottoman Empire, included lands where the sultan's power and authority were little more than nominal.[75] Most of these latter areas were inhabited by nomads and featured mountainous and inaccessible regions, Anatolia's eastern and southern frontiers being prime examples.[76] Although they were divided into the same administrative units as other territories and were ostensibly incorporated into the Ottoman system, these regions differed in that they were ruled by tribal authority and administrative roles were inherited.[77] Indeed hereditary *sancak*s (provinces) might almost be considered the property of the tribe, except that they were subject to taxation and provided valuable services to the state.

Rural land was divided into *timar*s, worked by peasant tenant farmers and administered by members of the Ottoman officer class (*sipahi*) in exchange for duties to the state. Timars were similar to the fiefs of feudal Europe, though the timar-holder technically was not a landowner.[78] The task of collecting taxes in the countryside fell to fiscal representatives, local administrators, or tax farmers (*mültezim*), who purchased from the state the right to collect taxes.[79] Tax farming was a risky business, as its agents were responsible for the delivery of taxes to the state regardless of what they managed to collect. Nonetheless, some profited greatly from this enterprise, extracting as much as they could from local peasants. In addition, provincial administrators and timar-holders drew part of their revenue from taxes, as well as from fines and other regional sources.[80] Thus, only a portion of what was collected made it back to the imperial treasury.

By the eighteenth century many of the timars of rural Anatolia had become hereditary tax farms—effectively private land that was no longer profiting the state.[81] At the same time, the short-term tax-farming system was facing a crisis; rural instability, decentralization, and other factors were making tax farms an increasingly difficult sell for the state. In part to combat these problems, the state launched the system of *malikane* revenue contracting, in which it sold tax farms granted to persons for the duration of their life. This system greatly increased the number of tax farms, and it provided a quick source of imperial profit.[82] In the long term, however, the central administration lost a vital source of revenue to speculators and other contractors (*malikane*) who purchased lifetime tax rights.

Through these developments, wealthy tax farmers had gained control of most of rural Anatolia by the early nineteenth century, and much potential state revenue disappeared into the hands of these intermediaries.[83] This system of tax collection ultimately proved detrimental to both the peasantry and the imperial treasury. During the Tanzimat era, the Ottoman administration made various efforts to reform it. Just months after the Gülhane proclamation in 1839, the timar system of landholding and tax farming was officially abolished and replaced by a bureaucratic system of tax collection based on the French model.[84] In 1830–31 and 1844 officials gathered demographic statistics designed to determine and evaluate the empire's taxable population.[85] In addition, the 1844 census was accompanied by surveys recording, classifying, and registering rural property and revenue (*temmetuat*), which the administration hoped would lead to a more efficient and profitable restructuring of the taxation process.[86] These and other measures culminated in the Land Code of 1858, which would become, for both Anatolia's sedentary and mobile populations, the most substantial and far-reaching initiative of this period.

The Land Code of 1858 (Arazi Kanunnamesi) represented an ambitious attempt by the Ottoman administration to expand its control over the countryside.[87] It was largely the work of the reform-minded official Ahmet Cevdet Pasha.[88] He constructed the law with Western principles in mind, though he opposed the direct application of European civil codes to the Ottoman case. In this vein the 1858 code became the first document to institute and regulate state control of forests following the arrival of the French forestry mission the previous year. At the same time, it promoted the privatization of agricultural land.[89] As the first Ottoman law to explicitly and generally identify land as property, the Land Code of 1858 aimed to clarify rural property divisions. It identified five distinct types of landed property: private property (*mülk*), state land (*miri*), public or communal land (*metruk*), foundation land (*vakıf*), and wastelands (*mawat*).[90] The law even distinguished between the ownership of actual land, meaning the earth or soil, and individual trees that stood on it.[91] In order to determine these distinctions, the code called for the surveying and registration of all rural property. Local inhabitants were required to formally register the land they exploited, backing up their property claims with support from legal documents (*tapu*).[92] In return they were recognized as owners and given much greater control over the land.[93] Those unable to produce the necessary documentation, or who neglected to register their land for other reasons, became landless and disadvantaged.

The 1858 code's system of registration succeeded in streamlining tax collection, and it absolved many rural peasants from chronic debts to tax farmers. Yet its merit for agriculturalists varied widely according to regional conditions, the extent of the code's application, the role and status of potential landowners, and a number of other factors.[94] The code's impact on most mobile pastoralists, in contrast, was uniformly negative whether they chose to continue to practice their traditional lifestyle or to settle and become farmers. Few nomads understood the terms of the code or were able to take advantage of the opportunity to register their property. Nomadic claims to pastureland were generally based on traditional usage rather than official ownership, and nomads often lacked the documentation necessary to legitimize their claims. In addition, the small, individual plots encouraged by the code did not meet the needs of mobile pastoralism. Although the law did make provisions for *yayla*s and *kışla*s (summer and winter pastures), its treatment of this type of traditional, communal property was vague.[95] As a result, many tribes were excluded from traditional pastureland or forced to pay ever-increasing rent for their pasture.[96] Those who settled faced a similar predicament, joining the ranks of impoverished, landless peasants.

Although the Land Code of 1858 aimed in part to protect forest resources, it ultimately had a detrimental effect on forests as well as nomads.[97] Its failure to address the problem of enforcement proved a major weakness. One of its provisions was a tax on the extraction of forest products in state and communal (village) lands.[98] In theory this tax was intended to generate revenue for the state and to limit forest exploitation. In practice, however, villagers found ways around the law. For example, one clause exempted inhabitants from the tax when they housed soldiers, and many used this exemption to take as much firewood as they wanted, whenever they pleased.[99] In addition, the multiple goals of the code worked against one another. The Ottoman administration hoped to use the legislation to promote agricultural expansion, which would benefit the state by increasing production and add to state revenue through registration. Thus, Article 19 of the code explicitly permitted new owners to burn down forests on their land, inviting owners of "such places as forests and woods [to] make them into arable land by opening them up for cultivation."[100] This clause effectively gave inhabitants a mandate to destroy forests indiscriminately. Finally, the law failed to regulate either forests or agriculture effectively, since its passage was immediately followed by a host of exceptions. The Land Code reflected the state's prioritization of agriculture over forest preservation, while the protection of pasture numbered among its lowest priorities. Yet French foresters and other Westerners held mobile pastoralists responsible for much of the deforestation that ensued. In this way the code helped to justify Ottoman efforts to settle nomadic tribes.

When the Land Code of 1858 was promulgated, the effort to subdue and sedentarize nomadic tribes was already a well-worn theme of Ottoman administration. Nineteenth-century sedentarization campaigns, however, differed critically from earlier initiatives in both their purpose and their methods. They no longer aimed merely to increase centralization and improve rural security; they also became a symbol of modernization and progress. Official attitudes toward mobile pastoralists began to echo Western biases, explicitly categorizing nomads as savage, primitive, and less advanced than their settled peers.[101] In addition, the sedentarization campaigns of the mid-nineteenth century exemplified Ottoman governmentality and its principles of reform. Cevdet Pasha, the mastermind of the Land Code of 1858, also contributed to some of the most ambitious sedentarization efforts, including the forced settlement of tribes in Adana from 1865 to 1869.[102] In another important shift, sedentarization orders finally began to enjoy real success in the mid- to late nineteenth century.[103] Developments in technology, transportation, communication, industry, and the military gave the central administration much greater and more immediate control over its vast periphery. The Ottoman state now had better tools to settle its mobile populations. Over the same period it had gained two significant sources of ethical justification for its sedentarization initiatives: the narratives of civilization and environmental conservation.

The Ottomans' shifting policy toward nomads paralleled the influx of Muslim migrants and refugees, resettled within the empire's shrinking borders as it lost territory in Europe.[104] The Crimean War was particularly significant, triggering the arrival of upwards of one million immigrants.[105] Approximately six hundred thousand of them were settled in Anatolia, many in the Adana region and along the peninsula's southwestern Mediterranean coast.[106] These new inhabitants presented serious problems of infrastructure, provisioning, and settlement. Both the Land Code of 1858 and the Forest Bill of 1861 appeared in this context and aimed in part to address attendant issues.

Nonetheless the government was unable to provide these immigrants with the services and support they needed, and many of them were simply given land and expected to cultivate it.[107] In a parallel to the convenient assumptions made by French colonists in Algeria, the Ottoman administration chose to identify territory occupied part time by nomadic pastoralists as vacant land. Thus, refugees frequently found themselves settled on the seasonal pastures of nomadic tribes.[108] To complicate the picture, some of these new inhabitants, such as the Circassians, were not farmers at all but nomads like the yörük. Their presence on the coastlands of southwestern Anatolia was no coincidence. The incoming sedentary population served as an effective check on the movements of nomadic tribes, supplementing the Sublime Porte's own efforts to control them. From a purely logistical perspective, the increased population density created by the influx of refugees left little room for mobile pastoralism. The practice of resettlement, however, could be detrimental for refugees and local tribes alike. In one case, the British consul in Konya warned that "the almost criminal action of the Turkish Government in sending thousands of refugees, a large proportion of them armed with modern weapons, to the country without making proper provision for their support or maintenance" was leading to deadly tensions between refugees and the local tribal populations.[109] In late nineteenth-century confrontations, immigrants usually had the Ottoman military on their side. Consequently, nomads ultimately had to choose whether to settle, to rent pasture from their new neighbors, or to relocate to less desirable land.[110]

The impact of immigration was acutely felt in southwestern Anatolia. The Mediterranean coastal regions surrounding Antalya, Alanya, Adana, and the nearby Taurus range had long been a last bastion of nomadic autonomy, largely because they were neglected by the rest of society.[111] In the early nineteenth century the Ottoman administration had attempted to gain control over these and other regions by resettling local dynastic rulers away from their power base. Through this initiative the entire Tekelioğlu clan, which had long dominated affairs in and beyond Antalya, was forcibly moved to Salonica. Yet the long-term impact of this campaign was limited.[112] It was only later in the century, when a steady stream of refugees began to settle in these regions, that the composition of these regions truly changed. The malarial swamps that had long acted as a deterrent to permanent residency now began to be drained for cultivation, and the mountain forests into which the yörük retreated annually experienced a flurry of timber and mining activity.[113] These changes heightened pressure on nomadic groups, limiting their movements, restricting pastureland, and subjecting them to ever-increasing fees, fines, and regulations. In this way nineteenth-century landed property transformations, including legislation and the influx of refugees, helped the Ottoman administration at last reckon with its mobile pastoral population. Through these measures the Ottomans gained control of these peripheral groups as well as the landscapes they inhabited.

In France, Algeria, and the Ottoman Empire the nineteenth-century transformation of property served to expand administrative control over Mediterranean pastoral populations. In Provence, this development took the form of the decline and disappearance of the commons. It favored large-scale, commercial agriculture and excluded most shepherds and small-scale pastoralists. In Algeria, property legislation became a way for the colonial population to exercise control over indigenous pastoralists, giving settlers legal recourse for policies of dispossession. The Ottoman Empire employed similar tactics, requiring a process of property registration that favored settled agriculturalists and left nomads landless. These measures did not pass uncontested, and the application of property legislation in these three contexts was tempered by protest, negotiation, and compromise. In Provence, pastoralists petitioned local and regional administrators, who were often sympathetic to their cause. Officials in the central administration were also aware of the vital role of the commons, and this acknowledgment led them to lighten the burden of administrative policies in certain cases. As a result, some communities were able to cling to the customary privilege of common pasture much longer than foresters, intellectuals, commercial farmers, entrepreneurs, and other critics would have liked.

The same trend of privatization had very different implications in Algeria and Anatolia. In Algeria, tribes fought back against colonial pressure, sometimes violently, as in the insurrection of 1871. Their resistance ultimately hardened settlers, foresters, and the colonial administration against them. As we will see in the next chapter, it also made them popular scapegoats for environmental damage, disasters, and decline. Yet their struggles also won them sympathy and support among progressive French intellectuals and officials. By the turn of the twentieth century a harsh critique of colonial practices had developed in mainland France, and the pendulum that had ushered in the settler colony era was poised to swing the other way. In Anatolia, tribes also voiced dissent toward administrative measures by staging uprisings and rebellions, which occurred throughout the early modern era.[114] In addition, nomads used their mobility to retreat into mountains and other peripheral regions, beyond the reach of Ottoman authorities.[115] These tactics taught the Ottoman state to treat its mobile pastoral population with care, taking advantage of the tribes' assets and doing its best to mitigate the administrative problems they posed. In the nineteenth century, however, improvements in technology, communication, and administrative power afforded the Sublime Porte better control over nomadic tribes, limiting their ability to evade authority. At the same time, the influx of immigrants augmented the agricultural population, expanded settlements, and decreased the number and size of inaccessible, remote regions. Boxed in by these ever-shrinking frontiers, nomads had nowhere to hide.

The transformation of landed property rights exercised a major impact on the Mediterranean environment as well as on mobile populations. Along with the application of scientific forestry, property reform was designed to streamline and modernize the process of land distribution, ownership, and management. Both were justified through the rhetoric of environmental stewardship, conservation, and sustainable exploitation. Both were driven by commercial interests and power politics, and both were indicative of trends in governmentality. In the late nineteenth century scientific forest legislation in the wider French Mediterranean continued to follow the path that the Forest Code of 1827, as well as nineteenth-century property regulations, had laid. It enriched the pockets of a select few at the expense of local populations and the environment.

CHAPTER 6

Nature's Scapegoats

Pastoralists and Natural Disasters

> Disasters without precedent are striking our country; it is not for me to investigate their causes here; I will say only that one might find the main cause, that which has generated all the others, in this practice of material exploitation . . . incompatible with the conservation of forests.
>
> —Louis Tassy, *Études sur l'aménagement des forêts*

In August 1863 a series of fires spread through the lucrative cork-producing forests of northeastern Algeria. Together they destroyed more than 162 square miles of woodlands, as well as fields and vineyards.[1] Wildfires were not at all uncommon in Algeria, but this outbreak was particularly extensive and destructive. At the time, it was the colony's largest fire to be attributed to arson. It also represented the first time the French government directed this charge collectively at the local indigenous population.[2] Following the blaze, cork concessionaires petitioned the government incessantly for compensation through the imposition of "collective tribal responsibility" (*responsabilité collective des tribus*), a measure that held all tribal members accountable for failing to identify the perpetrators of a crime. As the century progressed, the charge of arson grew steadily more common in cases of wildfire, and the application of collective responsibility, which an early colonial governor had called a "terrible" piece of legislation, gained widespread acceptance in Algeria.[3]

Fires were just one of many environmental hazards threatening nineteenth-century Mediterranean inhabitants. People also endured floods, droughts, locust invasions, and epidemics. Provence alone was hit by multiple earthquakes, famines, and a series of devastating floods over the course of the century. It also survived the worst phylloxera plague in history, which nearly destroyed France's reputation for fine wine.[4] Mobile pastoralists were acutely susceptible to natural disasters, especially certain weather-related phenomena, such as frosts and floods. While a major deluge might ruin a farmer's annual harvest, the toll it took on livestock—potentially drowning them or washing them downstream—could ruin a pastoralist for life. In addition, natural disasters compounded other strains on Mediterranean pastoralism. Even as they struggled to recover from such catastrophes, mobile pastoralists were often directly or indirectly implicated in the incidence of natural disasters in the Mediterranean world.

This chapter examines the nature and impact of natural disasters in Provence, Algeria, and Anatolia during the nineteenth century. These case studies reveal both connections and divergences within Mediterranean environmental history and policy. In Anatolia, the yörük struggled to survive a series of devastating droughts, famines, frosts, and other environmental calamities. These events coincided with the most ambitious and effective settlement campaigns of the Ottoman period. In France, scientific literature, drawing on the perceived link between Mediterranean pastoralism and deforestation, blamed pastoralists for a range of events, including floods, mudslides, droughts, frosts, and harvest failures.[5] The association of flooding with deforestation upstream led to a massive mountain reforestation campaign with far-reaching human and environmental consequences. French foresters and other European observers drew on such perspectives when evaluating mobile pastoralism in Algeria, Anatolia, and other Mediterranean contexts. Wildfires, endemic throughout the Mediterranean region, were also commonly associated with pastoralism, but the nature and extent of administrative responses varied. In Provence and Anatolia, foresters generally considered the relationship accidental;

they condemned the practice of occupational burning and worked to stamp it out. In French colonial Algeria, however, indigenous pastoralists were frequently accused of setting fires deliberately, as an act of protest. By the late nineteenth century the forest regime had all but ceased to distinguish between arson and accidental fires, subjecting indigenous perpetrators to the same harsh punishments regardless of the nature of their offense. In all three contexts mobile pastoralists suffered doubly in the wake of natural disasters. Even as shepherds reeled from environmental challenges, narratives associating their practices with such occurrences informed environmental policy and sedentarization efforts. While nineteenth-century natural disasters fueled the decline and transformation of their industry, Mediterranean pastoralists became veritable scapegoats for nature.

The Mountain Reforestation Project in Provence

Droughts, floods, and wildfires were all common visitors to nineteenth-century Provence. The region's susceptibility to drought intensified as its principal natural watercourses, the Durance and the Rhône Rivers, were reengineered over the course of the nineteenth century to serve an expanding agricultural population. As these rivers snaked through Basse-Provence, agriculturalists bled them with irrigation canals, so that the rivers themselves began to look more like empty riverbeds.[6] Those with access to irrigation systems benefited greatly from this innovation, but it exacerbated the impact of drought on Provence's remaining inhabitants, including most small-scale pastoralists. Meanwhile, climate change associated with the waning of the Little Ice Age brought glacier melt, wetter conditions, and periodic extreme weather to Provence in the early to mid-nineteenth century. As a result, when the region was not starved by drought, it was often inundated by floods. Major torrents heralded the new century in 1795 and 1802. In August 1806 a monstrous storm rained hail the size of eggs, drowning riverbeds, fields, and pastures and destroying forests, vines, and olive trees.[7] On 2 November 1840 the Rhône overflowed its banks, submerging the Crau in more than eight feet of water. A witness surveying the normally verdant pastures of Arles reported that "only the tops of trees are visible."[8] The region barely had time to recover before it was hit by another flood, in October 1841.[9] By one count the hinterland of the Rhône suffered twenty-six floods from 1800 to 1856.[10] In late spring of 1856 the Rhône valley was again submerged, and the Camargue was transformed into a huge lake.[11] The great flood of 1856 later gained notoriety as one of France's "worst floods in modern memory."[12]

These incidents profoundly affected the inhabitants of Provence, and pastoralists were among those hardest hit. Victims of the flood of 1856 actually watched their livestock float away and drown. Following a flood in 1886, sheep farmers in Arles claimed that their prize pastures had been transformed into "veritable swamps," which they feared would leave fifty thousand sheep without food for the winter season.[13] On the other hand, periods of prolonged dryness also put entire herds at risk of dying from hunger. During one drought in 1881 shepherds in Arles claimed that, unless given special permission to graze in forestlands, they could lose up to four hundred thousand sheep.[14]

Provence's environmental calamities garnered national interest and concern. As early as 1821 the minister of the interior sent a dispatch to France's eighty-six departments asking for statistics on "the extraordinary floods to which France seems to be becoming more and more subject."[15] Twenty years later Alexandre Surell, an engineer who spent much of his career managing watercourses and railroad construction in the Midi and the Alps, published a path-breaking study on the subject: *Étude sur les torrents des Hautes Alpes*.[16] Characterizing floods as "the most tragic scourge" of affected regions, Surell hypothesized that they were caused largely by erosion in mountain streams, creating "monstrous riverbeds, which grow constantly and threaten to wipe out everything."[17] His account linked erosion directly to mountain deforestation, which he called "the most disastrous of disturbances." While he recognized the need for cutting and clearing in an alpine economy, he reasoned that "in order for the mountains to be habitable, they must be forested."[18]

In the decade that followed, Surell's work inspired growing interest in mountain reforestation. In a treatise on mountain deforestation one of his contemporaries, Adolphe-Jérôme Blanqui, explained how, "under the influence of deforestation," ravaging floods "transform each day part of our four frontier departments into sterile solitudes."[19] French administrators also responded. In the summer of 1856, following another severe flood, Napoleon III personally visited Provence and oversaw the construction of levees to minimize future damage. He also vowed ambitiously that floodwaters would not rise during his reign.[20] The issue of alpine flooding also became a central topic of debate in the French Scientific Congress, as well as in the halls of Parliament.[21] Reforestation legislation was proposed in the 1840s and 1850s, though it failed to pass.[22]

When a series of particularly destructive floods inundated the Rhône delta and other lowlands in 1855 and 1856, these vivid environmental calamities seemed to validate Surell's words, and they heralded a fresh wave of support for the cause of afforestation and reforestation. Charles de Ribbe pioneered the movement with his 1857 publication *La Provence au point de vue des bois, des torrents, et des inondations avant et après 1789* (Provence with regard to woods, rivers, and floods). Building on Surell's work, Ribbe went even further in decrying the effects of alpine deforestation and promoting administrative reform. His introduction explains that the work was written "in the state of mourning produced by the latest floods," and its purpose was as much political as it was scientific.[23] Citing the "necessity of modifying many aspects of the Forest Code," Ribbe aimed to use his treatise to put the administration of forests "back on the path to a solution."[24] For him, together with a growing number of French scientists, foresters, policy makers, and environmental advocates, mountain deforestation represented a root cause of Mediterranean environmental decline.

The campaign for mountain reforestation won a major victory with the Mountain Reforestation Law, passed nearly unanimously by the Corps Législatif and approved by Napoleon III on 28 July 1860.[25] Framed as an effort to prevent alpine flooding and the formation of floodwaters, this law represented a direct response to recent environmental events.[26] As the first piece of French legislation focused exclusively on reforestation, it was also something entirely without precedent. The law adopted a two-pronged approach to reforestation. On the one hand, it encouraged voluntary efforts at re- and afforestation by providing growers with seeds, saplings, and subsidies.[27] On the other, it required the forest administration to institute projects wherever "the public interest demands that reforestation work be made mandatory, for reasons of the state of the soil and the dangers it poses to lands below."[28] The law's treatment of pastoralism was ambiguous. It authorized the practice in reforested areas once the new stands were mature enough to be *défensables*, but it did not specify an age for this status. As those who promoted the law explained, it was designed to employ the twin tactics of reward and coercion; it rewarded inhabitants for cultivating forests, and, when they did not, it forced them to do so.[29]

The implementation of the law of 1860 occurred in an atmosphere of heated opposition and debate.[30] Some contended that the costs of obligatory reforestation projects were too great a burden for local populations to bear. Indeed the law's reception in agropastoral communities was overwhelmingly negative. Others claimed that it was too soft on pastoralism, which they still considered a prime factor in deforestation. The press derided it as idealistic, infeasible, and impossible to enforce, while many in the French scientific elite argued that it did not go far enough. They also pointed out that a lack of local acceptance together with insufficient surveillance and staff were preventing the law from being effectively applied. Perhaps most important, because the law was instituted on a trial basis, it expired in 1870, and the turbulent political climate of this and subsequent years delayed discussion of a successor.

A few dedicated forest advocates sustained the mountain reforestation campaign in the wake of mounting criticism. One of the most renowned and enthusiastic voices for the 1860 law was that of the forester Louis Gabriel Prosper Demontzey. In his youth Demontzey had been counseled by Charles Vial, a family friend and fellow forest agent, to "create forests" everywhere he could.[31] A self-described "reforester" (*reboiseur*), Demontzey took these words to heart. After graduating from the Nancy school in 1852, Demontzey entered the forest service in Algeria, where he remained for ten years. He then spent the rest of his career

7. Prosper Demontzey, photograph, Bibliothèque nationale de France, accessed 7 October 2018, http://gallica.bnf.fr/ark:/12148/bpt6k6270993c. Originally published in *Revue des Eaux et Forêts* 38 (1898): 193.

in Provence, migrating from Nice, to Digne, and ultimately to Aix-en-Provence, where he became *conservateur* of the Bouches-du-Rhône in 1877. Throughout his life he made it his goal to have a measurable impact on the state of French forests. In his obituary a colleague recalled that "reforestation, for him, was everything!" and affirmed that the great forester had subordinated all other duties to this single cause.[32]

Demontzey's experience on both the northern and southern shores of the Mediterranean convinced him of the importance and urgency of mountain forest management. He worked eagerly to enforce the 1860 reforestation law in Provence and Algeria and defended it in publications. In 1878 he published *Étude sur les travaux de reboisement et de gazonnement des montagnes*.[33] Demontzey developed the text in response to the French forest director-general's call for proposals for practical approaches to mountain reforestation and replanting; it won first prize.[34] In his introduction Demontzey countered objections to the law by signaling its numerous and continuing successes: "In the many parts of France's mountain regions where reforestation, whether voluntary or obligatory, has been undertaken, the young forests that exist today present and maintain the most categorical refutation of the allegations of those miserable spirits who denied it the possibility of birth, life, and advancement."[35] Through his own efforts, Demontzey gave testimony to these words. Until 1877 he had directed efforts to minimize flooding in the Basses-Alpes, successfully limiting the impact of several major storms.[36] For the rest of the century he continued to promote mountain reforestation, both through his career in forestry and through a range of publications on this theme. These works embodied his faith in the practical application of law as well as his love of trees.[37]

If Demontzey's devotion to reforestation made him something of an idealist, he was also relatively pragmatic. As the title suggests, *Étude sur les travaux de reboisement et de gazonnement des montagnes* treats the project of both mountain reforestation and replanting or regenerating mountain pastures (*gazonnement*). His book acknowledges the importance of the pastoral industry to France's alpine regions and explores ways to minimize its damage to forests and its role in the formation of flood channels. In a departure from most of his colleagues, Demontzey maintained that transhumant pastoralism was less harmful than the non-migratory "indigenous" pastoralism practiced among some mountain populations. "The transhumant [herd]," he reasoned, "does not arrive in the mountains until the vegetation is full and the soil has regenerated . . . and leaves before the great autumn rains." He contrasted his character-ization with native animals, "which in winter leave no respite for the smallest patch uncovered from snow and roam different parts of the mountains, according to the season, in the most unfavorable conditions for soil stability and the conservation of grassy vegetation." In general Demontzey favored confining pastoralism to certain regions beyond the target zones of mountain reforestation initiatives. Even in this restricted sphere, he argued, the practice should be transhumant and limited to three sheep per hectare (about two and a half acres), a small number by any account. Demontzey's prescription was hardly music to pastoralists' ears, but it represented an important distinction from the forestry estab-lishment's long-standing efforts to eliminate pastoralism altogether.[38]

Louis Tassy, Demontzey's contemporary, shared his passion for mountain reforestation, if not his interest in accommodating local inhabitants. Tassy had gained prominence within the French forest administration for his role in establishing scientific forestry in the Ottoman Empire. Following his departure from Istanbul in 1868, he completed an extensive review of the forest regime in Algeria. In 1875 he retired to his native town of Aix-en-Provence, while Demontzey was stationed there as forest conservator of the Bouches-du-Rhône. Tassy rivaled Demontzey in both his familiarity with the Mediterranean region and his dedication to afforestation. These foresters both accepted the narrative that Mediterranean landscapes were tragically degraded and deforested, but Tassy went much further than his colleague in his criticism of current forest legislation and the impact of pastoralism. The zeal with which Tassy fought for mountain reforestation and admin-istrative reform following his retirement more than compensated for the discretion, caution, and compliance that his post in the Ottoman service had required.

In the debate that followed the termination of the Mountain Refor-estation Law of 1860, Tassy rebuked the program for its expense and for not going far enough. His treatise *La restauration des montagnes*, published in 1877, presents a direct assault on the forest regime's man-agement of the mountain reforestation project. He blamed the forest administration for softening its stance toward pastoralists and other mountain dwellers by financing reforestation and repasturing projects rather than penalizing them for the poor environmental practices that had degraded the land in the first place. He also chastised forest agents for handing the reins of reforestation to communities and thus limit-ing the scope of obligatory reforestation initiatives. Summarizing his assessment, he accused the French forest director-general of confusing local interests with the public good. In general Tassy used the text to inculpate mountain dwellers for unsustainable land use, erosion and flooding, and the destruction of France's dwindling forest resources. In this and subsequent publications his recommendations promoted the preservation of forest resources for future use, as well as the protection of communities downstream from the environmental consequences of deforestation, such as droughts and floods. He promoted these long-term environmental goals at the expense of contemporary local economies both in the mountains and on the plain. For Tassy, the mission of the forest service was ultimately to save forests, not people.[39]

The relentless lobbying of Tassy, Demontzey, and other prominent forest advocates throughout the 1870s eventually gained the attention of lawmakers. The result was a new piece of legislation, the Law on the Restoration and Conservation of Mountain Lands (Loi Relative à la Restauration et Conservation des Terrains en Montagne), passed 4 April 1882.[40] This text echoed many features of the 1860 law. It retained the idea of "public utility" as a prerequisite for maintenance, and it permitted the state to claim lands where such a need was identified. Like its predecessor, it was inspired largely by a fear of floods, and it aimed to protect lands from environmental catastrophes.[41] The 1882 law, however, differed on a number of points. In some ways it was more expansive. Whereas the Mountain Reforestation Law of 1860 had focused exclusively on planting and replanting trees, the new law also supported other projects promoting "conservation" and "restoration."[42] In addition, it governed not just forests but pastures and the pastoral industry, which it proposed to regulate rather than eliminate. Yet the law was in other ways more limited in its scope. It concentrated reforestation efforts in critical areas and encouraged agriculture, orchards, and the regrowth and fertilization of mountain pastures in others. Moreover, it no longer left the designation and implementation of restoration work to the discretion of forest agents. Instead it required that the "public utility" of each prospective project be determined by a vote of the local municipal council.[43] In this way the law reflected its political context. Promulgated in a new era of universal suffrage, it evinced the administration's concern for the support of rural constituents, who made up a significant portion of the voting population.[44] As its proponents explained, the law's liberal approach to environmental management made it much more practical and enforceable than the law of 1860.[45]

Nonetheless, the 1882 mountain restoration law did not escape criticism. Purists slandered it as a dirty compromise with local interests, particularly pastoralists.[46] In État des forêts en France, which appeared in 1887 and remained influential long after, Tassy demanded more intensive and extensive mountain forest restoration.[47] Dramatizing the pitfalls of the law, he claimed that if it was not reformed, "there will be, in ten years, no more question of mountain reforestation," since the French administration would have surrendered its ability to protect what few forests remained.[48] Yet Tassy's concerns proved unwarranted. In practice, reforestation and soil conservation programs continued to take precedence over the needs and livelihoods of local residents in the Alps and other regions. The law's promise to work with pastoralists rarely was achieved, as foresters and pastoralists disputed the meaning of sustainable practices. In addition, the law represented a trend away from commercialism toward conservationism.[49]

Foresters' evident preference for trees over people did not win them friends among the pastoralists of Provence. By empowering the forest administration to impose mandatory plantations, the laws of 1860 and

1882 generated antagonism between foresters and local populations, who rarely considered reforestation measures to be in their interest. Many continued to ignore new regulations and violate grazing restrictions. Fires of dubious origin frequently broke out in newly planted forests, suggesting that some inhabitants went even further to protest reforestation projects.[50] Local administrators generally shared their community's resentment toward reforestation projects. In 1886 the mayor of Pierrefeu objected that efforts at afforestation were having a detrimental effect on residents and robbing them of essential revenue. Characterizing the forest agents assigned to his village as "strangers to the country, its vegetation, and its forestlands," he maintained that their work, whether deliberately or through ignorance, had "completely failed . . . to respond to our wants, our needs, and our modest advice based on the evidence of experience."[51] Inhabitants of the alpine communities in Haute-Provence and the Basses-Alpes echoed the complaints of those on the plain and protested the loss of key pastures as well as access to other newly protected spaces.[52] Those accustomed to grazing sheep in forests destroyed by flooding or fire lost access to not only those spaces but a host of other newly protected areas as well. The process of mountain depopulation was already well under way when the mountain reforestation project reached these regions, and the law of 1882 effectively sealed the fate of an already diminishing mountain population. By the turn of the twentieth century most mountain villages supported only a small fraction of their former population.[53] The mountain reforestation project proved most successful in the Basses-Alpes, where foresters managed to expand forest cover by almost a hundred thousand acres, more than twice the amount in the Hautes-Alpes or other mountain departments.[54] In this way the mountains of Haute-Provence effectively became the empty forested spaces that foresters had long envisioned them to be.

Extinguishing Disasters in Algeria

In the late nineteenth century discussions surrounding mountain reforestation spread across the Mediterranean to French colonial

Algeria. Following the promulgation of the Law on the Restoration and Conservation of Mountain Lands in France, the settler-run Algerian Reforestation League called for similar legislation in Algeria. This campaign was spearheaded by the league's president, Jean Baptiste Paulin Trolard, a *pied-noir* who had spent his childhood in Algeria and, after completing his education in the metropole, returned to practice medicine in Algiers. Trolard argued that mountain reforestation served a public purpose so vital that "if this law did not already exist, it would have to be created expressly for Algeria."[55] He drew support from Louis Tassy's report on forestry in Algeria, describing his compatriot as "a high official whose competence is beyond doubt."[56] In particular Trolard pointed to Tassy's estimation of the vast economic potential of Algerian forests if only they were efficiently and effectively managed. Trolard's critics fired back that Algeria's environmental, political, and social features set it apart from continental France, rendering both the law of 1860 and its 1882 successor inapplicable. While acknowledging the importance of reforestation, another member of the league suggested that such regulations should cater to Algeria's specific nature, where "it is not the fear of floods and their disasters that counsel us to conserve forests and mountain pastures, it is the necessity of safeguarding easy and abundant access to springs." The league member added, "Limited exclusively to mountains, [the law] would not be able to address the main concerns of our colony."[57] Others branded reforestation as infeasible and suggested that the funds allocated for the project would be better employed in preventing the deforestation of massifs still "covered with trees" than in facing the challenges of a "difficult reforestation."[58] Although such recommendations for an Algerian-specific corollary to domestic mountain reforestation laws never came to fruition, detractors of the initiative successfully derailed it. Neither law was adopted in Algeria.

As colonial critics of the French mountain reforestation campaign keenly observed, Algeria's environment presented new types and degrees of hazards. In contrast to Provence, Algeria was generally spared the consequences of excess water, but the risk of drought was ever present. The colonial context presented additional environment-related challenges that were less prominent in Provence; they included famine, agricultural pests, plague, and epizootics. In the second half of the nineteenth century Algeria witnessed periodic shortages, and destructive insects regularly attacked fields and livestock. Particularly grave famines occurred from 1856 to 1857 and 1869 to 1870.[59] The department of Constantine was frequently beset by locusts, with full-scale invasions documented in 1845, 1866, 1874, 1877, and 1887, and 1889–91. Although floods were less frequent than in Provence, they were not unheard of. In the winter of 1856–57 torrential rains drenched the departments of Alger and Constantine.[60] Violent storms again pelted northern Algeria with rain and hail in June 1867, and storms returned in August, triggering major floods.[61]

Although colonists were not immune to such disasters, indigenous inhabitants generally suffered much more, a consequence intensified by France's "scorched earth" policy.[62] The years 1865–70 stand out for the range and intensity of environmental pressures, as well as for the severity of their impact on certain populations. The crisis began in 1863, which would become that decade's last good harvest year. The year 1865 witnessed a severe rash of wildfires, the worst since the French conquest, with damages calculated at 2 million francs.[63] It also saw famine and an outbreak of cholera. The situation grew steadily worse in the following years. In 1866 locusts devastated agricultural fields and pastures, and violent summer hail and rainstorms triggered floods. Epizootics attacked livestock, contributing to a 25 percent decline in wool stock and forcing many pastoralists to sell their remaining sheep. The following winter, which was exceptionally cold, decimated surviving herds.[64] Together these factors contributed to a decline in Algeria's indigenous population by as much as 20 percent.[65] The tragedy was instrumental in driving those who survived to revolt the following year.

While indigenous pastoralists reeled from a range of natural disasters, the settlers, foresters, and government officials focused on one in particular: wildfire. In the second half of the nineteenth century concerns over wildfire figured centrally in French colonial forest management as well as colonial administration in general. Wildfire has been a hallmark of Mediterranean lands since time immemorial. The thick, fire-resistant bark of cork and holm oak trees demonstrate how the very landscape has adapted to this ever-present force.[66] With its hot, dry summers, strong winds, and combustible vegetation, the region is ripe for wildfires. Likewise, fire has always been both friend and foe to Mediterranean inhabitants. Pastoralists and farmers around the inner sea regularly burned brushland, forests, and fields to regenerate pasture and expand farmland, but they also shrank from the inexorable force of uncontrolled wildfires. Throughout the Mediterranean region, pastoralists burned grasslands regularly, sometimes as often as every few years. Alfred Grove and Oliver Rackham have called burning for pasture "as necessary to shepherding as ploughing is to farming" because it encourages the regeneration of edible plants and shrubs, removes distasteful or poisonous vegetation, and hinders the growth of trees.[67] In the nineteenth century, however, few outsiders saw it this way.

French officials typically viewed fire in the Mediterranean region in purely negative terms. In the words of one scholar, it was "the eternal enemy, the most terrible scourge of the forest lands of Basse-Provence."[68] Another characterized it as "a plague that brings shame on our civilization."[69] A third source called it a "calamitous curse."[70] Accounts of wildfires in Mediterranean France described "murderous flames" of "frightening energy" that "would not cease to expand their ravages."[71] Some scholars' antagonism ran deeper, linking fire to long-term environmental effects. A common refrain implied that fire encouraged erosion and flooding by dislodging rocks and soil and destroying forests.[72] Others cited fire as a force of desiccation, arguing that the long-term practice of controlled burning in the Mediterranean had gradually transformed much of the landscape from timber forests into *maquis*.[73] Such claims implied that fire, if allowed to persist, ultimately would turn the region into a desert. In addition, nineteenth-century French foresters tended to view Mediterranean wildfires as anything but natural, attributing them to human agents and mobile pastoralists in particular. This perspective influenced policy and practices in Provence and Anatolia, but it was especially pronounced in Algeria, where it became emblematic of the divide between colonizers and colonized. French fire management was developed as a useful and necessary corollary to forest administration in the driest, most fire-prone regions of continental France; in French colonial Algeria it became a tool for indigenous suppression, resource exploitation, and the legitimization of colonialism.

Fires played a unique role in Algerian colonial society and administration. On the opposite Mediterranean shore, fires were more frequent and costly than in continental France, leading French colonial foresters to view them with particular dread. The year 1865 was particularly devastating, with a total surface area of almost 400,000 acres burned and damages calculated at 2 million francs.[74] The worst rash of fires of the colonial era, however, occurred in 1881, when a total of 244 fires destroyed nearly 420,000 acres at a cost of more than 9 million francs. The Algerian case was also distinguished by the range of civilian responses to antifire legislation, which often pitted the settler population against indigenous groups. Indeed European settlers often expressed greater anxiety and alarm over the prospect of wildfires than the forest officials who fought them. Some even complained that the forest regime did not go far enough in its restrictions where indigenous Algerians were concerned. François Trottier, a Frenchman who settled in Algeria in the 1870s and became a prominent forest advocate, listed pastoralism, fire, and cork exploitation as "the three great scourges that we must fight" in Algeria's forests.[75] Another settler put it this way: "We have the choice . . . between seeing the forest disappear immediately through fire or in fifty years by the abuses of pastoralism."[76] Whether they rated pastoralism or fire as the greater threat to the future of the colony, European settlers generally agreed that both must be extinguished; so much the better if they could be beaten with a single blow.

Nineteenth-century observers had compelling reasons to seek human culprits for Mediterranean wildfires. Because the occupations and lifestyles of the Mediterranean world's human inhabitants have long depended on the controlled use of fire, a Mediterranean wildfire, particularly in the summer months, is much more likely to have originated from a human source than from the rare instance of lightning. In the nineteenth-century context finding someone to blame was often the first step in obtaining compensation for the losses occasioned by a fire and the costs of fighting it. Attributing fire to human causes in cases lacking a suspect or clear explanation, moreover, allowed foresters, officials, and other investigators to accept and perpetuate common declensionist narratives. By blaming pastoralists for the proliferation of wildfires in the Mediterranean region, administrators used fire management to fight the practice of Mediterranean pastoralism.

Nineteenth-century French foresters usually ascribed Mediterranean wildfires to either imprudence or arson (*malveillance*), but the meaning and frequency of these designations varied widely in applications of French forestry around the Mediterranean. "Imprudence" referred to any wildfire caused accidentally by human actions. Throughout the nineteenth century imprudence was and remained the most common cause cited for wildfires in Provence, while the incidence of arson steadily declined.[77] Out of seven fires reported in the Var Department during the summer of 1867, six were attributed to imprudence, including one major burn, while only one was believed to be caused by arson.[78] These proportions changed little over the remainder of the nineteenth century. The only major shift occurred with the development and spread of the railroad, which posed a significant new fire hazard because of sparks flying from trains. In 1935 imprudence remained by far the most common cause of fires in southwestern France, at 60 percent, but it was followed by railroads at 25 percent, while malveillance came in third at 9 percent.[79] Military exercises accounted for half of the remaining 6 percent, and wildfires attributed to natural causes continued to be virtually nonexistent.

In Provence, fires attributed to imprudence were attributed to a range of actors, but pastoralists tended to top the list. As recently as the eighteenth century some Provençal communities were known to hold local shepherds collectively responsible when the perpetrator of a wildfire could not be identified.[80] Although this practice went out of fashion in the nineteenth century, villagers and local officials continued to view pastoralists with suspicion when wildfires occurred. Charles de Ribbe's 1869 study on wildfires in the Var Department rates pastoralists second in his list of possible perpetrators, just after arsonists, even though he admits that the "savage process" of burning for pasture was virtually extinct in Provence.[81] Indeed scholars were still listing shepherds as prime suspects in the early twentieth century.[82] Why did pastoralists continue to make the cut, long after they had ceased to perform the actions that had implicated them? Part of the reason may lie in developments taking place across the Mediterranean, where French forest agents faced both mobile pastoralism and wildfire in more formidable forms.

Despite similarities in the social and environmental conditions surrounding forest fires in Provence and Algeria, the French forest administration treated fires in these two contexts in dramatically different ways. In June 1840 the French director of finance sent a letter to the governor of Algeria in which he expressed his concern for Algeria's lucrative forest resources. "We are entering the season when the Arabs choose to set fire to brushland," he wrote, addressing this tradition with reluctant resignation as "always and everywhere disastrous."[83] In Algeria, foresters' reports blamed indigenous groups for the vast majority of fires, charging European settlers rarely and only in clear-cut cases. Accusations took multiple forms. Initially most outbreaks of fire were attributed to occupational burning, a practice that colonial agents worked relentlessly to stamp out.[84] A colonial report on Algerian forests published in 1845 cited the local tradition of burning woodlands to create pasture as the principal cause of deforestation in Algeria up to that time and the greatest threat to forests in the future. It urged the administration to take immediate "measures to prevent the system of devastation by fire used by the Arabs, a system that results in denuding immense spaces of land of all vegetation."[85]

In the second half of the nineteenth century the French inclination to blame "natives" for forest fires gained a more sinister aspect. In the wake of the rebellion of 1871, foresters increasingly suggested that the fires they investigated had been set deliberately, not for any practical purpose but as acts of protest. Through such charges, stock explanations the French offered for wildfires in Algeria ultimately crossed the line from so-called imprudence to malveillance. This is a tricky word to translate, particularly in the colonial context. Although its official use corresponds roughly to the English word *arson*, the meaning of the latter is narrower and fails to convey the combination of deliberate destruction, hatred, and vengeance evoked by *malevolence*, the English cognate of the French term. In Algeria, moreover, the term "malveillance" gained new connotations. While it was very rarely used to describe fires set by European colonists, it became a standard accusation against indigenous inhabitants. Its association soon reached beyond clear cases of protest and rebellion to cover all wildfires. As the century progressed, charges of malveillance began to include fires caused by imprudence or arson, without distinction. The year 1871, which witnessed both Napoleon III's fall from power and the greatest rebellion of the colonial era, represented a turning point. French investigators associated nearly all of the year's wildfires with the indigenous rebellion. Reports of the event described the rebels setting "fire to forests, harvests, farms, and villages."[86] Indeed one observer interpreted the revolt almost entirely in terms of nomads' greed for pastureland, claiming, "In order to gain pastureland and to distance themselves from the European element, which they fear, the natives took advantage of these conditions to set fire to forests."[87] Such descriptions helped to solidify the narrative of indigenous responsibility for fire. At the same time, these descriptions nurtured stereotypes casting indigenous Algerians, and "Arab" pastoralists in particular, as enemies of the earth.

By the late nineteenth century malveillance had become the most common explanation for forest fires in Algeria. Under pressure to identify a cause, foresters who lacked clear evidence for the cause of a fire tended to cite malveillance. In a typical case, a fire that occurred on 30 July 1902 in the province of Tébessa was described as cause "unknown, probably due to malveillance."[88] Indeed forest agents were so eager to attribute fires to indigenous malevolence that they could become quite baffled, and even frustrated, when there was no local tribe to blame. At the same time, fires attributed to Europeans, whether through imprudence or malveillance, became the rare exception in Algeria.[89] Yet the few cases against Europeans confirm that they too interacted with Algerian woodlands and were equally capable of burning them. Some intrepid officials commented on this imbalance. In an 1887 letter to the governor-general, the prefect of Oran challenged the status quo by remarking that fires "should not necessarily be attributed exclusively" to indigenous pastoralists. He reminded his correspondent that fires could equally be due to the imprudence of hunters or passersby and added meaningfully, "You are certainly aware, Mr. Governor, that forests have often been affected by fires originating in the agricultural fields of Europeans."[90] Such arguments, however, fell mostly on deaf ears, and most officials took for granted indigenous associations with fire. In the words of a 1902 report on a forest fire in the vicinity of Bône, fire was "one of the particular forms of native banditry, a protest against France's possession of the forests, revenge for the trouble caused to their predatory habits."[91]

The French forest administration took a number of steps to prevent Mediterranean wildfires. Fire bans were implemented in all Mediterranean forests under the French forest regime, which included most communal and state forests in Provence, as well as all forests in Algeria. Such legislation customarily outlawed any use of fire in forest regime land and often limited access and other forms of exploitation. Fire bans usually remained in effect for the entire hot season and sometimes longer. A typical fire ban implemented in Constantine in the late nineteenth century lasted from the beginning of July until November. Notices publicizing the ban were printed in French and Arabic, in what may have been a genuine attempt to raise awareness among indigenous Algerians or, conversely, to hold them accountable.[92] The forest administration also fought fires through legislation, punishments, and fines. In continental France, fires attributed to imprudence generally resulted in fines for the guilty parties (when their identity could be determined) according to the nature of their offense and the scale of the fire. Penalties for fires caused by imprudence were always lighter than those for cases of arson (malveillance). In contrast, punishments for indigenous Algerians linked to cases of fire were always harsh, but they grew progressively worse in the late nineteenth century. In the 1860s Napoleon III made some attempts to lessen the burden of the indigenous population and promote more objectivity in assigning blame.[93] Yet the rebellion of 1871, which coincided with the emperor's fall from power, silenced any conciliation toward Algeria's indigenous population. Blaming the revolt on indigenous overconfidence resulting from "an impunity as bizarre as it is habitual," colonial observers sought to ensure that "all the instigators of the revolt, regardless of distinction, receive a punishment proportional to their crimes."[94] In the period that followed, the administration saddled indigenous groups with fines they could never afford to pay, as well as further dispossession of ancestral lands. In addition, this period witnessed an intensification of forestry legislation designed ostensibly to help prevent wildfires but that in practice almost completely excluded indigenous Algerians from forests.

Emblematic of this shift was a law passed on 17 July 1874, at the height of the settler colony era, which reframed the principle of collective tribal responsibility. The idea of collective punishment of tribes first arose in Algeria in 1844.[95] Its original intent was to deal with cases in which one or more of the members of a tribe had committed an offense and the tribe had refused to identify the guilty party. Originally such punishment was considered only in cases of theft and murder and then only as a last resort. Through the middle decades of the nineteenth century Algerian settlers frequently petitioned the government to apply collective responsibility in cases of wildfire, but with little success.[96] That pattern changed drastically following the passage of the Law of 17 July 1874. This law broadened the applicability of the principle of collective responsibility and made it the default ruling for fires of undetermined origin.[97] Officials applied this charge liberally in the late nineteenth century, though no tribe ever fully paid off its fine.[98] The Law of 17 July 1874 restricted indigenous activities in other ways as well. It aimed to address settlers' concerns about fire danger and to gratify outrage against indigenous groups following the 1871 rebellion. Consequently, the law placed even greater restrictions on indigenous access to forests and increased penalties for infractions.[99] It also targeted pastoralists explicitly. It banned grazing in burned forests for at least six years, apparently to discourage occupational burning.[100] Indeed the law proved much more successful in limiting indigenous rights than in preventing fires. In fact the number of forest fires continued to rise in the late nineteenth and early twentieth centuries.[101]

In Algeria the aims and impact of fire prevention policies extended far beyond forest management. Such policies expanded state control of subject populations through new regulations and the criminalization of traditional practices. They also promoted governmentality in a deeper ideological sense, by providing social and environmental justification for such control. Just as fire had long been a tool of shepherds and farmers around the Mediterranean and around the world, fire control became a tool of empire.

Natural Disasters in Anatolia

In the account of his mid-nineteenth-century journey through Anatolia, Baptistin Poujoulat recites a local proverb, joking that "the three great scourges of the Orient are the plague, fire, and dragomans."[102] Although southwestern Anatolia was as prone to wildfires as Provence and Algeria, official responses to them set this region starkly apart. Prior to the arrival of French forest experts in the mid-nineteenth century and the implementation of scientific forestry, the Ottoman state expressed little to no concern about the incidence of wildfires in its territories, unless they occurred in forests reserved for the imperial shipyard. In the mid- to late nineteenth century the Ottoman administration began efforts to better regulate activities that could lead to wildfires and to punish perpetrators. For example, it prohibited open fires in many contexts, and the Ottoman penal code of 1858 threatened arsonists with a life-long sentence of forced labor.[103] In practice, however, fire remained relatively low on the nineteenth-century forest regime's agenda in Anatolia, and such limited measures proved ineffective in checking the widespread incidence of fires. Indeed some historians of Ottoman forestry have suggested that the use of fire for forest clearing actually increased in the second half of the nineteenth century, as the population expanded and sought additional land for pasture, agriculture, and charcoal production.[104] In any case, wildfires resulting from natural causes, imprudence, and arson continued unabated well into the twentieth century. In 1903 a British consul complained of "the havoc played by the frequent fires due to the ill-will of shepherds and woodcutters," which he ascribed to the "negligence" of local authorities and the lax enforcement of forestry legislation.[105]

Meanwhile, the Ottoman administration was busy addressing a host of other environmental challenges plaguing Anatolia during the nineteenth century. Three major earthquakes shook the peninsula during the course of the century, with at least twenty tremors of an intensity of 5.0 or above in its Mediterranean coastlands alone. In 1831 an epidemic in Izmir marked the arrival of cholera in the Ottoman Empire. At the same time, urban centers throughout the empire continued to suffer from outbreaks of plague, even as it receded from western Europe.[106] For the rural population, however, the impact of droughts, frosts, locusts, epidemics, and wildfires was undoubtedly worse. Locusts made a regular appearance in fields and pasturelands, with particularly devastating visitations in the 1850s, 1860s, and 1880s.[107] In 1845 officials claimed to have exterminated two hundred thousand *kıyya* (*okka*), or nearly three hundred tons, of the insects.[108] In 1880 the British consul in Antalya reported that locusts had destroyed harvests in the *vilayet* of Antalya for at least five years.[109] The region, which had formerly exported grain, was forced to import it. When environmental factors such as pests, drought, flooding, frosts, and hailstorms appeared in rapid sequence, or when heavy snowfall hampered transportation, widespread famine could result. Famine often occurred in the wake of periodic droughts, and Ottoman reports documented droughts and famine in Anatolia in nearly every decade of the nineteenth century.[110] In the early 1870s two years of brutal dryness precipitated a particularly devastating famine in rural Anatolia. Inhabitants were still "bewail[ing] the disasters" when British traveler Frederick Burnaby visited the region several years later.[111]

In such desperate times nomadic tribes suffered alongside their sedentary neighbors, since they depended on agricultural success for their own nourishment as well as that of their herds. Food shortages typically hurt livestock more than human inhabitants. During the famine of 1873–74 many pastoralists lost their herds to starvation or were forced to slaughter them to survive. One community whose chief industry was goat (angora) hair lost 60 percent of its goats, sheep, and cattle.[112] In the vilayet of Konya, where mobile pastoralism made up a major part of the economy, the devastation of livestock was even more dramatic. The famine reduced one village's population of seventy thousand cattle, sheep, and goats to no more than four thousand.[113] It also temporarily halted the manufacture and trade of carpets, along with transportation and communication between Konya and Antalya and other Mediterranean centers south of the Taurus range. The famine, moreover, occurred in the wake of the Ottoman state's most ambitious forced settlement campaign, Fırka-i İslahiye, and greatly intensified its impact on Anatolian nomads.[114]

Nomadic victims of natural disasters had little to hope for in terms of administrative aid. Traditionally, Ottoman disaster victims fell under the realm of charity and consequently outside of the government's purview, but the Tanzimat era inspired new initiatives in disaster management and relief.[115] Following the famine of 1873–74, Sultan Abdulaziz invested considerable expense in alleviating its effects through provisioning and resettlement.[116] The state also periodically provided relief for victims of earthquakes.[117] Likewise, the Ottoman state fought locust invasions through mass pest-extermination campaigns, though it did little else to prevent these occurrences or deal with their social and economic ramifications. As a rule, however, nomads received few benefits from such programs. In the wake of the disastrous events of the 1870s, the British consul-general in Anatolia, Col. Charles Wilson, hypothesized that their exclusion was probably deliberate, part of the effort to compel nomads to settle.[118]

Nomads responded in various ways to environmental stress. For many, environmental disasters provided additional motivation to relinquish their peripatetic lifestyle. Beginning in the 1870s, agricultural production skyrocketed along Anatolia's Mediterranean coastlands, while nomadism dwindled. Many of the region's new farmers were tribal members who had responded to government incentives—or pressure—to settle.[119] However, the combination of environmental and political pressure proved all the more irksome for Anatolia's remaining nomads. Years scarred by famine, drought, plagues, and other calamities witnessed an increase in brigandage, destruction, and open rebellion among members of the rural Anatolian population, especially nomadic tribes. In the late 1870s and 1880s lawlessness erupted along Anatolia's southwestern Mediterranean coast, ending the relative stability of preceding decades. Incidents of tribal resistance and rebellion occurred periodically through the end of the century.[120] This region, where nomads and farmers had coexisted relatively peacefully for centuries, began to reflect the stereotypes of nomadic banditry with which Europeans had long associated it.

European observers ultimately tended to view Anatolia's political and environmental challenges as casualties of the Ottoman administrative system. Amid the crises of the 1870s, Colonel Wilson fumed, "At no time in the past 25 years has the provincial government been so weak, outrage so rife and unchecked, anarchy so rampant, or the reins of authority so slack."[121] Although he shared the state's preference for sedentarization, the British consul-general maintained that its neglect of nomadic groups affected by natural disasters was "not the way to obtain the desired result." He surmised, "It is by such ill-endured measures that the Porte either ruins the people or forces them to become robbers."[122] Similarly, Burnaby blamed the inability of Anatolia's population to recover from years of famine on an oppressive system of taxation as well as on rural communities' limited economic options.[123] Others made even broader connections between Ottoman administration and environmental phenomena. Baptistin Poujoulat linked Ottoman "decadence" to desertification, claiming that the lands of the Ottoman rulers "have become arid deserts in their hands."[124] In the eyes of another mid-nineteenth-century French traveler, Georges Perrot, the connection between Ottoman administration and environmental degradation was even more explicit. In a memoir of his visit to Anatolia he exclaimed, "Nothing is sadder, in Asia Minor, than these vast deserts where it seems that the earth is waiting and calling the plow that fertilized it for so long," and he lamented the woes of a rural population forced to fight constantly against "poverty" and "bad government."[125]

Such comments expose critical differences in perceptions of natural disasters. Nineteenth-century Ottoman subjects did not, as a rule, view natural disasters as the result of destructive human actions. Well into the twentieth century, both Muslim and non-Muslim inhabitants continued to interpret major calamities in religious terms. The Ottoman state, for its part, was less concerned with the causes of natural disasters than with their effects and the needs of affected communities.[126] In contrast, Europeans tended to emphasize the human role in natural disasters. On the one hand, they blamed nomads for contributing to threatening environmental conditions; on the other, they faulted the Ottoman government for neglecting its environment and its population.

The human element is and always has been implicit in standard uses of the term "natural disasters." Newspapers do not report on hurricanes, earthquakes, avalanches, or floods unless they ravage coastal towns, destroy buildings, trap people under snow, rocks, or rubble, or at least exact some human toll. In addition, human actions can influence the incidence and intensity of certain environmental calamities. Scholars have even suggested that in our Anthropocene world, humans are not just the victims of natural disasters but the cause.[127] Nineteenth-century French scientists, scholars, officials, and the broader public also tended to view natural disasters through the lens of human actions, but for different reasons. In that context, human associations with environmental phenomena helped observers to make sense of them and to find someone to blame. The high number of floods, droughts, fires, and other environmental catastrophes reported throughout the Mediterranean world in the nineteenth century provided undeniable benefits for the forest administration. Such events, along with the widespread belief that they were occurring more frequently, served to justify stricter legislation and to promote the image of foresters and other governmental officials as protectors of the earth and its human populations.[128] In addition, natural disasters supported the foresters' mission by legitimizing the marginalization of Mediterranean mobile pastoralism. They also expedited the decline of this practice directly—by wiping out livestock and ruining pastures. Fire belongs to this broader history of Mediterranean natural disasters, but it is also distinct. Nineteenth-century French scholars, foresters, officials, and colonists tended to view Mediterranean wildfires as the direct result of human actions, and they focused blame on mobile pastoralists. In the Algerian case, accusations took on a racial dimension, providing additional fodder for the demonization of indigenous tribes and the justification of the French imperial mission. Through their association with fire and other environmental catastrophes, Algerian pastoralists—and to a lesser extent mobile pastoralists in other Mediterranean contexts—effectively became scapegoats for nature.

CHAPTER 7

Sheep to the Slaughter

Mediterranean Pastoralism and Forestry at
the Turn of the Twentieth Century

Today nomadism—which no longer exists around the Mediterranean in
its residual state, it is true—consists of the knot of about ten people who
might be seen round a fire at nightfall . . . the relic of a tradition which is
slowly disappearing.

—Fernand Braudel, *The Mediterranean and the Mediterranean World in the Age of
Philip II*

In the early twentieth century the French forest service stepped back
to evaluate its mountain restoration project. The project originally had
aimed to extend France's forest cover by nearly three million acres.
By 1925, however, it had reforested little more than one-tenth of that
amount.[1] Still, it had succeeded in turning the tide of deforestation;
forests were advancing slowly but steadily across France. The campaign
against Mediterranean pastoralism had also seen results. The mountain
reforestation project and more broadly the French forest administration had largely eliminated pastoralism from Provence's woodlands,
as well as from many other parts of the Mediterranean landscape. This
initiative succeeded in part because it complemented other pressures
facing mobile pastoralists. In addition to the environment-related challenges of forest legislation, property designation, and natural disasters,
nineteenth-century Mediterranean pastoralists increasingly had to
cope with the social and political challenges of modern society. In

Provence, Algeria, and Anatolia, the arrival of new communication and transportation networks made it easier for the state to control mobile populations and harder for the latter to evade authority or to remain mobile. Similarly, population growth in Mediterranean coastlands forced pastoralists to share space with agriculturalists, industrialists, and officials, preventing the extensive access to pastures that their tradition required. While they struggled to adapt to such circumstances, Mediterranean pastoralists witnessed the virtual collapse of the market for their wool. Together such factors drove the transformation of Mediterranean pastoralism over the course of the nineteenth century. This tradition had entered the nineteenth century as a unifying feature of Mediterranean society. By the turn of the twentieth century its representatives around the inner sea were virtually unrecognizable from their predecessors and from one another.

In the process, the field of French forestry had also changed. Whereas early scientific foresters' efforts were aimed at achieving sustainable exploitation, their successors pursued the more ambitious goal of environmental preservation. They strove to transform forests into constructed "natural" spaces, to reserve the forest for the trees. Through these actions they became true conservationists *avant la lettre*. At the same time, the field developed a largely antithetical concern for the impact of forest science on human society. In short, French forestry gained a soul. Mediterranean pastoralists played an integral role in both of these developments. On the one hand, those convinced of pastoralism's environmental toll pursued policies of preservation and reforestation with ever more enthusiasm, intensity, and success. On the other hand, the marginalization of mobile pastoralists through these very measures led others to contest dominant narratives and to advocate for this age-old Mediterranean tradition. Thus, as mobile pastoralism receded from Mediterranean landscapes, it won new champions for its cause.

This chapter updates the story of Mediterranean pastoralism to the twentieth century by examining the interplay of such developments. It shows how new perspectives on French scientific forestry, complemented by social and political factors, led to the divergence of pastoral industries in Provence, Algeria, and Anatolia. Although these changes fueled the marginalization of both mobile pastoralists and forests around the Mediterranean, they ultimately failed to destroy either.

New Perspectives on Pastoralism in Provence

In the second half of the nineteenth century a small but influential faction of foresters began to express newfound sympathy and support for Mediterranean pastoralists. This perspective owed much to the mid-nineteenth-century forester-sociologist Frédéric Le Play, one of the first French foresters to study the relationship between forests and their social milieu.[2] Although he shared his contemporaries' anxieties about deforestation, Le Play blamed the government and the aristocracy for much of the damage.[3] The forest conservator Félix Briot echoed this view in *Les Alpes françaises*, first published in 1896. In this critical study of forest administration and the Alpine economy, Briot contested prevailing declensionist narratives, branding claims of "abuses of exploitation committed by the population" as "certainly exaggerated."[4] In stark contrast to other forest agents of the time, moreover, Briot painted mountain peoples as "in reality so wise, so intelligent and diligent," and he advocated working with them to recreate the agro-sylvo-pastoral society that for three-quarters of a century his colleagues had been trying to destroy.[5]

The geographer Lucien-Albert Fabre went even further in his support for pastoralists. His work includes more than seventy-six volumes devoted to mountain economy, rural life, and botany.[6] His choice of titles, including *La fuite des populations pastorales françaises* (The flight of French pastoral populations) and *L'exode du montagnard et la transhumance du mouton en France* (The exodus of the mountain dweller and sheep transhumance in France), suggest where his sympathies lay. Fabre became one of the most ardent proponents of a sylvopastoral economy, as well as one of the greatest critics of reforestation efforts that disregarded local populations. "We spend more than three million per year to bandage the earth's wounds without ever attacking the harm at its source," he wrote, referring to the desperate economic state of mountain peoples.[7] Fabre's perspective presented difficulties for his career as a forest agent, but it won him considerable distinction in the scientific community, and it helped spread interest in a more humanistic approach to forest science.[8] A handful of other foresters, geographers, and scholars joined Briot and Fabre in reassessing the environmental impact of pastoralism in Provence.[9] These forester-sociologists formulated the ideal of a practical, harmonious agro-sylvo-pastoral economy. They contested the perception of mobile pastoralism as a devastating force, and they introduced the suggestion that it might be, on the contrary, the most economically viable and environmentally sustainable use of certain Mediterranean ecosystems.

A similar reconceptualization of mobile pastoralism infiltrated the popular literature of Provence during this period. No sooner had shepherds and sheep retreated from the realm of common daily experience than they began to reappear in romanticized form. Frédéric Mistral and other prominent Provençal authors and poets of the mid-nineteenth to mid-twentieth centuries, including Alphonse Daudet, Marie Mauron, Émile Zola, and Jean Giono, took up the pen in support and celebration of this tradition. These writers recast pastoralists as gentle, noble, solitary, and wise, praising their innocence and simplicity of life. Their characterizations often echoed the image of the "noble savage" that had influenced European colonial perceptions more than a century earlier. In other ways as well, their works depicted a bygone, idealized provincial society predating the imposition of modern institutions and the state. By treating Provençal shepherds as noble savages and integrating them into a broader regionalist campaign, these literary figures indirectly confirmed the triumph of governmentality and empire. Their work is evidence that by the turn of the twentieth century France's internal colonization of Provence was complete.

The French state facilitated the decline of Provence's mobile pastoral industry not only through state forestry and other environmental policies but also through social, economic, and technological means. In the second half of the nineteenth century the national trend of urbanization reduced the potential number of forest users—and abusers—in rural areas, but in other places increasing population density made mobile pastoralism progressively less feasible. At the same time, the construction of railroads and other transportation networks improved communication and access to previously remote regions. Such developments created new obstacles for migrating herds and made it harder for shepherds to elude forest regulations. Meanwhile, French pastoralists were reeling from an even greater blow: the wool crisis. In 1860 France signed the Cobden-Chevalier Commerce Agreement with England, lowering import duties. Following implementation of the agreement, French imports of English wool doubled. Although French exports of wool also rose, the price of wool plummeted. During this time, rising competition from Australia, Spain, and the internal market of Algeria were already placing substantial pressure on French sheep farmers.[10] The reentry of American cotton following the conclusion of the Civil War in the United States dealt another crushing blow to the French wool trade.

TABLE 2. Prices of French wool, 1850–1910

Year	Price per kilo of washed wool (francs)
1850–54	8.20
1860–64	8.91
1870–74	6.80
1875	7
1891	4
1894	3.40
1899	4.50
1910	3

SOURCE: A. Orange and M. Amalbert, *Le mérinos d'Arles* (Antibes: F. Genre & Cie, 1924), 35.

These circumstances precipitated a crisis that would haunt French wool producers for the rest of the nineteenth century. Across France the practice of pastoralism declined swiftly or disappeared altogether. In the second half of the nineteenth century the number of merino sheep declined 70 percent throughout France.[11] Pastoralists suddenly found themselves forced to sell fleeces at a fraction of the cost of maintaining their herds, and these circumstances put many out of business.[12] Mountain dwellers were hit particularly hard, and such economic troubles helped to drive many inhabitants toward commercial centers on the plain.[13] As a rule, only large commercial enterprises were able to survive. They did so by shifting their focus from wool to meat production, a trend that has continued to characterize Provence's sheep farming industry to the present day.[14] The wool crisis, together with privatization and other nineteenth-century developments, ultimately made forest legislation considerably easier to enforce. At the turn of the twentieth century the numbers of sheep around Arles again reached pre–wool crisis levels, but most of these animals belonged to large-scale farms and were destined for slaughter. Provençal sheep farmers still complain that the cost of shearing their sheep outweighs the profit they gain from it. Clearly, the age of the "golden fleece" has passed.

Fin-de-Siècle Algerian Society and the Forest Code of 1903

In Algeria as well the fin de siècle brought new environmental and political perspectives to the fore. The debate over mountain reforestation had appeared against the backdrop of the settler colony era, when colonial rights triumphed at the expense of the indigenous population and the environment. In the final decades of the nineteenth century settlers still tended to view mobile pastoralism as "the worst enemy of [Algeria's] forests."[15] At the same time, both the colonial forest administration and the settler colony era came increasingly under fire. Drawing on new perspectives on colonialism and France's domestic pastoral industry, critics in Algeria as well as in the metropole began to assert that the French colonial mission had taken a grievous wrong turn, and they worked actively to put it back on track. As the commander general of Setif, a town in northern Constantine Province, explained in a letter to the regional forest inspector in 1895, the challenge was to "safeguard native rights without compromising the interests of the state and the future of its forests."[16] Even for a true *indigenophile*, this was no easy task. A few influential allies, however, helped raise the profile of resistance to settler politics.

In the late nineteenth century revisionist perceptions of Algerian pastoralism spread within French scientific circles and forest administration. Louis Trabut, a respected botanist, professor of natural history, and doctor practicing in Algiers, was one of the first scientists to challenge prevailing views on pastoralism in colonial Maghreb. In 1891 he and Auguste Mathieu, the French forest conservator of Oran, conducted an official investigation of that province's Hauts-Plateaux.[17] Their report painted the region as a model "pastoral steppe" unsuitable for agriculture or other industries.[18] They countered common characterizations of tribes as ignorant, noting, "Native pastoralists don't have any of the science one learns in school, but, living in daily contact with nature, they benefit from personal observations and tradition."[19] The authors even suggested that indigenous inhabitants could make better use of this environment than European settlers. They agreed that the region had suffered from environmental decline, but they blamed colonists for the majority of the damage.

Such assessments drew support from the central government in France. In the spring of 1892 a commission headed by former French prime minister Jules Ferry arrived to investigate accounts of colonial exploitation of the environment and the indigenous population.[20] Ferry's report provided a harsh critique of the colonial government and forest administration. He branded the government's approach to indigenous affairs a "policy of oppression," and he called for more enlightened administration and oversight. "The interests of indigenous people," he wrote, "must to no degree be abandoned to the European element."[21] As a result of this investigation, the governor-general was forced to resign, and the settler-friendly Warnier Law was finally suspended after holding sway for twenty years.[22] Needless to say, such prescriptions by the metropole only sharpened the lines of debate about colonial government. Paulin Trolard fired back with a forecast that the Senate's initiatives would ultimately doom Algeria's indigenous population, claiming, "With your own hands, senators, you will have dug their grave."[23]

As if to prove Trolard right, the colonial lobby obstinately blocked every effort by the new governor-general to change the status quo, and the years following the Jules Ferry Commission consequently witnessed little improvement in the state of indigenous affairs. But if Ferry's efforts to reform colonial politics ultimately fell flat, his commission did succeed in gaining serious political consideration for the subject of Algerian forest legislation. In his report Ferry had argued strongly in favor of an independent Algerian forest code that would take into account Algeria's distinct environmental and social characteristics.[24] Ferry was particularly concerned about the regulation of forest grazing, which he maintained was "for the forest dweller, one of the necessities of life."[25] He and the other members of the 1892 senatorial commission agreed that neither the French Forest Code of 1827 nor the law of 1885 was appropriate for the Algerian case. In the fall of 1892 they created a committee within the colonial administration to develop a new forest bill specific to Algeria.[26] This development was greeted with enthusiasm by multiple colonial factions representing a wide range of notions about what such a law would entail.

The idea of creating an Algerian forest code had been voiced at various times since the beginning of the colonial era, but it had never been pursued.[27] In the 1870s, however, a number of developments had served to place this subject in the public eye and to rally support among the colonial population. The 1871 revolt was instrumental in this process. The ensuing rash of forest fires had been particularly devastating for cork concessionaires. Convinced of the culpability of "malevolent" indigenous tribes, they reproached the forest administration for failing to take punitive action. Many officials agreed. A report published in 1874 cited the vast cost and extent of the damages as evidence of the urgent need for an Algerian forest code. Repeating a common complaint, it argued, "The forest code . . . was not written for Algerian forests. A special law is necessary."[28] Interest in an independent forest regime gained speed following Louis Tassy's investigation of the Algerian forest administration. Although he did not recommend the detachment of the Algerian forest service from French supervision,

European settlers cited the problems he described as support for this cause. Following the publication and public distribution of his report in 1882, Tassy's name was used freely to endorse the liberation of the Algerian forest regime, along with various other perspectives that the forester had never intended. Indeed by 1885 the governor-general of Algeria was using the report, which he characterized as an example of "declarations made by very competent men," to support his claim that the forest service was jealously guarding nonforested land that should be opened up to colonial exploitation.[29]

By the late nineteenth century most French Algerians agreed that no number of amendments or modifications to France's Forest Code of 1827 would be sufficient to adapt it to the Algerian case, but they disagreed over the form that an independent code should take. Many French settlers and colonial administrators supported the perspective of the Ligue du Reboisement, which argued strongly in favor of a code that would impose strict forest regulations, especially with respect to pastoralism and other indigenous practices. Others, including numerous officials from metropolitan France, promoted the relaxation of forest laws and greater latitude toward local traditions. After a tedious process of development and review, Algeria finally received an independent forest code in 1903. Given the debate that had preceded its development, public opinion on the code was surprisingly favorable. For French settlers and *pieds-noirs* it represented a powerful symbol of freedom from the French central government. Internationally, forest experts judged the code to be well suited to its environment while maintaining an exemplary balance between conservation and commercial exploitation. The American forester Theodore Woolsey was so impressed with the code that he took the trouble to translate it into English so that it could be appreciated by a wider audience.[30]

The Algerian Forest Code maintained the ethic of its French model, tailoring it to the Algerian case. Superficially the code seemed to recognize and accept the practice of indigenous pastoralism. It tempered some of the previous legislation's provisions to account for local circumstances. It reauthorized grazing in woodlands, a practice that had persisted throughout the colonial era despite being banned by the French Forest Code of 1827, and it referred specifically to livestock typical of Algeria: cattle, camels, sheep, and goats.[31] Yet the overall impact of the new code on nomadic pastoralists was resoundingly negative.[32] Its conditions for grazing were even more complex than those in effect in France. The right of pasture could be granted only where both the civil administration and the forest service agreed that it was "an absolute necessity to the inhabitants" and then only in certain designated areas.[33] In addition, the passage of the law definitively stripped pastoralists of free access to communal pasture. Instead sheep and goats were banned from communal woods, and shepherds were compelled to rent pastures.[34] The law also targeted indigenous Algerians by maintaining the state's right to expropriation for "public utility."[35] It imposed penalties, including fines, imprisonment, and forced labor, for illegal grazing as well as for the broadly defined crime of "deforestation," which could be committed through "excessive exploitation."[36] Through such measures the law presented new challenges for Algerian pastoral tribes.

The situation of indigenous Algerians in the early twentieth century made the new legislation's impact all the more dramatic. By this time many pastoralists had already lost their land, their livestock, and their livelihood. In the final decade of the nineteenth century the number of Algerian-owned sheep had fallen by 25 percent.[37] In 1911 the number of sheep owned by indigenous Algerians was half of what it had been in 1870, while totals of European-owned sheep had doubled.[38] Moreover, indigenous pastoralists had all but disappeared from the more fertile regions of Algeria. Through forest and property legislation as well as the persistent efforts of the colonial population, they were relegated to the sparsely vegetated Hauts-Plateaux, the Atlas, and the Sahara. This shift was paradoxically due in part to the very efforts of sympathetic scientists to promote an indigenous agro-sylvo-pastoral economy. Inspired by the findings of Trabut, Mathieu, and other scholars, a growing body of scientists and colonial officials promoted the Hauts-Plateaux as a haven for indigenous pastoralism. From an administrative perspective, this region seemed to promise an outlet for population pressure, indigenous unrest, and environmental stress. Accordingly, the government began systematically relocating indigenous groups from across Algeria to these dry, sparsely vegetated plateaus.[39]

As early as 1906 one critic recognized the potential of the Algerian Forest Code to "impede the movements of the natives."[40] Indeed the law had sought successfully to relegate indigenous pastoral groups to places where the forest administration believed that they would do the least environmental harm. The close of the nineteenth century thus marked the definitive retreat of pastoralism from the fertile Tell.[41] Looking back on this transformation, a Bedouin poet later wrote, "From our spacious tents, we have made miserable huts; from our immense herds, we now have a few beasts."[42] In contrast to Provence, where mobile pastoralism's scientific and intellectual paladins won it fresh appreciation, advocacy for indigenous pastoralists in Algeria hastened their retreat from Mediterranean coastlands. In this way Algerian nomads had finally been "civilized."

The resettlement process had environmental consequences as well. Unlike the fertile Tell region, the Hauts-Plateaux area has only meager grasslands, and water sources are scarce. Pastoralism in this region requires more space and effort, and it yields less profit. Once relocated to the Hauts-Plateaux, pastoralists found their herds needing nearly four times the pastureland that had sufficed on the Tell.[43] Crowded onto less-productive terrain and denied the possibility of transhumance, indigenous pastoralists quickly depleted the ecological potential of their new home. One scholar has summarized this story by remarking ironically, "The degradation of the best pastures by the expansion of agriculture led to a greater overexploitation of the poorest pastures."[44] As in Provence, nineteenth-century French environmental policies, together with social and political developments, effectively made claims about Algerian pastoralism's environmental impact a reality.

The Legacy of the French Forest Mission in Anatolia

The lives of pastoral tribes in southwestern Anatolia also changed dramatically during the second half of the nineteenth century, but their fate differed from that of mobile pastoralists in Provence, Algeria, and other parts of the Mediterranean. By the 1880s a combination of factors had greatly reduced the number, status, and independence of nomadic tribes. The role of environmental administration, though not insignificant, was less prominent in this transformation than property legislation, settlement campaigns, the arrival of refugees, and other changes associated with the modern Ottoman state. Ironically, it was in Anatolia, where forest legislation exercised much less of a direct effect on pastoral groups than in Provence or Algeria, that the marginalization and sedentarization of Mediterranean mobile pastoralists enjoyed the greatest success.

In the nineteenth century Ottoman policy toward nomads changed swiftly as the government bureaucracy developed the means and the justification to extend control over the mobile population. Thus, in the late 1850s the Ottoman administration dropped its conciliatory tone toward nomads and stepped up sedentarization efforts.[45] Over the course of the second half of the nineteenth century tribes throughout Anatolia were settled through a combination of voluntary submission, coercion, and occasional violent confrontations. In 1865 the state launched its most expansive military operation of this campaign (Fırka-i İslahiye), which drove the submission and sedentarization of tribes throughout the empire.[46] In the 1870s this initiative finally began to yield results, in part because it targeted populations already weakened by famine and other environmental stresses.[47] Its impact on the province of Adana, where nomads previously had made up nine-tenths of the population, was particularly profound. In general, tribes seen as dangerous or politically threatening were prime candidates for sedentarization, as were those who ranged on or near trade routes, areas affected by agricultural expansion, or valued resources such as forests.[48] Many members of such groups suffered a fate much worse than the loss of their peripatetic lifestyle. By the turn of the twentieth century approximately twenty thousand nomads had died in Ottoman settlement campaigns.[49]

Tribal inhabitants who had not been settled by force often chose to settle voluntarily for economic reasons or government incentives. By the early twentieth century yörük tribes counted both mobile and sedentary members, and the Anatolian tradition of nomadism was quickly disappearing. Nonetheless, much of Anatolia remained sparsely populated well into the twentieth century, and those who chose to remain mobile gradually retreated into these peripheries. In some cases even regions with a substantial agricultural population continued to provide seasonal pasture for mobile pastoralists. Thus, multiple yörük tribes still inhabited the hills and mountains surrounding Antalya in the late nineteenth century, despite improvements in transportation and communication, the expansion of agriculture, and an influx of refugees.[50] They remained, though in much smaller numbers, three-quarters of a century later, when anthropologists and human geographers arrived to record their daily habits and traditions.[51] As these scholars recognized, their subjects were the last representatives of a way of life on the eve of extinction.

A number of forces reshaped Anatolian nomads' relationship to their environment and assisted sedentarization efforts in the second half of the nineteenth century. The development of transportation contributed to rural population growth and agricultural expansion in rural Anatolia, and it exerted growing pressure on nomads and on forests.[52] During this period the camel's reign of more than a thousand years was challenged by the train. The decline in camel transport dealt a particularly significant blow to nomads who supplemented their income by raising camels for overland transportation and by leading caravans. The construction of railroads required forest clearing and timber, and trees became locomotive fuel during coal shortages.[53] The improvement of transportation led to increased security and administration of the countryside, which in turn fueled rural agricultural and industrial settlement.[54] Communities that quickly appeared along railway lines also exploited the surrounding forests for local use or trade, and as security increased, people grew progressively less wary of settling in regions previously dominated by nomadic tribes.[55]

In addition, the railroad facilitated the exploitation of Anatolia's resources. Transportation systems brought agricultural, commercial, and industrial enterprises to the remote countryside and the mountains. Advances in irrigation and technology allowed peasant farmers to extend the range of agriculture above and beyond the plains into areas previously frequented only by nomads. Mining operations also expanded during this period, placing an additional strain on forest resources and on nomads. The proximity of forests was a major consideration in the location of mines, which required timber for mineshaft construction and charcoal for smelting.[56] As mines burrowed ever deeper into the mountains of Anatolia, they began to appear along traditional migration routes and to draw from forests long used by nomads. At the same time, irrigation and drainage efforts began to open the marshy coastal plains of southwestern Anatolia to increasing cultivation, while the mass production of cheap quinine reduced the threat of malaria. Land previously abandoned to pasture was requisitioned for agriculture. By the 1970s such developments had made the region south of Adana, at the eastern edge of the Mediterranean, "one of the most productive and densely populated parts of Turkey," according to one account.[57] As cultivation increased, the practice of nomadism all but ceased.

In late nineteenth-century Anatolia, moreover, forestry and sedentarization worked hand in hand. As in metropolitan France and French colonial Algeria, scientific forest management in the Ottoman Empire gained assistance from the new tools of the modern state, which helped to counterbalance legislative and administrative weaknesses. The Ottomans' most ambitious settlement campaign appeared in the same year as Tassy's reappointment to the Ottoman Council of Public Works.[58] Likewise, the passage of the Land Code of 1858, despite its negative impact on forests, gave the forest regime important advantages in dealing with pastoral tribes. Such measures were complemented by the arrival of refugees, which continued in the 1870s in the wake of the Russo-Turkish War and unrest in the Balkans.[59] Together these factors progressively marginalized the empire's nomadic population. In the face of such pressures many chose to settle, and those who remained mobile lost the power to evade and resist administration.

The forest administration, meanwhile, was still struggling to gain solid footing in the final years of the nineteenth century. By 1878 the French mission had served its term, and the remaining French foresters returned to France. For the rest of the century the management of Anatolian forests fell to Ottoman hands.[60] Although French foresters continued to exercise a prominent role in Ottoman forestry through the 1870s, the empire's forest administration gradually became authentically Turkish. The forest school began to furnish more—and more competent—graduates to work under and alongside French engineers. In 1869 the forest service gained its first Turkish general director, Aristidi Baltacı. According to a Turkish source, Bricogne praised him as not only the first to hold this office but "possibly the best."[61] During the same period, the organization and responsibilities of this growing cadre of Turkish forestry personnel were codified in a series of formal instructions and legislative acts.

On 14 January 1870 the Ottoman state instituted forest legislation (Orman Nizamnamesi) designed to provide a legal basis for the forest regime. The 1870 law reflected the influence of the French foresters who helped to craft it. It incorporated elements of scientific forestry and in many ways echoed the French Forest Code of 1827. It formalized the establishment of a bureaucratic forest administration under the Ministry of Trade, presented guidelines for the forest school, specified restrictions on forest use and access, and standardized penalties for infractions.[62] It distinguished four types of forests: state (*miri*), *evkaf* (pious endowments), communal, and private.[63] Certain forests were set in reserve for commercial, naval, and other state use; customary rights were enumerated in other forests. The rights and privileges allowed in Ottoman forests, however, were generally much more extensive than in France. Under certain conditions the regulation allowed communities to exploit their local woodlands to extract timber for municipal construction projects, to burn for charcoal, to allow for the grazing livestock, and to gather firewood, deadwood, and other forest products.[64] One of the principal goals of the 1870 legislation was to discourage the use of forests as pasture.[65] The 1870 document continued to allow grazing in woodlands, but it placed a number of restrictions on the practice. Indeed five of the document's fifty-two articles dealt with grazing regulations, and five others enumerated penalties for infractions.[66] Like the French code, it limited the number of beasts admitted as well as the permitted locations, and it required the registration and taxation of herd owners.[67]

The 1870 Orman Nizamnamesi remained the basis of Ottoman forest legislation and the administration of pastoralism well into the twentieth century.[68] In the years that followed, the state supplemented this document with a number of amendments designed to clarify or modify certain clauses, and it established a special commission, headed by Yusuf Ziya Pasha, the general director of forests from 1873 to 1874, to investigate further possible improvements. By 1876 the commission had added six additional regulations to the legislation of 1870 and published instructions reorganizing the Ottoman forest administration into a complex, hierarchical bureaucracy.[69] For the first time, forest inspectors and guards were assigned to provincial posts throughout the empire. They represented a first step in bringing environmental exploitation in far-flung regions under state supervision and control. Finally, the forest directorate was transferred from the Council of Public Works to the Ministry of Finance, a move that reflected the Ottoman state's continued interest in the commercial potential of its forests, rather than conservation.[70] Back in France observers cheered the promulgation of Ottoman forest legislation, claiming that it would "offer a solid basis for agents' operations and commercial transactions from now on."[71]

In reality, however, these measures proved far from sufficient for the effective implementation of forest management.[72] The legislation of the 1870s failed to change traditional practices in part because of its relatively tolerant stance. In contrast to the French forest code, Ottoman forest administration chose to regulate and restrict customary rights, rather than banning them completely. Equally problematic was the forest service's perpetual understaffing, which prevented even these mild restrictions from being enforced. The ordinance of 1876 divided the entire empire among four head forest inspectors. Just one of these inspectors was expected to survey and oversee the provinces of Aydın, Konya, Adana, and Syria, as well as the Aegean Islands.[73] Even with a staff of sub-inspectors, guards, and day laborers, this was a daunting task. When the century reached its final decade, there were only 60 officials in the forest service assigned to the entire province of Konya, 46 in Adana, and a mere 17 in Syria.[74] In comparison, the French forest service in Algeria included by this time 355 forest guards in the province of Constantine alone, and most still considered it understaffed.[75] In addition, the forest service was forced to operate on a meager budget, a common weakness among Tanzimat reforms, and this underfunding crippled efforts to expand its operations, authority, and influence.[76]

These complications quickly became glaringly apparent. Members of the French mission were, as a rule, hesitant to criticize Ottoman forest administration. As one agent delicately explained, "Our scruples restrain us from judging that of which we are a part."[77] Still, although French observers publicly praised the 1870 regulation, they also condemned its limitations in private. Citing the rights and privileges it gave local communities, they worried that its implementation might actually expedite deforestation.[78] Turkish foresters also joined in the criticism. Looking back from the early twentieth century, the forest inspector Ali Rıza commented that the law lacked the elements necessary for "good protection and orderly administration."[79] The pitfalls of its legislation and the limitations of its administration would plague Ottoman forestry for the rest of the century.

For the Ottoman state, moreover, the late nineteenth century was rife with administrative problems. The empire suffered a severe economic crisis, a growing national debt, wars, rebellions, and nationalist unrest, not to mention the fallout from natural disasters. The oppressive and paranoid sultan Abdülhamid II, who reigned from 1876 to 1909, met such challenges with callous repression and autocratic rule. Even as he sought to tighten his grip, however, the administration of this vast, cosmopolitan empire was slipping steadily out of his control, and the forest regime was no exception. Efforts to create stricter forest legislation seemed to lead only to more violations.[80] Court records adjudicating forest abuses indicate that forest regulations were to some extent enforced but more often disregarded.[81] Foresters were censured for mismanagement or outright corruption.[82] As members of the Ottoman elite debated new initiatives, directions, and solutions for forest administration, the empire's forest cover declined steadily.

In the twentieth century Turkish forestry finally found its stride. The forest administration was revised extensively through sweeping legislation in 1917, 1920, and 1937. In 2011 the Turkish Forest Law (no. 6831) was nominated for the prestigious International Policy Award for Visionary Forest Policies, given by the World Future Council. In 2018 Turkey's forests covered nearly fifteen thousand square miles, equivalent to 27 percent of the country's surface area, a figure just 4 percent lower than France's forest cover.[83] By contrast, only 2 percent of Algeria is forested.[84] Antalya holds the most extensive old growth forest area of any province of Turkey, at about forty-two hundred square miles. Adana's forest cover, while less extensive, is impressive nonetheless. With nearly thirty-five hundred square miles of high and coppice forests combined, it reflects that area's agricultural character. Although Turkish forest administration has matured greatly since the nineteenth century, it has not forgotten its debt to France. Turkish scholarship still refers to the Ottoman forest institute established in 1857 as the Tassy School.[85] This relationship also remained a source of pride in France long after its official end. In the late nineteenth century the general director of forests in France recalled how fondly Louis Tassy's Turkish colleague, the minister of public works, had remembered him. "You would have been proud like me," he declared, "to have heard with what expressions of respect and affection this foreigner, while retaining his oriental solemnity, displayed regret to have lost Monsieur Tassy."[86]

The forests of Anatolia suffered immense strain during the nineteenth century, though their greatest antagonists were not goats or nomads but agriculture, industry, and the railroad.[87] Late nineteenth-century European observers were correct to consider the nomads' treatment of forests to be destructive, but they were wrong to assume that such treatment was standard and entrenched. What they saw was not any long-term tradition but the final attempt of a desperate, dwindling nomadic population to exploit the marginal lands remaining as pasture. Although nomadism did not disappear entirely from Anatolia in the nineteenth century, it began an irreversible course of decline. The ultimate success of the Ottoman campaign against nomads owed much to European biases against pastoralism. Such attitudes allowed Ottoman officials to cast sedentarization as environmental progress and a step toward modernity, despite the detrimental effects it had on a major portion of its population.

By the early twentieth century mobile pastoral traditions had faded from many parts of the Mediterranean. In Provence, flocks of sheep had surrendered to orchards, vineyards, and forests of Aleppo pine. For a handful of large-scale commercial farmers in the vicinity of Arles the pastoral industry was thriving. Yet many of them kept their livestock on the plain year round, and nearly all of those who still engaged in transhumance had begun shipping their herds to summer pastures by train. In Algeria, most indigenous pastoralists had been relegated to the Sahara and the Hauts-Plateaux, where they lost not only their ancestral lands but also the option to migrate seasonally.[88] Westerners who encountered them dismissed them as "malnourished and poorly dressed."[89] European settlers arrived in increasing numbers to compete with them, sometimes occupying the very territories that indigenous pastoralists had been forced to relinquish. Meanwhile, mobile pastoralism was also disappearing from the eastern Mediterranean. In southwestern Anatolia, the nomadic tribes had lost much of the freedom and fortune that had once made their lifestyle appealing. Required to rent pastures for amounts nearly equal to their total profits, limited to ever more marginal lands and sinuous migration routes, and under increasing supervision by suspicious state officials, many nomads chose to settle.[90] By many accounts, those who maintained their mobile habits had become, like their Algerian counterparts, a pitiful bunch indeed. "Roving over barren districts," wrote an American observer of the yörük in 1905, "the members of this group are true half-starved human products bred in areas of defective food supply."[91]

Still, mobile pastoralists have persevered all around the Mediterranean. In Anatolia and Algeria, tribes still range in the periphery, resisting administration and sedentarization.[92] In Provence, the practice of transhumance by foot has actually regained appeal in recent years. Locals seem to enjoy the sight of passing herds enough to tolerate the resulting traffic jams, and they value the meat of transhumant over stabled sheep.[93] Today's shepherds are often urban dwellers seeking a return to nature or captivated by the romanticized past of their trade.[94] To this day Provence continues to celebrate its past. Pastoral festivals are held seasonally along traditional migration routes, and images of sheep and shepherds grace postcards, placards, sculptures, and other art throughout the region. Through these vestiges, Mediterranean pastoralism remains an immemorial tradition. Even where pastoralism disappeared completely, this process was not simply a tale of exploitation and expropriation, of the triumph of the state. Instead it involved the confluence of a number of factors, including developments in transportation, technology, communication, population expansion, and the transformation of the global wool market, as well as the agency of Mediterranean pastoralists themselves. In Provence, Algeria, and Anatolia the nineteenth-century transformation of pastoralism involved much more than the simple imposition of imperial values, science, politics, and culture. It was a process of negotiation whose effects on French forestry and administration were equally profound.

Conclusion

Planting Politics

It is not part of a true culture to tame tigers, any more than it is to make sheep ferocious.

—Henry David Thoreau

The nineteenth-century encounters of mobile pastoral groups and foresters in Provence, northern Algeria, and southwestern Anatolia belong to a Mediterranean story. The connections and continuities among their experiences originated in part through their relation to this inner sea. It connected their economies through trade and brought similar weather to their shores. Today Mediterranean societies continue to display unifying elements in diet, architecture, industry, economy, lifestyle, language, and more. In the past such ties were even stronger. Moreover, the very variability of the Mediterranean world—its dramatic topography, diverse vegetation, capricious climate systems and rainfall patterns, multitude of states and societies—might itself be characterized as a common feature. The intersection of mobile pastoralism and French scientific forestry in the nineteenth century added new dimensions to this Mediterranean connection. At the turn of the nineteenth century most of the inhabitants of these three places were active participants in local agro-sylvo-pastoral societies. Over the following decades all witnessed the development and implementation of "scientific" forestry as well as the propagation, evolution, and solidification of narratives condemning their traditional lifestyle. They responded to these challenges in similar ways: through petitions, infractions, protest, arson, vandalism, violence, and ultimately adaptation.

The foresters who led the campaign against mobile pastoralists in these three contexts also shared certain features. First and foremost they were French, and they had graduated from a school that taught them to believe in the ideal of lush, dense woodlands and the evils of sheep and goats. Through the lens of such narratives the environments and economies of the Mediterranean world appeared especially impoverished. In Provence, Algeria, and Anatolia the French foresters proudly defended this cause by promoting reforestation and antipastoral legislation. Although the success of their war on deforestation varied considerably, by the turn of the twentieth century one of their main goals had come to pass. In all three places pastoralism had lost much of its value, and it no longer played a central role in society. Many pastoralists had given up their peripatetic ways, while those who clung to their mobility had retreated into the periphery. This long-standing bastion of the Mediterranean economy had finally broken down.

Yet this study is not simply a tale of two factions. Pastoralists and foresters were players in a broader web of governmentality. The nineteenth-century French government used forest administration to regulate people as well as forests, and it helped to bring France's mobile populations in Provence and Algeria under the control of the state. Perhaps even more significantly, French scientific forestry provided a vehicle for the legitimization of state initiatives against mobile pastoral groups in the French Empire and in other applications of French forestry around the Mediterranean. Moreover, foresters and states drew considerable benefit from trends that fueled the transformation of Mediterranean pastoralism and the marginalization of mobile pastoral groups, such as urbanization and population growth.

This story also involves critical distinctions in the cases of Provence, Algeria, and Anatolia. Each context represents a unique complex of relations among local pastoralists, foresters, and the state. In geographical terms French forest agents were closest to home in Provence, but even there local inhabitants often regarded them as foreigners, while foresters responded by branding local customs as primitive and environmentally destructive. Through legislation, rhetoric, enforcement, and reforestation, the forest regime played a crucial role in the gradual consolidation of French central authority and French nationality in Provence. The French state used forest administration as a more explicit tool of empire across the sea in Algeria, where scientific forestry served both to subjugate indigenous groups and to validate their subjugation. At the same time, French colonists took advantage of their distance from Paris to exercise greater influence over the colonial forest regime, manipulating it to serve their ends. Foresters for their part struggled to balance their understanding of the environment with this powerful colonial lobby and with state demands. In contrast to Provence, where their perspective was typically representative of the central administration, forest agents in Algeria often found themselves at odds with both settlers and the state. They ultimately succeeded in driving indigenous pastoralists from more fertile regions, only to watch the country's forests disappear at the hands of cork farmers and other colonial entrepreneurs.

Anatolia represents a third model. French imperialist interests there were only vaguely defined. The forestry mission allowed the French state to exercise a form of soft power over the Ottoman Empire as well as to increase its international reputation as a torchbearer of modern science. In practice, however, French foresters found themselves virtually powerless within the Ottoman government apparatus. In their view their hosts lacked the interest, flexibility, and finances to successfully implement forest management reform. However, under their watch mobile pastoralism retreated much more quickly in southwestern Anatolia than in other Mediterranean contexts. French foresters could hardly take credit for this change, but they could celebrate its results. At the turn of the twentieth century, however, French scientific forestry's social and environmental legacy in the Mediterranean world was anything but clear.

In recent years ecologists have begun to reassess the nature of environmental change in the Mediterranean region. Land degradation is difficult to measure, but many scientists argue that it is now taking place at a much faster pace than ever before.[1] In any case, the Mediterranean environment is clearly in flux, and humans have a major hand in its ongoing changes. At the same time, scientists have developed new perspectives on the impact of sheep and goat grazing in Mediterranean ecosystems. Although they continue to recognize the potential of grazing to stimulate soil erosion and land degradation, most specialists now agree that traditional pastoral systems were not a major factor in past environmental decline.[2] Moreover, many ecologists have argued convincingly for the essential role of grazing in the life cycle of Mediterranean vegetation, which has adapted to the presence of sheep and goats.[3] In general, scientists today recognize the value of grazing and controlled burns in limiting wildfires. As recent studies have shown, grazing exercises a pruning effect on vegetation that increases its nutritive value while limiting the extent of fire-prone biomass.[4] In places where the current environmental impact of sheep and goats is significant, it is usually due not to grazing but to overgrazing. In these regions, often having marginal or already degraded soils, the density and intensity of pastoral consumption exceeds the vegetation's rate of regeneration. Notably, these are the very characteristics of the grazing regimes that nineteenth-century administrators created when they banished pastoralists to the Hauts-Plateaux in Algeria, crowded them onto the Crau in Provence, or forced them to settle in Anatolia. In these and other contexts contemporary scientists, engineers, and officials are working to undo or mitigate the environmental damage of systems that their predecessors put into place more than a century ago.

Ecologists also have begun to develop new management strategies that take into account the inherent instability of ecosystems. Rather than viewing livestock populations in terms of a rigid carrying capacity, as they did in the past, contemporary scholars now view ecosystems as being in a state of "nonequilibrium," one that is continually reshaped by human and natural disturbances.[5] From this perspective the relative adaptability of mobile pastoralism makes it a more appropriate and practical use of certain dynamic environments than unirrigated agriculture or other economic systems. Its environmental impact is also minor compared to that of sedentary pastoralism or agriculture.[6] Likewise, common property regimes have proven more effective than private landownership in regions of limited annual precipitation or of high annual rainfall variability, where the risk burden of environmental disturbances is too great for a single individual to bear.[7] All of these factors made mobile pastoralism particularly well suited to the Mediterranean environment, especially before the increase in population density in the mid- to late nineteenth century.[8] Thus, contemporary science has resurrected and redeemed many of the very practices that nineteenth-century French forest science fought so bitterly to destroy.

✻

While nineteenth-century French foresters were cultivating trees, they also were effectively planting politics. In the application of scientific forestry to Provence, Algeria, and Anatolia, French administrators were influenced by standards of environmental legislation set in other states, especially Prussia and other German principalities. The French forest mission to the Ottoman Empire provided an ideal opportunity for France to maintain its international reputation as a Great Power and to secure economic and intellectual assets abroad. Likewise, growing concerns about deforestation throughout Europe spurred the French state to capitalize on the forest resources of continental France as well as its colonies. Even in the Ottoman Empire, a prime exporter of timber, French foresters were motivated in part by the prospect of provisioning. Throughout its Mediterranean application, moreover, French scientific forestry embodied the interests of empire. It helped to extend French power and authority into new realms, and it facilitated the expansion of the French state in the lives and minds of its constituents. In Algeria the process of dispossession and oppression of indigenous groups represented a deliberate effort by the colonial administration to bury autochthonous culture and society under a new French colonial state.[9] Even at home in Provence the imposition of national forest administration formed a type of internal colonization.

The institutions and individuals who promoted and carried out this mission were motivated not solely by self-interest but also by the ideals of progress, civilization, and environmental preservation. Foresters, officials, and nonspecialists alike firmly believed in narratives blaming pastoralists for Mediterranean deforestation and other forms of environmental destruction. They accepted these narratives in part because they stemmed from observable facts: forests were disappearing at a staggering pace, and sheep and goats do exercise a visible environmental impact. Yet these narratives surfaced, spread, and solidified largely because they proved necessary in the achievement and validation of the imperial mission. The systematic subjugation and sedentarization of Mediterranean pastoralists confirm the power of the antipastoral narratives that legitimized this process. As in other contexts of soft and hard imperialism, as well as other applications of French forestry around the Mediterranean, European environmental perceptions ultimately sealed the fate of mobile pastoralists in the nineteenth century.

The nineteenth-century transformation of Mediterranean pastoralism owed much to environmental policy. In Provence the forest regime gradually tightened its grip on communal grazing over the course of the nineteenth century. During the same period, changes in landed property rights restricted the extent of common land available, encouraging privatization and the rental of pastures. Together these factors helped to drive Provence's age-old custom of forest grazing to the point of extinction by the turn of the twentieth century. Likewise, forest administration in late nineteenth-century Anatolia was also targeting the traditional practices of nomadic tribes, though with limited results. Meanwhile, the sedentarization campaigns of the 1860s and 1870s succeeded in breaking down the regional autonomy of nomadic tribes and either compelling them to settle or forcing them toward the empire's shrinking frontiers.

Legislation exercised a particularly marked impact on nomadic pastoralists in Algeria, where it was relatively unencumbered by political resistance from these groups. Through the application of the French Forest Code of 1827 and its subsequent amendments, the Warnier Law of 1873, the Fire Law of 1874, and the Algerian Forest Code of 1903, indigenous rights and access to pastureland were slowly extinguished. During this period Algeria's sheep population remained relatively stable, but that was due largely to the entrance of European settlers into the market. The broader picture nonetheless still suggests the gradual decline of this industry. By 1960 Algeria's total sheep population had fallen to five million, less than half its size in 1885.[10] Colonial pressures, moreover, succeeded in transforming not only the lifestyle of mobile pastoralists but also indigenous peoples' perceptions of it. By the early decades of the twentieth century Algerians tended to view the profession of shepherd "as a social abasement . . . exercised by the poorest members [*les plus misérables*] of the tribe."[11] Some indigenous farmers expressed this sentiment by joining the Ligue du Reboisement, an institution that had been instrumental in their oppression.[12] Yet nineteenth-century transformation of pastoralism was not simply a tale of exploitation and expropriation, of the triumph of the state. Even Augustin Bernard and Napoléon Lacroix, two enthusiastic proponents of sedentarization, acknowledged in their 1906 study *L'évolution du nomadisme en Algérie* (The evolution of nomadism in Algeria), "No more than the English Parliament can, as the proverb goes, change a man into a woman, a legislative measure cannot make a barbarian civilized, a nomad sedentary."[13] Instead this process also required the confluence of other factors, including developments in transportation, technology, communication, population expansion, natural disasters and climate change, and the transformation of the global wool market, as well as the agency of Mediterranean pastoralists themselves.

The comparison of pastoral traditions in Provence, Anatolia, and Algeria highlights key connections and distinctions. Considering these three cases together within a Mediterranean framework provides valuable insight into the relations of people, states, and economies around

8. Sculpture celebrating Provence's pastoral tradition, Sisteron, France. Photo from the author's collection.

this inner sea. In addition, this transnational perspective sheds light on the history and practice of pastoralism worldwide and offers new approaches to the investigation of this subject. The fate of mobile pastoralists, French officials, and forestry in the Mediterranean is not unique. It is representative of the political implications of science and environmental policy around the world. It provides just one example of the ways in which, throughout history, power relationships have been determined and contested on environmental terms. Although this condition is particularly evident in imperial and colonial contexts, it is much more universal. It is indicative of the pressures and challenges that nomads and other subaltern peoples around the world have faced in the past and continue to face today.

NOTES

ABBREVIATIONS FOR ARCHIVES

ANOM   Archives Nationales d'Outre-Mer, Aix-en-Provence

BDR   Archives Départementales des Bouches-du-Rhône, Aix-en-Provence and Marseille

BOA   Başbakanlık Osmanlı Arşivi, Istanbul

CADN   Centre des Archives Diplomatiques, Nantes

CARAN   Centre d'Accueil et de Recherche des Archives Nationales, Paris

TNA   The National Archives of the United Kingdom, Kew, London

INTRODUCTION

1. Planhol, *De la plaine pamphylienne aux lacs pisidiens*, 194, cited in Braudel, *Mediterranean and the Mediterranean World*, 88.

2. Braudel, *Mediterranean and the Mediterranean World*, 14.

3. Horden and Purcell, *Corrupting Sea*, 13. For a critique of traditional conceptualizations of the Mediterranean environment, see Horden and Purcell, *Corrupting Sea*, 9–49.

4. See, for example, Herzfeld, *Anthropology through the Looking-Glass*; and Pina-Cabral, "Mediterranean as a Category of Regional Comparison," 399–406.

5. Pina-Cabral, "Mediterranean as a Category of Regional Comparison," 399.

6. See, for example, A. Grove and Rackham, *Nature of Mediterranean Europe*; M. Williams, *Deforesting the Earth*; and J. Hughes, *Mediterranean*.

7. Seigue, *La forêt circumméditerranéenne et ses problèmes*, 315–30; Thornes, "Land Degradation," 575–77; Lloret, Piñol, and Castellnou, "Wildfires," 553; A. Grove and Rackham, *Nature of Mediterranean Europe*, 227, 240.

8. A thorough presentation of this history is found in Ford, *Natural Interests*.

9. See, for example, the works of Lucien-Albert Fabre, Philippe Arbos, René Baehrel, and Xavier de Planhol. For more recent examples, see Nedonsel, "Contribution à l'étude de l'élevage ovin transhumant dans les Bouches-du-Rhône"; Duclos, *L'homme et le mouton dans l'espace de la transhumance*; and Laffont, *Transhumance et estivage en Occident des origines aux enjeux actuels*.

10. Typical of this genre are Musset, *Histoire et actualité de la transhumance en Provence*; *Le mouton en Provence*; Mauron, *La transhumance du pays d'Arles aux Grandes Alpes*; Annequin, *Aux origines de la transhumance*; and Moyal, *Transhumance*. Notable exceptions include Madeline and Moriceau, *Acteurs et espaces de l'élevage*; and Moriceau, *Histoire et géographie de l'élevage français*.

11. The development of this field owes much to Andrée Corvol, who has made a number of contributions in addition to her classic study *L'homme aux bois*. For the connection between forestry and pastoralism, see Dumoulin, *La forêt provençale au XIXe siècle*; Dumoulin, "Communes et pâturages forestiers en Provence au XIXe siècle"; and Whited, *Forests and Peasant Politics in Modern France*.

12. Key contributions to scholarship on pastoralism in the Ottoman Empire include Refik, *Anadolu'da Türk Aşiretleri*; Özbayri, *Tahtacılar ve Yörükler*; Gündüz, *Anadolu'da ve Rumeli'de Yörükler ve Türkmenler*; Beşirli and Erdal, *Osmanlı'dan Cumhuriyet'e Yörükler ve Türkmenler*; Kasaba, *Moveable Empire*; and the works of Cengiz Orhonlu, Rudi Lindner, and Yusuf Halaçoğlu. Most of these works address the early modern era, the administration of nomads, or tribal folklore. A few scholars have examined Ottoman nomadism in the context of environmental legislation. See S. White, *Climate of Rebellion in the Early Modern Ottoman Empire*; and Toksöz, *Nomads, Migrants and Cotton in the Eastern Mediterranean*. For Ottoman forestry see Dursun, "Forest and the State"; Özdönmez and Ekizoğlu, "Tanzimat ve Meşrutiyet Dönemleri Ormancılığında Katkıları Olan Yabancı Uzmanlar"; Keskin, "Osmanlı Ormancılığı'nın Gelişiminde Fransız Uzmanların Rolü"; and various works by Bekir Koç and Halil Kutluk.

13. Davis has made a number of other contributions to this subject, including "Desert 'Wastes' of the Maghreb"; "Potential Forests"; and "Neoliberalism, Environmentalism, and Agricultural Restructuring in Morocco." A number of other scholars have studied the connections between French colonial environmental policy and the indigenous population in Algeria. See in particular Cutler, "Evoking the State"; Ford, "Reforestation, Landscape Conservation, and the Anxieties of Empire"; Prochaska, "Fire on the Mountain"; and Sivak, "Law, Territory, and the Legal Geography of French Rule in Algeria."

14. Davis, *Resurrecting the Granary of Rome*, 166.

15. Phillips and Phillips, *Spain's Golden Fleece*, ix.

1. LAND OF THE GOLDEN FLEECE

1. Royer, "Les transhumants du Roi Rene," 13–14.
2. See, for example, Labrouche, *Ariane Mnouchkine*, 66.
3. Khazanov, *Nomads and the Outside World*, 16, 21; Braudel, *Mediterranean and the Mediterranean World*, 1:101; Ryder, *Sheep and Men*, 210; Bonnet, "Aspects de la transhumance ovine provençale," 2: 11; Cleary, "Contemporary Transhumance in Languedoc and Provence," 107; Corbier, "La transhumance dans les pays de la Méditerranée antique," 67.
4. Horden and Purcell represent an important exception. They avoid the problem of regional terminology by describing all Mediterranean mobile pastoralism as transhumance. See Horden and Purcell, *Corrupting Sea*, 63.
5. Harding, Palutikof, and Holt, "Climate System," 69.
6. Harding, Palutikof, and Holt, "Climate System," 69.
7. Harding, Palutikof, and Holt, "Climate System," 77–78.
8. McNeill, *Mountains of the Mediterranean World*, 12. McNeill notes that if one travels along the Mediterranean coast, except along the lowlands between Tunisia and Sinai, one is almost always in sight of both mountains and sea.
9. Harding, Palutikof, and Holt, "Climate System," 69–70; Davis, *Resurrecting the Granary of Rome*, 178.
10. Harding, Palutikof, and Holt, "Climate System," 69.
11. Wainwright, "Weathering, Soils, and Slope Processes," 179.
12. Wainwright, "Weathering, Soils, and Slope Processes," 179.
13. Wainwright, "Weathering, Soils, and Slope Processes," 178; Allen, "Vegetation and Ecosystem Dynamics," 205–6; Blondel, "Nature and Origin of the Vertebrate Fauna," 157–58.
14. Allen, "Vegetation and Ecosystem Dynamics," 203. See also Wainwright, "Weathering, Soils, and Slope Processes," 173; Davis, *Resurrecting the Granary of Rome*, 184; and Pyne, *Vestal Fire*, 20–21.
15. Allen, "Vegetation and Ecosystem Dynamics," 204–6.
16. Niamir-Fuller and Turner, "Review of Recent Literature on Pastoralism and Transhumance in Africa," 32–33.
17. See, for example, Blondel, "'Design' of Mediterranean Landscapes."
18. Allen, "Vegetation and Ecosystem Dynamics," 220, 225; Thornes, "Land Degradation," 575; Davis, *Resurrecting the Granary of Rome*, 184–86.
19. Wainwright, "Weathering, Soils, and Slope Processes," 179.
20. See, for example, Allen, "Vegetation and Ecosystem Dynamics," 224–25; Thornes, "Land Degradation" 575; Leach and Mearns, *Lie of the Land*; Niamir-Fuller, *Managing Mobility in African Rangelands*; and Davis, *Resurrecting the Granary of Rome*, 184–86.

21. Louis Gachon, in *Nouvelles Littéraires*, 10 February 1940, cited in Braudel, *Mediterranean and the Mediterranean World*, 1:399; Livet, *Habitat rural et structures agraires en Basse-Provence*.

22. Marcelin, "Contribution à l'étude géographique de la garrigue nîmoise," 45–46; J. Hughes, *Mediterranean*, 9–10; Allen, "Vegetation and Ecosystem Dynamics," 203–27.

23. Allen, "Vegetation and Ecosystem Dynamics," 212.

24. Marcelin, "Contribution à l'étude géographique de la garrigue nîmoise," 45–46; A. Grove and Rackham, *Nature of Mediterranean Europe*, 46.

25. *Le mouton en Provence*.

26. For historical descriptions of coussouls, see Michel, *Observations sur le commerce des bêtes à laine*, 1–3; and Briot, *Les Alpes françaises*, 74.

27. Davis, *Resurrecting the Granary of Rome*, 177. "Maghreb" is technically an anglicization of the Arabic word for "the West," literally "the sunset," used to describe Islam's westernmost provinces, المغرب or al-Maghrib.

28. Ruedy, *Modern Algeria*, 5.

29. See Woolsey, *French Forests and Forestry*, 47.

30. Woolsey, *French Forests and Forestry*, 47.

31. Woolsey, *French Forests and Forestry*, 54–55.

32. Statistique Général de l'Algérie (30 June 1911), 296–300, cited in Woolsey, *French Forests and Forestry*, 57.

33. Lucas, *Exploration scientifique de l'Algérie*, 8.

34. Lucas, *Exploration scientifique de l'Algérie*, 54, 88. For a map showing the distribution of commercial forests in Algeria, see Marc, *Notes sur les forêts de l'Algérie*, appendix.

35. Dewdney, *Turkey*, 25.

36. McNeill, *Mountains of the Mediterranean World*, 22–24.

37. McNeill, *Mountains of the Mediterranean World*, 20; Dewdney, *Turkey*, 36.

38. Gould, "Pashas and Brigands," 12; Kasaba, *Moveable Empire*.

39. Chang and Koster, "Beyond Bones"; Ryder, *Sheep and Men*, 18–27; Cribb, *Nomads in Archaeology*.

40. See Guthrie, *World History of Sheep and Wool*, 4–6; Ryder, *Sheep and Men*, 117–22.

41. Pliny the Younger, *Complete Letters*, trans. Philemon Holland, quoted in Ritchie, *Influence of Man on Animal Life in Scotland*, 45.

42. See, for example, Ladoucette, *Histoire, topographie, antiquités, usages, dialectes des Hautes-Alpes*, 544; and Belleval, "Réflexions sur la transhumance des troupeaux en Provence," 23. The mayor of Allauch also used this phrase in an appeal to the forest administrative office regarding the acceptance of goats

on communal land. See BDR 2 O 3 1, "Allauch: Le Conseil municipal décision à faire autoriser le parcours des chèvres" (1817). For more modern use, see Sclafert, "À propos du déboisement des Alpes du Sud," 127; and Cointat, "La dégradation des forêts dans le Département du Gard," 110.

43. *Le mouton en Provence*, 10; Brun, Congés, and Badan, "Les bergeries romaines de la Crau d'Arles." See also Leveau, "Entre le delta du Rhône, la Crau et les Alpes," 83; and Butzer, "Environmental History in the Mediterranean World," 1787.

44. Leveau, "Entre le delta du Rhône, la Crau et les Alpes," 88; Brun, Congés, and Badan, "Les bergeries romaines de la Crau d'Arles."

45. Pliny 31.57, quoted in and translated by Leveau, "Entre le delta du Rhône, la Crau et les Alpes," 93.

46. Villeneuve, *Statistique du département des Bouches-du-Rhône*, 4:86–89.

47. "Terre en friche, lieu inculte et sauvage." "Gâtine / Gastine," *Dictionnaire du Moyen Français (1330–1500)*, accessed 2 February 2013, http://www.atilf.fr/dmf/.

48. Villeneuve, *Statistique du département des Bouches-du-Rhône*, 4:87.

49. Leveau, "Entre le delta du Rhône, la Crau et les Alpes," 90.

50. Villeneuve, *Voyage dans la vallée de Barcelonette*, 67–68.

51. Villeneuve, *Voyage dans la vallée de Barcelonette*, 72; Tolley, "Qui sont les bergers?," 75.

52. Archiloque, "D'hier à aujourd'hui, la passion d'être berger," 97–98.

53. Tolley, "Qui sont les bergers?," 75.

54. Archiloque, "D'hier à aujourd'hui, la passion d'être berger," 100.

55. Villeneuve, *Voyage dans la vallée de Barcelonette*, 68–69.

56. Royer, "Les transhumants du Roi Rene," 17; Villeneuve, *Voyage dans la vallée de Barcelonette*, 70.

57. Villeneuve, *Voyage dans la vallée de Barcelonette*, 69–70; Ladoucette, *Histoire, topographie, antiquités, usages, dialectes des Hautes-Alpes*, 545; Archiloque, "D'hier à aujourd'hui, la passion d'être berger," 100.

58. Royer, "Les transhumants du Roi Rene," 19.

59. Moriceau, *Histoire du méchant loup*, 13.

60. Mistral, *Mirèio*, canto IV.

61. Villeneuve, *Voyage dans la vallée de Barcelonette*, 68.

62. Shippers, "Le cycle annuel d'un berger transhumant," 63.

63. Bloch, *French Rural History*, 204.

64. Ladoucette, *Histoire, topographie, antiquités, usages, dialectes des Hautes-Alpes*, 545. Rented alpages were either private or communal, and use was regulated by a head tax (546).

65. Sclafert, "À propos du déboisement des Alpes du Sud," 138.

66. Royer, "Les transhumants du Roi Rene," 17–18.

67. Ladoucette, *Histoire, topographie, antiquités, usages, dialectes des Hautes-Alpes*, 545.

68. Royer, "Les transhumants du Roi Rene," 20.

69. Ryder, *Sheep and Men*, 251.

70. Ageron, *Modern Algeria*, 66.

71. Ruedy, *Modern Algeria*, 32, 34–35.

72. Ruedy, *Modern Algeria*, 33–34, 36–37.

73. Ruedy, *Modern Algeria*, 36.

74. Hirtz, *L'Algérie nomade et ksourienne*, 51–58; Boukhobza, "Nomadisme et colonisation," 45–70.

75. Berque, *Maghreb*, 32, cited in Boukhobza, "Nomadisme et colonisation," 3.

76. Lehuraux, *Le nomadisme et la colonisation dans les Hauts Plateaux de l'Algérie*, 6.

77. Quoted in Yacono, *La colonisation des plaines du Chelif*, 2:77.

78. Boukhobza, "Nomadisme et colonisation," 28.

79. Lehuraux, *Le nomadisme et la colonisation dans les Hauts Plateaux de l'Algérie*, 6.

80. Boukhobza, "Nomadisme et colonisation," 32; Lehuraux, *Le nomadisme et la colonisation dans les Hauts Plateaux de l'Algérie*, 57–58, 94.

81. Boukhobza, "Nomadisme et colonisation," 35.

82. Boukhobza, "Nomadisme et colonisation," 19–21.

83. Lehuraux, *Le nomadisme et la colonisation dans les Hauts Plateaux de l'Algérie*, 72.

84. Vryonis, "Nomadization and Islamization in Asia Minor"; İnalcık and Quataert, *Economic and Social History of the Ottoman Empire*, 98; İnalcık, "Yuruks."

85. Ryder, *Sheep and Men*, 57. Juliet Clutton-Brock cites earlier evidence of sheep under human control, though not necessarily domesticated, in Iraq. She hypothesizes that sheep were first domesticated in western Asia, either in or near Anatolia. See Clutton-Brock, *Natural History of Domesticated Mammals*, 74. See also Chang and Koster, "Beyond Bones."

86. Ryder, *Sheep and Men*, 56–57.

87. Planhol, *De la plaine pamphylienne aux lacs pisidiens*, 79.

88. Aristotle, *Historia animalium*, 3.12, cited in Ryder, *Sheep and Men*, 143. Ryder also notes that Strabo remarked in the first century CE that some people had amassed great wealth by grazing their flocks on the Konya Plain (57).

89. Planhol, "Geography, Politics and Nomadism in Anatolia"; Planhol, *De la plaine pamphylienne aux lacs pisidiens*, 93–94; Johnson, *Nature of Nomadism*, 20l; Lindner, *Nomads and Ottomans in Medieval Anatolia*.

90. S. White, "Ecology, Climate, and Crisis in the Ottoman Near East," 151; Frödin; "Les formes de la vie pastorale en Turquie," 223; Ş. Çelik, "Yüzyılda İçel Yörükleri Hakkında Bazı Değerlendirmeler"; Armağan, "Yüzyıllarda Teke Sancağı'nda Konar-Göçerler," 78–88.

91. S. White, *Ecology, Climate, and Crisis in the Ottoman Near East*, 151; Murphey, "Some Features of Nomadism in the Ottoman Empire."

92. Karpat and Zens, *Ottoman Borderlands*; Peacock, *Frontiers of the Ottoman World*.

93. Ibn Battuta and Gibb, *Travels of Ibn Battūta*, 415–16.

94. Crane, "Evliya Çelebi's Journey through the Pamphylian Plain."

95. Şahin, *Osmanlı Döneminde Konar-Göçerler/Nomads in the Ottoman Empire*, 34–35. See also Johnson, *Nature of Nomadism*, 20–21. I use the general term "yörük" except when referring to specific tribes.

96. Şahin, *Osmanlı Döneminde Konar-Göçerler / Nomads in the Ottoman Empire*, 98.

97. McNeill, *Mountains of the Mediterranean World*, 113; Johnson, *Nature of Nomadism*, 29.

98. Johnson, *Nature of Nomadism*, 29; Kolars, *Tradition, Season, and Change in a Turkish Village*, 18–19; McNeill, *Mountains of the Mediterranean World*, 158; Sarre, *Reise in Kleinasien, Sommer 1895*, 116.

99. See Bulliet, *Camel and the Wheel*.

100. Ryder, *Sheep and Men*, 212.

101. McNeill, *Mountains of the Mediterranean World*, 113; Roux, *Les traditions des nomades de la Turquie méridionale*, 68.

102. Ryder, *Sheep and Men*, 212; Aksoy, "Konar-Göçer Yörük Alt-Kültüründe Kadın Kimliği."

103. Van Lennep, *Travels in Little-Known Parts of Asia Minor*, 293.

104. Van Lennep, *Travels in Little-Known Parts of Asia Minor*, 136.

105. Lindner, *Nomads and Ottomans in Medieval Anatolia*; Gould, "Pashas and Brigands," 23; McCarthy, *Death and Exile*, 43–44. See also Findley, *Turkey, Islam, Nationalism, and Modernity*, 117.

106. A. Grove and Rackham, *Nature of Mediterranean Europe*, 191.

107. Whited, *Forests and Peasant Politics*, 15–16.

108. Wagstaff, *Evolution of Middle Eastern Landscapes*, 159.

109. Planhol, "Les nomades, la steppe, et la forêt en Anatolie," 106.

110. Wagstaff, *Evolution of Middle Eastern Landscapes*, 159.

111. Özbayri, *Tahtacılar ve yörükler*.

112. Bates, *Nomads and Farmers*, 19; Kolars, *Tradition, Season, and Change in a Turkish Village*; Rowton, "Woodlands of Ancient Asia," cited in Wagstaff, *Evolution of Middle Eastern Landscapes*, 123.

113. Boukhobza, "Nomadisme et colonisation," 28; Ageron, *Les Algériens musulmans et la France*, 1:103–6; Ford, "Reforestation, Landscape Conservation, and the Anxieties of Empire," 344–45.

114. Livet, *Habitat rural et structures agraires en Basse-Provence*, 97.

115. Louis Gachon, in *Nouvelles Littéraires*, 10 February 1940, quoted in Braudel, *Mediterranean and the Mediterranean World*, 1:399.

116. Livet, *Habitat rural et structures agraires en Basse-Provence*, 23.

117. Livet, *Habitat rural et structures agraires en Basse-Provence*, 86; Villeneuve, *Voyage dans la vallée de Barcelonette*, 70–71; Chang and Koster, "Beyond Bones," 103.

118. Villeneuve, *Voyage dans la vallée de Barcelonette*, 70.

119. Villeneuve, *Voyage dans la vallée de Barcelonette*, 70–71; Klein, *Mesta*, 144; Royer, "Les transhumants du Roi Rene," 16.

120. Hirtz, *L'Algérie nomade et ksourienne*, 17–18.

121. Ruedy, *Modern Algeria*, 25; Boukhobza, "Nomadisme et colonisation," 32.

122. Ruedy, *Modern Algeria*, 26.

123. Leeuwen, *Nomads in Central Asia*, 67.

124. Boukhobza, "Nomadisme et colonisation," 3.

125. Boukhobza, "Nomadisme et colonisation," 19; Lehuraux, *Le nomadisme et la colonisation dans les Hauts Plateaux de l'Algérie*, 74.

126. Boukhobza, "Nomadisme et colonisation," 35; Lehuraux, *Le nomadisme et la colonisation dans les Hauts Plateaux de l'Algérie*, 1–4.

127. İnalcık and Quataert, *Economic and Social History of the Ottoman Empire*, 38; Dewdney, *Turkey*, 109; Ryder, *Sheep and Men*, 222.

128. McNeill, *Mountains of the Mediterranean World*, 157; İnalcık, "Yuruks," 49; İnalcık and Quataert, *Economic and Social History of the Ottoman Empire*, 36–38, 40.

129. Murphey, "Some Features of Nomadism in the Ottoman Empire," 190.

130. McNeill, *Mountains of the Mediterranean World*, 113; Kolars, *Tradition, Season, and Change in a Turkish Village*, 22; Murphey, "Some Features of Nomadism in the Ottoman Empire," 190.

131. Planhol, *De la plaine pamphylienne aux lacs pisidiens*, 115, 120.

132. Planhol, *De la plaine pamphylienne aux lacs pisidiens*, 120.

133. McNeill, *Mountains of the Mediterranean World*, 156; Armağan, "Yüzyıllarda Teke Sancağı'nda Konar-Göçerler," 78–79, 81, 89–102.

134. McNeill, *Mountains of the Mediterranean World*, 157.

135. See, for example, Appleby, "Epidemics and Famine in the Little Ice Age"; J. Grove, "Little Ice Age in the Massif of Mont Blanc;" Swan, "Mexico in the Little Ice Age"; and Parker and Smith, *General Crisis of the Seventeenth Century*.

136. Jorda and Roditis, "Les épisodes de gel du Rhône depuis l'an mil," 21.

137. Macklin and Woodward, "Rivers and Environmental Change," 344.

138. Lachiver, *Les années de misère*, 207; Emmanuelli, *L'intendance de Provence à la fin du XVIIe siècle*; Collins, *State in Early Modern France*, 150.

139. Tabak, *Waning of the Mediterranean*, 17, 205; Gould, "Pashas and Brigands," 12.

140. Armağan, "Yüzyıllarda Teke Sancağı'nda Konar-Göçerler," 88–89; S. White, *Climate of Rebellion in the Early Modern Ottoman Empire*, 229.

141. Barkan, "Essai sur les données statistiques des registres de recensement dans l'Empire ottoman," 30.

142. Çelebi, *Seyahatname*, 3:102–4; S. White, *Climate of Rebellion in the Early Modern Ottoman Empire*, 240–41; Hütteroth, "Influence of Social Structure on Land Division and Settlement in Inner Anatolia"; Türkiye Diyanet Vakfı Islam, *Türkiye Diyanet Vakfı İslâm Ansiklopedisi*, s.v. "Salnâme-i Vilayet-i Konya."

143. Lapidus, "Tribes and State Formation in Islamic History," 43.

144. A few other notable cases include the Mestas of Spain; the shepherds of Italy, Greece, and the Balkans; the Bedouin of the Levant; and the Berbers of Morocco. For Spain, see Phillips and Phillips, *Spain's Golden Fleece*. For Italy, see Marino, *Pastoral Economics in the Kingdom of Naples*. For Greece, see McNeill, *Mountains of the Mediterranean World*. For the Levant, see Lewis, *Nomads and Settlers in Syria and Jordan*.

2. BLACK SHEEP

1. Ammer, *American Heritage Dictionary of Idioms*, 64.

2. Sykes, *Black Sheep*, 11.

3. Spary, *Utopia's Garden*, 101–17; Gay, *Age of Enlightenment*, 22; Matteson, "Masters of Their Woods," 113.

4. Almond, Chodorow, and Pearce, *Progress and Its Discontents*.

5. Prominent French orientalists were Silvestre de Sacy, his student William de Slane, and Joseph de Hammer, among others. For a critical survey of these scholars and their interpretations, see Hannoum, *Violent Modernity*, 60–69; and Burke, "Sociology of Islam." For a general description of the orientalist renaissance in European literature, see Lockman, *Contending Visions of the Middle East*, 67–69.

6. Slane, *Histoire des Berbères et des dynasties musulmans de l'Afrique septentrionale par Ibn Khaldoun*, i, quoted in and translated by Hannoum, *Violent Modernity*, 64.

7. Boukhobza, "Nomadisme et colonisation," 2; Lacoste, *Ibn Khaldun*, 146, 150. Ibn Khaldun's ةمدقم (Muqaddimah) was originally published in 1377.

8. Slane, *Les prolégomènes d'Ibn Khaldoun*, 1:85.

9. Slane, *Les prolégomènes d'Ibn Khaldoun*, 1:310.

10. Slane, *Les prolégomènes d'Ibn Khaldoun*, 1:cxi.

11. Ibn Khaldun, *Muqaddimah*, 93.

12. Ibn Khaldun, *Muqaddimah*, 92.

13. Lorcin, *Imperial Identities*, 2; Davis, *Resurrecting the Granary of Rome*, 57–58; Ford, "Reforestation, Landscape Conservation, and the Anxieties of Empire,"

347–48, 354; Hannoum, *Violent Modernity*, 66–67; Lacoste, *Ibn Khaldun*; Ageron, *Histoire de l'Algérie contemporaine*, 138–40. For an example of the Kabyle myth, see Lehuraux, *Le nomadisme et la colonisation dans les Hauts Plateaux de l'Algérie*, 11–23.

14. Tocqueville, *Writings on Empire and Slavery*, 21.
15. Tocqueville, *Writings on Empire and Slavery*, 9–10.
16. ANOM F8010, Ministère de la Guerre, Rapport sur la colonisation de l'ex-Régence d'Alger par Mr de la P. (November 1833), quoted in and translated by Abi-Mershed, *Apostles of Modernity*, 13.
17. *Exploration scientifique de l'Algérie* comprises a multivolume series of works on the science and natural history of the region. The volumes were published in various forms and at various times in the mid-nineteenth century.
18. Duveyrier, *Exploration du Sahara*. See also Asher, *Death in the Sahara*, 18–20.
19. Ageron, *Modern Algeria*, 23.
20. Ageron, *Modern Algeria*, 23.
21. See, for example, ANOM ALG GGA/P2, Direction provinciale des affaires arabes, Province d'Oran, letter to Gouverneur Général (12 July 1868). See also Ford, "Reforestation, Landscape Conservation, and the Anxieties of Empire," 358–59.
22. Davis, *Resurrecting the Granary of Rome*, 52–53; Heffernan and Sutton, "Landscape of Colonialism"; Abi-Mershed, *Apostles of Modernity*.
23. Bugeaud, quoted in Enfantin, *Colonisation de l'Algérie*, 481.
24. Enfantin, *Colonisation de l'Algérie*, 42.
25. Enfantin, *Colonisation de l'Algérie*, 481.
26. Kasaba, *Moveable Empire*, 4–7; Kasaba, "Do States Always Favor Stasis?," 27–48; Murphey, "Some Features of Nomadism in the Ottoman Empire," 191, 193–95.
27. For this terminology, see Ágoston, "Flexible Empire."
28. Kasaba, *Moveable Empire*, 59–61; İnalcık and Quataert, *Economic and Social History of the Ottoman Empire*, 438, 646. See also Faroqhi, "Towns, Agriculture and the State in Sixteenth-Century Ottoman Anatolia"; Issawi, *Economic History of Turkey*; Tankut, "Urban Transformation in the Eighteenth-Century Ottoman City"; and İnalcık and Quataert, *Economic and Social History of the Ottoman Empire*, 781.
29. Barkey, *Bandits and Bureaucrats*, 199–201; Murphey, "Some Features of Nomadism in the Ottoman Empire," 191.
30. Kasaba, *Moveable Empire*, 8–9.
31. Ágoston, "Flexible Empire," 17.
32. Murphey, "Some Features of Nomadism in the Ottoman Empire," 194; S. White, "Ecology, Climate, and Crisis in the Ottoman Near East," 64–65.

33. Kasaba, *Moveable Empire*, 54–58. For the Ottomans' early flexibility, see Ágoston, "Flexible Empire," 15–31.

34. See "Part II: The Little Ice Age and the Celali Rebellion," in S. White, "Ecology, Climate, and Crisis in the Ottoman Near East," 171–267; S. White, *Climate of Rebellion in the Early Modern Ottoman Empire*, 123–225; Griswold, "Climatic Change"; Griswold, *Great Anatolian Rebellion*, 983; and Barkey, *Bandits and Bureaucrats*, 141–228.

35. S. Shaw, *History of the Ottoman Empire and Modern Turkey*, 1:174–75.

36. Wright, *Ottoman Statecraft*, 89, 93, 117, 119, 126–27; Barkey, *Bandits and Bureaucrats*, 200–201. See also Akdağ, *Türk Halkının Dirlik ve Düzenlik Kavgası*; and Akdağ, "Celâli İsyanlarından Büyük Kaçgunluk," 1–49.

37. S. White, *Climate of Rebellion in the Early Modern Ottoman Empire*, 243–44; Kasaba, *Moveable Empire*, 65.

38. Refik, *Anadolu'da Türk Aşiretleri*; Orhonlu, *Osmanlı İmparatorluğu'nda Aşiretlerin İskâni*, 57–65, 107–9; Kasaba, *Moveable Empire*, 66; S. White, *Climate of Rebellion in the Early Modern Ottoman Empire*, 243–47.

39. Kasaba, *Moveable Empire*, 29–30; Refik, *Anadolu'da Türk Aşiretleri*.

40. Refik, quoted in and translated by Kasaba, *Moveable Empire*, 67.

41. Orhonlu, *Osmanlı İmparatorluğu'nda Aşiretlerin İskâni*; Halaçoğlu, *XVIII. Yüzyılda Osmanlı İmparatorluğu'nun İskân Siyaseti ve Aşiretlerin Yerleştirilmesi*; Saydam, "XIX. Yüzyılın İlk Yarısında Aşiretlerin İskânına Dair Gözlemler," 217–29; Planhol, *De la plaine pamphylienne aux lacs pisidiens*, 117; McNeill, *Mountains of the Mediterranean World*, 156.

42. Kasaba, *Moveable Empire*, 5; Fleischer, "Royal Authority," 199–203; Broadbridge, "Royal Authority, Justice, and Order in Society."

43. Fleischer, "Royal Authority," 198–219; Kasaba, *Moveable Empire*, 5.

44. See Kasaba, *Moveable Empire*, 7.

45. CADN 166PO/D76/1, Mémoire Concernant Satalie (18 April 1810).

46. TNA FO 195/55, Letter book, Consulate General (1857).

47. Poujoulat, *Voyage dans l'Asie Mineure*, 70.

48. Poujoulat, *Voyage dans l'Asie Mineure*, 203.

49. Finkel, *Osman's Dream*, 457. Tsar Nicholas I became the first to use this phrase, in 1853.

50. Hogarth, *Nearer East*, 199; Locher, *With Star and Crescent*, 450; Van Lennep, *Travels in Little-Known Parts of Asia Minor*, 136–37, 294. For examples of such language among British consuls, see TNA FO 78/2987, FO 222/1 (1879), FO 222/2 (180), and FO 196/55 (1857).

51. Van Lennep, *Travels in Little-Known Parts of Asia Minor*, 136–37.

52. Burnaby, *On Horseback through Asia Minor*, 55–56.

53. Burnaby, *On Horseback through Asia Minor*, 83.

54. Burnaby, *On Horseback through Asia Minor*, 85.

55. Lindner, *Nomads and Ottomans in Medieval Anatolia*; Köprülü, *Les origines de l'Empire ottoman*; Fleischer, *Bureaucrat and Intellectual in the Ottoman Empire*, 274–76.

56. Wittek, "Le rôle des tribus turques dans l'empire ottoman"; Lindner, *Nomads and Ottomans in Medieval Anatolia*, 105.

57. Deringil, *Well-Protected Domains*, 41.

58. Lewis, *Nomads and Settlers in Syria and Jordan*, 24, cited in Kasaba, *Moveable Empire*, 114.

59. BOA I.DH 4 Gurre-i, Ramazan 1313 (February 1896); Mardin, *Religion and Social Change in Modern Turkey*, 169, cited in Deringil, *Well-Protected Domains*, 41.

60. The *Oxford English Dictionary* (2012) gives the usage history of the word *nomad*, including the usage history over five hundred years.

61. Laffont, *Transhumance et estivage en Occident*, 9.

62. Laffont, *Transhumance et estivage en Occident*, 14.

63. Gautier et al., *Histoire et historiens de l'Algérie*, 103, cited in Boukhobza, "Nomadisme et colonisation," 7.

64. A. Grove and Rackham, *Nature of Mediterranean Europe*, chap. 1; Ford, "Reforestation, Landscape Conservation, and the Anxieties of Empire," 346; Davis, "Desert 'Wastes' of the Maghreb," 362.

65. Davis, *Resurrecting the Granary of Rome*, 8.

66. A. Grove and Rackham, *Nature of Mediterranean Europe*, 9, neatly summarizes one iteration of this declensionist narrative.

67. Buffon, *Les époques de la nature*, cited in Larrère, "Rauch ou Rougier de la Bergerie," 248; Matteson, *Forests in Revolutionary France*, 166.

68. Poivre, *Voyages d'un philosophe*, 31.

69. Ladoucette, *Histoire, topographie, antiquités, usages, dialectes des Hautes-Alpes*, c–ci.

70. BDR 7 M 163: Eaux et Forêts, "Reboisement, 1822, Première Arrondissement."

71. Dussard, *Le Journal des Économistes*, July 1842, cited in Ribbe, *La Provence au point de vue des bois*, 16.

72. Bouche, *Essai sur l'histoire de Provence*, 2:535, in a chapter titled "Contrées abandonnées," quoted in Ribbe, *La Provence au point de vue des bois*, 23.

73. Bouche, *Essai sur l'histoire de Provence*, 2:536, quoted in Ribbe, *La Provence au point de vue des bois*, 23.

74. Boyer de Fonscolombe, "Mémoire sur la destruction et le rétablissement des bois dans les départemens qui composoient la Provence," 6. This treatise was first published in 1803.

75. Ladoucette, *Histoire, topographie, antiquités, usages, dialectes des Hautes-Alpes*, 464–65.

76. Corvol, *L'homme aux bois*, 65. For example, see Demontzey, *Étude sur les travaux de reboisement et de gazonnement des montagnes*, 253; BDR 7 M 101, Letter: Comte de Villeneuve, to Préfet des Bouches-du-Rhône (Paris, 17 December 1817).

77. *Forêts perdues*, 37.

78. Quoted in Woronoff, *Révolution et espaces forestiers*, 49.

79. Gibelin, "Observations sur les chèvres," 134.

80. Boyer de Fonscolombe, "Mémoire sur la destruction et le rétablissement des bois dans les départemens qui composoient la Provence," 7.

81. Ribbe, *La Provence au point de vue des bois*, 130.

82. Michelet, *Histoire de la Révolution française*; Woronoff, *Révolution et espaces forestiers*, 49. Andrée Corvol estimates that the number of goats increased 15 percent in the Var Department in 1795–1805. Corvol, *L'homme aux bois*, 27.

83. Lecugy, "A mort les chèvres!," 37–39.

84. See, for example, Fillias, *État actuel de l'Algérie*, 65; and Carteron, *Voyages en Algérie*, 104, 114, 182, 283–84, 290.

85. Carbonnieres, "De l'économie pastorale dans les Hautes-Pyrénées, de ses vices et des moyens d'y porter remède," cited in Métailie, "Lutter contre l'érosion," 97–98.

86. Guénot, "Le déboisement des Pyrénées," 67, quoted in Puyo, "La science forestière vue par les géographes français," 626.

87. Gibelin, "Observations sur les chèvres," 134.

88. Charles Ambroise de Caffarelli du Falga (Ardèche prefect, 1800), quoted in Corvol, *L'homme aux bois*, 27.

89. Bory de Saint-Vincent, *Note sur la commission exploratrice et scientifique d'Algérie*, 6. For a brief biography of Bory de Saint-Vincent, see Broc, "Les grandes missions scientifiques françaises," 321.

90. Broc, "Les grandes missions scientifiques françaises," 327; Davis, *Resurrecting the Granary of Rome*, 36.

91. Broc, "Les grandes missions scientifiques françaises," 327.

92. Lucas, *Exploration scientifique de l'Algérie*. See also Broc, "Les grandes missions scientifiques françaises," 326–31.

93. Lucas, *Exploration scientifique de l'Algérie*, 5.

94. ANOM FM F80981, Copy of letter from Conseiller d'État (signed Baude), to Ministre Secrétaire d'État de la Guerre (signed Bernard), Paris (24 April 1838).

95. The term "granary of Rome" is present in French colonial literature, but I owe Diana K. Davis for this usage. See Davis, *Resurrecting the Granary of Rome*.

96. Bernard and Lacroix, *L'évolution du nomadisme en Algérie*, 40.

97. ANOM ALG GGA/P59, Ligue du Reboisement de l'Algérie, Comité Central: Veu pour l'Organisation d'un Service de l'Agriculture (1882), 233; ANOM ALG GGA/ P59, *Bulletin de la Ligue du Reboisement de l'Algérie* 2–3 (February–March 1882): 28.

98. Davis, *Resurrecting the Granary of Rome*, 63 and passim.

99. Slane, *Les prolégomènes d'Ibn Khaldoun*, 45, 164. See also Davis, *Resurrecting the Granary of Rome*, 56.

100. Slane, *Les prolégomènes d'Ibn Khaldoun*, 310; Ibn Khaldun, *Muqaddimah*, 26.

101. Davis, *Resurrecting the Granary of Rome*, 97; Hannoum, *Violent Modernity*, 64.

102. Warnier, *L'Algérie devant l'Empereur*, 9.

103. ANOM FM F80 1785, M. de Bassano, "Examen de la question forestière dans la Subdivision de Bône" (8 February 1846).

104. Bernard and Lacroix, *L'évolution du nomadisme en Algérie*, 44.

105. İnalcık "Yuruks," 39–65; Murphey, "Some Features of Nomadism in the Ottoman Empire"; White, "Ecology, Climate, and Crisis in the Ottoman Near East," 64–65, 92.

106. White, "Ecology, Climate, and Crisis in the Ottoman Near East," 65. See also Kasaba, *Moveable Empire*, 53.

107. Dursun, "Forest and the State," 41; İnalcık, "Yuruks," 52–53.

108. Baudrillart, *Traité général des eaux et forêts*, 3.

109. Hamilton, *Researches in Asia Minor*, 1:70.

110. Chateaubriand, *Itinéraire de Paris à Jérusalem*, 195, quoted in Planhol, "Les nomades, la steppe, et la forêt en Anatolie."

111. Bernard and Lacroix, *L'évolution du nomadisme en Algérie*, 5.

112. Bent, "Exploration in Cilicia Tracheia," 435, 455.

113. TNA FO 222/2 (1880).

114. Dursun, *Forest and the State*, 101–14.

115. Meyer, "*Turquerie* and Eighteenth-Century Music"; Göçek, *East Encounters West*, 72–73.

116. Dursun, "Forest and the State," 106–9.

3. COUNTING SHEEP

1. Ladoucette, *Histoire, topographie, antiquités, usages, dialectes des Hautes-Alpes*, 464 (also 5–6, 91, 257, 303, 396, 464, and 484).

2. Duma, *Les Bourbon-Penthièvre*, 98.

3. See Matteson, *Forests in Revolutionary France*, 159. Planhol locates the nadir of French forests slightly later, in 1840. See Planhol, *Historical Geography of France*, 359.

4. Foucault, "Governmentality," 102.

5. "Ordonnance sur les eaux et forêts du 29 mai 1346," printed in Jourdan et al., *Recueil général des anciennes lois françaises depuis l'an 420 jusqu'à la Révolution de*

1789, 4:522–29. For the role of the 1346 edict in early French forest legislation, see Dumoulin, *La forêt provençale au XIXe siècle*, 1; Devèze, *La vie de la forêt française au XVIe siècle*, 1:64–72; and Matteson, *Forests in Revolutionary France*, 34.

6. Edict on forests (1516), Article 72, printed in Jourdan et al., *Recueil général des anciennes lois françaises depuis l'an 420 jusqu'à la Révolution de 1789*, 12:68.

7. Dumoulin, *La forêt provençale au XIXe siècle*, 1; Baudrillart, *Traité général des eaux et forêts*, 15–16.

8. Deposition of Elisabeth Fricon at the divorce of Louis XII (Amboise, 27 September, 1498), cited in Maulde La Clavière, *Procédures politiques du règne de Louis XII*, 978.

9. *Ordonnance de Louis XIV*.

10. Bamford, *Forests and French Sea Power*, 20–24.

11. Cameron, "Policing of Forests in Eighteenth-Century France"; Graham, "Policing the Forests of Pre-Industrial France"; Blais, "Contribution à une histoire des gardes forestiers au XVIII siècle," 17–26; and Matteson, *Forests in Revolutionary France*, 121.

12. Corvol, *L'homme aux bois*, 185, 189.

13. Dumoulin, *La forêt provençale au XIXe siècle*, 48; Blais, "Contribution à une histoire des gardes forestiers au XVIII siècle," 21.

14. Matteson, *Forests in Revolutionary France*, 121; Blais, "Contribution à une histoire des gardes forestiers au XVIII siècle," 21; Dumoulin, *La forêt provençale au XIXe siècle*, 41; Graham, "Policing the Forests of Pre-Industrial France"; BDR 7 M 163, Third File: "Reboisement, 1822, Première Arrondissement."

15. BDR 2 U 2 2827: "Analyse procédure contre François Tessère et Lazare Ambroise, bergers de Gémenos, pour avoir fait paître un troupeau de chèvres dans la forêt communale." For the lack of enforcement in general, see Larrère and Nougarede, *L'homme et la forêt*, 58.

16. Corvol, *L'homme aux bois*, 184–85.

17. Corvol, *L'homme aux bois*; Whited, *Forests and Peasant Politics*, 21–51; Dumoulin, *La forêt provençale au XIXe siècle*, 5.

18. The "Arrête sur la dévastation des bois" (BDR 14 J 227), passed in 1815 under the prefecture of Jean-Baptiste-Suzanne, marquis d'Albertas, provides a typical example.

19. *Forêts perdues*, 20.

20. *Forêts perdues*, 19; Corvol, "Région Rhône, Alpes, Provence et Côte d'Azur."

21. BDR 7 M 163, Third File: "Reboisement, 1822, Première Arrondissement."

22. BDR 7 M 163, Third File: "Reboisement, 1822, Première Arrondissement."

23. BDR 7 M 163, Third File: "Reboisement, 1822, Première Arrondissement."

24. "Antoine Claire Thibaudeau," in Robert, Bourloton, and Cougny, *Dictionnaire des parlementaires français*, 4:396–97.

25. *Forêts perdues*, 8.

26. BDR 7 M 163, Eighth File, Eaux et Forêts.

27. BDR 7 M 163, Eighth File, Eaux et Forêts.

28. BDR 14 J 173, printed in *Forêts perdues*, 18.

29. *Forêts perdues*, 19.

30. R. Grove, *Green Imperialism*, chap. 1. For other perspectives on the origins of conservationism, see Drayton, *Nature's Government*; Beinart and Hughes, *Environment and Empire*, 14–18; Merchant, *Reinventing Eden*; Radkau, *Nature and Power*, 167; and Ford, *Natural Interests*.

31. Braudel, *Capitalism and Material Life*, 2:106, quoted in R. Grove, *Green Imperialism*, 55.

32. R. Grove, *Green Imperialism*, chap. 3.

33. R. Grove, *Green Imperialism*, 120.

34. R. Grove, *Green Imperialism*, 5–6.

35. R. Grove, *Green Imperialism*, 60.

36. Albion, *Forests and Sea Power*, cited in Pincetl, "Some Origins of French Environmentalism," 81. See also Bamford, *Forests and French Sea Power*, esp. 157.

37. Matteson, *Forests in Revolutionary France*, 93; Devèze, "La crise forestière en France dans la première moitié du XVIIIe siècle."

38. Buffon, *Mémoire sur la conservation et le rétablissement des forêts*, cited in Matteson, *Forests in Revolutionary France*, 114.

39. Matteson, *Forests in Revolutionary France*, 65–66.

40. Duhamel du Monceau, *Du transport, de la conservation et de la force des bois*, viii, quoted in and translated by Matteson, *Forests in Revolutionary France*, 65.

41. Larrère and Nougarede, *L'homme et la forêt*, 59–60.

42. Tocqueville, *Old Regime and the French Revolution*, 16.

43. Tassy, *Études sur l'aménagement*, quoted in Larrère and Nougarede, *L'homme et la forêt*, 79. The higher figure is cited in Allard, *Les forêts et le régime forestier en Provence*, 51.

44. Pincetl, "Some Origins of French Environmentalism," 81; Matteson, *Forests in Revolutionary France*, 213–89.

45. Festy, *Les animaux ruraux de l'an III*, cited in Solakian, "De la multiplication des chèvres sous la Révolution," 55.

46. Rougier de La Bergerie, *Traité d'agriculture pratique, ou Annuaire de cultivateurs*, 190, quoted in Solakian, "De la multiplication des chèvres sous la Révolution," 53.

47. Solakian, "De la multiplication des chèvres sous la Révolution," 53.

48. Quoted in Corvol, *L'homme aux bois*, 27.

49. Michelet, *Tableau de la France*, 33, quoted in Woronoff, "La 'dévastation révolutionnaire' des forêts," 49.

50. Woronoff, "La 'dévastation révolutionnaire' des forêts," 45.

51. Moriceau, *Histoire et géographie de l'élevage français*, 164; Sclafert, "À propos du déboisement des Alpes du Sud," 135; Solakian, "De la multiplication des chèvres sous la Révolution," 58.

52. The Var (1800–1806) provides one example.

53. Allard, *Les forêts et le régime forestier en Provence*, 53–54; Corvol, *L'homme aux bois*, 27; Michelet, *Histoire de la Révolution française*; Larrère and Nougarede, *L'homme et la forêt*, 44; Solakian, "De la multiplication des chèvres sous la Révolution."

54. Larrère and Nougarede, *L'homme et la forêt*, 74.

55. Allard, *Les forêts et le régime forestier en Provence*, 52.

56. BDR 7 M 163, "Renseignements recueillis pour la solution des questions proposées par la circulaire, No. 18, de S.E. le ministre de l'intérieur, sur le changement du système météorologique du département des Bouches-du-Rhône" (1842).

57. M. Rougier de La Bergerie (1817), cited in Baudrillart, *Traité général des eaux et forêts*, 9, 30.

58. Michelet, *Tableau de la France*, 73, quoted in Trottier, *Boisement et colonisation*, 6.

59. Trottier, *Boisement et colonisation*, 6.

60. Horn, *Path Not Taken*.

61. Merchant, *Death of Nature*.

62. Corvol, "Les partenaires de l'environnementalisme," 39.

63. Kalaora, *La forêt pacifiée*, 18.

64. See Radkau, *Wood*, 149–50; Lowood, "Calculating Forester"; Whited, *Forests and Peasant Politics*, 27; and Scott, *Seeing Like a State*, chap. 1.

65. Lowood, "Calculating Forester," 316.

66. Foucault, "Governmentality," 96. See also Drayton, *Nature's Government*, 69–74.

67. Drayton, *Nature's Government*, 70; Wakefield, *Disordered Police State*; Vann, *Making of a State*.

68. Lowood, "Calculating Forester," 318; Radkau, *Wood*, 149–50.

69. Whited, *Forests and Peasant Politics*, 27; Lowood, "Calculating Forester," 319–20; Dombrowski, *Allgemeine Encyklopädie der gesammten Forst-und Jagdwissenschaften*, 510. The journal first appeared in 1763.

70. Lowood, "Calculating Forester," 319–20.

71. Hundeshagen, *Encyklopädie der Forstwissenschaft, systematisch abgefasst*, 1:6, quoted in and translated by Lowood, "Calculating Forester," 317.

72. "Chronique forestière" (1870), 68.

73. *Ordonnance du Roi pour l'exécution du Code forestier (du 21 mai 1827)*, Article 41: "L'enseignement dans l'école royale," printed in Baudrillart, *Code forestier*, 75.

74. Hartig, *Grundsätze der Forst-Direction*, 64, quoted in Lowood, "Calculating Forester," 338. See also Radkau, "Wood and Forestry in German History," 66; Steen et al., *History of Sustained-Yield Forestry*; and Smith, *Practice of Silviculture*, 344–52.

75. See "Loi du 3 mai 1844, sur la police de la chasse," in Bourdeaux, *Code forestier*, 174–95; Radkau, "Wood and Forestry in German History."

76. CARAN, F10 1743: Écoles forestières (1871–83).

77. Radkau, *Wood*, 151–52; Radkau, "Wood and Forestry in German History," 70–71; Badré, *Histoire de la forêt française*, 141; Smith, *Practice of Silviculture*, 330, 474–85; Matteson, *Forests in Revolutionary France*, 75.

78. Jean-Baptiste Sylvère Gaye, vicomte de Martignac, to the Chamber of Deputies (Paris, c. 1827), quoted in Allard, *Les forêts et le régime forestier en Provence*, 60.

79. Baudrillart, *Code forestier (du 21 mai 1827)*, Article 1.

80. Baudrillart, *Code forestier (du 21 mai 1827)*, Title III, section VIII. See also Corvol, "La privatisation des forêts nationales aux XVIIIe et XIXe siècles," 219–20; Whited, *Forests and Peasant Politics*, 36–37; Matteson, *Forests in Revolutionary France*, 334; and *Les eaux et forêts du XIIe au XXe siècle*, 222–23.

81. *Ordonnance du Roi pour l'exécution du Code forestier (du 21 mai 1827)*, Articles 40–56.

82. Dumoulin, *La forêt provençale au XIXe siècle*, 7.

83. CARAN F10 1724: "Commission du réorganisation du service forestier" (1878), Rapport de M. Charles Geraud; CARAN F10 1725: "Organisation militaire"; *Les eaux et forêts du XIIe au XXe siècle*, 483, 495, cited in Matteson, *Forests in Revolutionary France*, 365–66.

84. Baudrillart, *Code forestier (du 21 mai 1827)*, Article 119.

85. Baudrillart, *Code forestier (du 21 mai 1827)*, Article 119. See also Articles 63, 64, 90, 110, and 118.

86. BDR 150 E IN 2; *Ordonnance du Roi* (7 June 1829), Article 1.

87. Simon, Clément, and Pech, "Forestry Disputes in Provincial France," 343.

88. Baudrillart, *Code forestier (du 21 mai 1827)*, Article 76; Simon, Clément, and Pech, "Forestry Disputes in Provincial France," 343.

89. "Forêt," Centre national de ressources textuelles et lexicales (CNRTL), accessed 29 August 2017, www.cnrtl.fr/definition/forêt. It is also in this general sense that Olivier de Serres uses the term, though he also distinguishes between high-growth forests (*bois de haute futaie*) and coppices (*bois, taillis*). See Serres, *Le théâtre d'agriculture et mesnage des champs*, 558–60.

90. *Les eaux et forêts du XIIe au XXe siècle*, 12–13, 48, cited in Matteson, *Forests in Revolutionary France*, 59n78; Gadant, *L'atlas des forêts de France*. Specifically, this is the designation in the law of 1219, from which the phrase *eaux et forêts* originated.

91. Panckoucke, *Encyclopédie méthodique*, 4:144.

92. Edict of 1669, Article I.ii, in *Ordonnance de Louis XIV*, 9.

93. Furetière, *Dictionnaire universel, contenant généralement tous les mots françois*, vol. 1, s.v. "bois."

94. Bazire and Gadant, *La forêt en France*, 25–26, cited in Whited, *Forests and Peasant Politics*, 9.

95. *Ordonnance du Roi pour l'exécution du Code forestier (du 21 mai 1827)*, Article 59.

96. Baudrillart, *Code forestier (du 21 mai 1827)*, Article 90.

97. Lowood, "Calculating Forester," 320.

98. "Chronique forestière" (1890), 37–38, 137, 424, 564.

99. ANOM ALG GGA/P1, Sixth File: Documents sur forêts; Seventh File: Législation forestière étrangère.

100. ANOM ALG GGA/P1, Seventh File: Législation forestière étrangère.

101. "Chronique forestière" (1867), 60.

102. Miller, *Gifford Pinchot and the Making of Modern Environmentalism*, 77–97.

4. THE FOREST FOR THE TREES

1. BDR 103 E 2N1: "Statistique des Bouches-du-Rhône" (Allauch, 1816). See also BDR 8 F 30/2 (Arles); and BDR 8 F 30/1 (Aix-en-Provence), cited in *Forêts perdues*, 24–25.

2. Villeneuve, *Statistique du département des Bouches-du-Rhône*, 4:102, cited in *Forêts perdues*, 20. The French terms are *pins, chênes blancs, chênes verts*, and *bois blanc*.

3. BDR 14 J 173 (1811); *Forêts perdues*, 18–19. See also chapter 3.

4. Woronoff, *Révolution et espaces forestiers*, 62, 58; *Forêts perdues*, 36; Dumoulin, *La forêt provençale au XIXe siècle*, 130, 133–34.

5. BDR 7 M 250, Letter: Sous-préfet d'Aix-en-Provence to Sénateur (Aix-en-Provence, 22 January 1866).

6. See, for example, BDR 167 E 2N 1 (Les Baux-de-Provence); BDR 150 E 1N 2 (Roquefort-la-Bédoule); BDR 7 M 249 (Ceyreste and Gémenos); and BDR 174 E N 3 (Egyalières).

7. Baudrillart, *Code forestier (du 21 mai 1827)*, Articles 64 and 110. See also Dumoulin, *La forêt provençale au XIXe siècle*, 23.

8. Corvol, *L'homme aux bois*, 341.

9. Simon, Clément, and Pech, "Forestry Disputes in Provincial France," 343–44; Dumoulin, *La forêt provençale au XIXe siècle*, 113–15; BDR P5 32: Logging reserves (1849–60), cited in *Forêts perdues*, 35. See also the example of Ceyreste (BDR 7 M 249: Ceyreste).

10. Baudrillart, *Code forestier (du 21 mai 1827)*, Article 90.

11. BDR 150 E IN 2: "Roquefort-la-Bédoule," Extrait des registres des délibérations du Conseil Municipal de la Commune de Roquefort (28 March 1830).

12. Dumoulin, *La forêt provençale au XIXe siècle*, 14–15; Villeneuve, *Statistique du département des Bouches-du-Rhône*, 4:88.

13. Baudrillart, *Code forestier (du 21 mai 1827)*, Article 67.

14. Dumoulin, *La forêt provençale au XIXe siècle*, 196; Baudrillart, *Code forestier (du 21 mai 1827)*, Article 199.

15. Baudrillart, *Code forestier (du 21 mai 1827)*, Articles 78 and 110.

16. BDR 150 E IN 2: Roquefort-la-Bédoule: Redevance sur les troupeaux paissant sur les terres gastes (1807–1882); BDR 7 M 250: Orgon à Roquevaire: Herbages et pacages des troupeaux, titres de recettes des droits de pâturage, rôles de redevances (1857–1917).

17. BDR 150 E IN 2, Letter: Préfet to Maire de Roquefort (Marseille, 11 March 1830).

18. For a few examples, see BDR 150 E IN 2, Letter: Préfet to Maire de Roquefort (Marseille, 23 January 1835); and BDR 150 E IN 2, Letter: Préfet to Maire de Roquefort (Marseille, 12 August 1842).

19. Christophe, comte de Villeneuve, is one example.

20. BDR 7 M 163: Forêt: Reboisement, Correspondance, 1842. The community mentioned was Roquefort.

21. BDR P 5–22, Gémenos, délibération du Conseil municipal, 5 décembre 1858, quoted in Dumoulin, *La forêt provençale au XIXe siècle*, 21.

22. Simon, Clément, and Pech, "Forestry Disputes in Provincial France," 338–39.

23. BDR 7 M 163, Seventh File (1840).

24. BDR 7 M 163, Seventh File (1840).

25. Corvol, *L'homme aux bois*, 185.

26. Dumoulin, *La forêt provençale au XIXe siècle*, 19.

27. Dumoulin, *La forêt provençale au XIXe siècle*, 63.

28. BDR 174 E N 3, Second File: "Pâturages; Eaux et Forêts, 1872–1873" (Marseille, 12 September 1872), Ministère des Finances, République française, Sous-préfecture d'Arles.

29. BDR 174 E N 3, Third File: "Pâturages; Eaux et Forêts, 1872–1873," Direction générale des forêts (7 November 1871).

30. BDR, Marseille P5–18: Ceyreste, délibération du conseil municipal du 2 août 1854, lettre du préfet au maire du 20 février 1855, lettre de la direction générale

des forêts au préfet du 10 mars 1855, et délibération du conseil municipal du 9 mai 1855, cited in Dumoulin, *La forêt provençale au XIXe siècle*, 19.

31. Tassy, *Études sur l'aménagement des forêts*, ix–x.

32. BDR 7 M 249, Letter: Ministre des Finances to Sous-Préfet d'Arles (Paris, 9 November 1869).

33. Baudrillart, *Code forestier (du 21 mai 1827)*, Article 78.

34. BDR PER 124, Tome III (1874): "Circulaires 82–145 (1868–1873)," Circulaire 82 (23 March 1868).

35. *Loi du 18 juin–19 novembre 1859*, printed in Duvergier, *Collection complète des lois, décrets, ordonnances, règlemens, avis du Conseil-d'État*, 386; Dumoulin, *La forêt provençale au XIXe siècle*, 177.

36. Duvergier, *Collection complète des lois, décrets, ordonnances, règlemens, avis du Conseil-d'État*, 394–97.

37. Baudrillart, *Code forestier (du 21 mai 1827)*, 61.

38. Ageron, *Modern Algeria*, 5; Julien, *Histoire de l'Algérie contemporaine*, chap. 1.

39. Quoted in Ageron, *Modern Algeria*, 5.

40. Ford, "Reforestation, Landscape Conservation, and the Anxieties of Empire," 344; ANOM FM F80 981, "Note relative à la lettre adresse par Monsieur l'Intendant Militaire de la division d'Alger à Monsieur le Gouverneur Général" (13 March 1843).

41. See, for example, ANOM ALG GGA/P9: *Statistique générale*; and ANOM FM F80 1785, M. de Bassano, "Examen de la question forestière dans la Subdivision de Bône" (8 February 1846).

42. Carette, *Études sur la Kabilie*, 245, quoted in and translated by Davis, *Resurrecting the Granary of Rome*, 37.

43. ANOM ALG GGA/P9a, Report (14 March 1845).

44. Reynaud, *Mémoires historiques et géographiques sur l'Algérie*, 307.

45. ANOM FM F80 981, Letter: Inspecteur Général Directeur des Finances to Monsieur le Maréchal Gouverneur Général de l'Algérie (Algiers, 16 June 1840); ANOM FM F80 981, Letter: Conseiller d'État Baudé to Ministre Secrétaire d'État de la Guerre (Paris, 24 April 1838).

46. ANOM ALG GGA/P14, Minute de la lettre écrite du Ministre de la Guerre (Algérie) to Gouverneur Général (26 August 1842).

47. Trehonnais, *Rapport: L'Agriculture en Algérie*, 3.

48. Sautayra, *Législation de l'Algérie*, 286.

49. ANOM FM F80 981, Note relative à la lettre adressée par M. l'Intendant Militaire de la division d'Alger à M. le Gouverneur Général du 13 mars 1843.

50. Prochaska, "Fire on the Mountain," 234–35; Ford "Reforestation, Landscape Conservation, and the Anxieties of Empire," 344–45.

51. For eucalyptus, see ANOM ALG GGA/P59, Reboisements: Ligue du Reboisement. For viticulture, see ANOM ALG GGA/P12, Sol forestier: Bois et forets, 1845–1897, and ANOM FM F80 1788, "Viticulture." For other "exotic" tree species, see ANOM FM F80 1788, "Acclimatisation des plantes et arbres exotiques en Algérie."

52. See *Décret relatif à la centralisation et à l'organisation du service des forêts, et à leur soumission au régime forestier du 27 septembre 1873*, Article 2, printed in Estoublon and Lefébure, *Code de l'Algérie annoté*, 421.

53. See, for example, ANOM FM F80 175, Statistique Forestière 1854–1857; ANOM FM F80 971 (1858); and ANOM ALG GGA/P9a (1868–88).

54. Ford, "Reforestation, Landscape Conservation, and the Anxieties of Empire," 345.

55. ANOM FM F80 1786, Notice sur le Cercle et la Ville de la Calle (Algérie), Doucement fourni par M Dubouchage, concessionnaire des forêts de chêne liège de La Calle, confiées à M Duval 23 mars 1855.

56. ANOM ALG GGA/P9a, Algérie, Province de Constantine, Service des Forêts, Conservation de Constantine, "Forêts de l'Algérie, Statistique Générale," Fait et Dressé par l'Inspecteur Hors de Conservateur des Forêts, Constantine le 5 juin 1868.

57. ANOM FM F80 981, Letter: Direction des Finances, Service des Forêts, to Monsieur le Lieutenant Général Gouverneur Général, par intérim (Alger, 7 January 1844).

58. ANOM FM F80 981, Letter: Direction des Finances, Service des Forêts, to Monsieur le Lieutenant Général Gouverneur Général, par intérim (Alger, 7 January 1844), 106–8.

59. ANOM ALG GGA/22K/10, "Forêts: Correspondances, 1896–1897."

60. ANOM FM F80 971, M. le Secrétaire Général du Gouvernement, Alger, de l'Algérie, département de Constantine, Service des Forêts (l'Inspecteur chef du Service des Forêts), Constantine, le 12 avril 1854.

61. Woolsey, *French Forests and Forestry*, 106.

62. ANOM ALG GGA/P60, Letter: Conservateur des Forêts to Gouverneur-Général de l'Algérie (Constantine, 5 June 1889).

63. Tassy, *Service forestier de l'Algérie*, 15.

64. ANOM ALG GGA/P60, Letter: Conservateur des Forêts to Gouverneur-Général de l'Algérie (Constantine, 5 June 1889). See also Fillias, *État actuel de l'Algérie*, 62; and Carteron, *Voyages en Algérie*, 104, 114, 182, 283–84, 290.

65. ANOM FM F80 982, Circulaire, Marcotte (Ministre des Finances) to forest conservators (Paris, 23 June 1834).

66. See, for example, ANOM FM F80 730, Letter: M. Poussin to Secrétariat-Général du Gouvernement, Direction des Affaires civiles de l'Algérie, 3e Bureau (18 April 1857).

67. ANOM FM F80 1785, Forêts du Kaidat de l'Edough (Province de Bône), reconnues par M. de Klopestein, Sous Inspecteur des forêts en Algérie, pendant l'année 1845 (Ministère de la Guerre).

68. ANOM ALG /GGA/P1, "Propositions faites le 14 Jan 1848 par le chef du service des forêts pour servir à la rédaction d'un règlement forestier, spécial à l'Algérie."

69. BOA A Amd. no. 78/28 (1856–57); BOA I.MV, no. 16327; BOA I.MV, no. 16518, all printed in Batmaz and Koç, *Osmanlı Ormancılığı ile İlgili Belgeler*, 1:155–58, 161–65, 175. See also Dursun, "Forest and the State," 179.

70. BOA A Amd. no. 78/28, printed in Batmaz and Koç, *Osmanlı Ormancılığı ile İlgili Belgeler*, 1:155–58. See also Dursun, "Forest and the State," 179–80.

71. Imber, *Ottoman Empire, 1300–1600*, 294.

72. Mikhail, *Nature and Empire in Ottoman Egypt*, 99–101.

73. Imber, *Ottoman Empire, 1300–1600*, 294; [L. C.] M., "Recherches sur le mouvement," 33; Aydın, *Sultanın Kalyonları*; Bostan, *Osmanlı Bahriye Teskilâtı*, 102.

74. Baytop, *Türkiye'de Botanik Tarihi Araştirmaları*, 38, cited in Dursun, "Forest and the State," 27.

75. Dursun, "Forest and the State," 33.

76. Batmaz and Koç, *Osmanlı Ormancılığı ile İlgili Belgeler*, 1:xv; Koç, "Tanzimat Sonrası Hukuk Metinlerinde Çevre Bilincinin," 274; Dursun, "Forest and the State," 185.

77. Dursun, "Forest and the State," 37–38; Imber, *Ottoman Empire, 1300–1600*, 295; Imber, "Navy of Suleyman the Magnificent"; Aydın, *Sultanın Kalyonları*; Bostan, *Osmanlı Bahriye Teskilâtı*.

78. Bricogne, "Les forêts de l'Empire ottoman," 273.

79. Dursun, "Forest and the State," 39.

80. Hanioğlu, *Brief History of the Late Ottoman Empire*, 72–73; Karpat, "Transformation of the Ottoman State," 258–59.

81. Hanioğlu, *Brief History of the Late Ottoman Empire*, 73.

82. Batmaz and Koç, *Osmanlı Ormancılığı ile İlgili Belgeler*, xiv.

83. Dursun, "Forest and the State," 177; BOA I.MV, no. 1637 (1857), printed in Batmaz and Koç, *Osmanlı Ormancılığı ile İlgili Belgeler*, 1:156–59.

84. Bernhard, *Türkiye Ormancılığının Mevzuatı*, 109; BOA I.MV, no. 16327 (1857), both cited in Dursun, "Forest and the State," 179.

85. Takeda, *Between Crown and Commerce*, 81–82; Brewer, "Gold, Frankincense, and Myrrh."

86. Göçek, *East Encounters West*, 72. See also Pettet, "Veritable Bedouin," 41.

87. Hanioğlu, *Brief History of the Late Ottoman Empire*, 34.

88. Hanioğlu, *Brief History of the Late Ottoman Empire*, 44.

89. Hanioğlu, *Brief History of the Late Ottoman Empire*, 48–49.

90. Finkel, *Osman's Dream*, 457.

91. CADN 2 Mi 1209, Correspondance du départ: Sublime Porte (Pera), Ministère des Affaires Étrangères (Ottoman) à M. Chauvenel, Ambassadeur de sa majesté l'Empereur des Français (29 August 1857).

92. Dursun "Forest and the State," 183.

93. [L. C.] M., "Des forêts de la Turquie," 405.

94. [L. C.] M., "Des forêts de la Turquie," 405.

95. Bricogne, "La mission forestière en Turquie" (1876): 362.

96. [L. C.] M., "Mission forestière de MM Tassy et Sthème en Turquie," 113.

97. "Notice sur Louis Tassy," 6.

98. "Notice sur Louis Tassy," 7.

99. Buttoud, *L'état forestier*, s.v. "Tassy, Louis,"; "Chronique forestière" (1865), 150–51.

100. "Chronique forestière" (1865), 150–51.

101. Özdönmez and Ekizoğlu, "Tanzimat ve Meşrutiyet Dönemleri Ormancılığında Katkıları Olan Yabancı Uzmanlar," 60.

102. Bricogne, "Les fôrets de l'Empire ottoman," 425.

103. Bricogne, "Les fôrets de l'Empire ottoman," 430.

104. Bricogne, "Les fôrets de l'Empire ottoman," 382–83, 427–29.

105. Bricogne, "Les fôrets de l'Empire ottoman," 425.

106. "Chronique forestière" (1865), 150–51; Dursun, "Forest and the State," 180.

107. BOA I.DH 38044 (26 L 1282/14 March 1866), supplement 4; Dursun, "Forest and the State," 184–89.

108. Bricogne, "Les fôrets de l'Empire ottoman," 382–83.

109. Bricogne, "Les fôrets de l'Empire ottoman," 430.

110. Bricogne, "La mission forestière en Turquie" (1870), 223.

111. Eraslan, *Türkiye'de Ormancılık Öğretim*, 3–4, cited in Dursun, "Forest and the State," 180.

112. *Journal de Constantinople*, 8 January 1860, quoted in "Chronique forestière: École forestière de Constantinople," *Annales Forestières* 20 (1861).

113. Bricogne, "La mission forestière en Turquie" (1870), 223; Yund, "100 Yıllık Türk Ormancılık Öğretimine Bakış," 22, cited in Dursun, "Forest and the State," 182.

114. Tassy, *Études sur l'aménagement des forêts*, xi–xvi; Dursun, "Forest and the State," 181, 184.

115. *Ottoman Forest Bill of 1861*, Article 39, printed in Dursun, "Forest and the State," 413–14.

116. *Ottoman Forest Bill of 1861*, Articles 50, 51, and 53, printed in Dursun, "Forest and the State," 414.

117. Dursun, "Forest and the State," 198–200.

118. Dursun, "Forest and the State," 216.

119. Bricogne, "Les fôrets de l'Empire ottoman," 7; Dursun, "Forest and the State," 189–97.

120. Dursun, "Forest and the State," 195.

121. "Tarifs et douanes," 259.

122. Bricogne, "La mission forestière en Turquie" (1876), 363; Dursun, "Forest and the State," 182.

123. Bricogne, "La mission forestière en Turquie" (1870), 223; Keskin, "Osmanlı Ormancılığı'nın Gelişiminde Fransız Uzmanların Rolü," 128.

124. BOA A.MKT.UM no. 326/55 (1858), printed in Batmaz and Koç, *Osmanlı Ormancılığı ile İlgili Belgeler*, 1:177.

5. AGAINST THE GRAIN

1. Ladoucette, *Histoire, topographie, antiquités, usages, dialectes des Hautes-Alpes*, 544–45.

2. Fassin, "Le droit d'esplèche dans la Crau d'Arles," 7.

3. *Forêts perdues*, 21.

4. *Forêts perdues*, 32.

5. Dumoulin, *La forêt provençale au XIXe siècle*, 32; Ladoucette, *Histoire, topographie, antiquités, usages, dialectes des Hautes-Alpes*, 546.

6. Fassin, *Le droit d'esplèche dans la Crau d'Arles*; Shippers, "Le cycle annuel d'un berger transhumant," 64.

7. BDR 7 M 248 (Les Baux-de-Provence, 1890–1900).

8. BDR 2 U 2 2827 (Gémenos, an XII); BDR 2 U 2 2862 (Draguignan, 1849).

9. *Loi du 10–11 juin 1793: Décret concernant le mode de partage des biens communaux*, Sec. IV, Article Ier, printed in Jay, *Bulletin des lois des justices de paix*, 1:120–21.

10. Matteson, *Forests in Revolutionary France*, 130–34; McPhee, *Revolution and Environment in Southern France*, chap. 6; Rosenthal, *Fruits of Revolution*.

11. *Le nouveau code rural*, 10–11; Fassin, *Le droit d'esplèche dans la Crau d'Arles*, 37.

12. Fassin, *Le droit d'esplèche dans la Crau d'Arles*, 36. The right of *vaine pâture* was abolished through the *Loi du 9 juillet 1889*, Article 2.

13. Matteson, "Masters of Their Woods," 114–15.

14. Freeman, "Forest Conservancy in the Alps of Dauphiné," 173–74.

15. Ferrand, *De la propriété communale en France*, 20.

16. Quirot, Commissioner of the Doubs, *Compte de la situation politique pendant le mois de Messidor an 6* (June–July 1798), quoted in and translated by Matteson, *Forests in Revolutionary France*, 106.

17. Allard, *Les forêts et le régime forestier en Provence*, 51.

18. Godechot, *Les institutions de la France sous la Révolution et l'Empire*, 663.

19. *Code civil* (1804), Book II, Chap. 3, Article 537, in *Recueil des lois composant le code civil.*

20. Godechot, *Les institutions de la France sous la Révolution et l'Empire,* 570.

21. These forest lands totaled about 870,000 acres. *Les eaux et forêts du XIIe au XXe siècle,* 523, cited in Matteson, *Forests in Revolutionary France,* 256; Husson, *Les forêts françaises,* 161.

22. Weber, *Peasants into Frenchmen,* 128.

23. *Forêts perdues,* 20–21.

24. See, for example, BDR 150 E 1N 2, Extrait des registres des délibérations du Conseil municipal de la commune de Roquefort (28 March 1830).

25. BDR 2 U 2 2821: Cour d'appel, 25 avril 1825; BDR 2 U 2 2862: "Analyse procédure contre Eléonore Astier, pour introduction de chèvres dans une forêt communale"; BDR 2 U 2 2827: "Analyse procédure contre François Tessère et Lazare Ambroise, bergers de Gémenos, pour avoir fait paître un troupeau de chèvres dans la forêt communale"; BDR 7 M 192: "Livrets: Rapports des gardes forestiers"; and BDR 7 M 240: "Délits forestiers: Amendes et transactions."

26. BDR 2 O 3 1: "Pâturage: Allauch. Préfecture (1808–1824), Marseille."

27. BDR 2 U 2 2821: Cour d'appel, 25 avril 1825.

28. BDR 2 O 3 1, Letter: Mayor of Allauch to Préfet, Marseille (17 February 1818); Dumoulin, *La forêt provençale au XIXe siècle,* 32, 35; *Forêts perdues,* 21.

29. Dumoulin, *La forêt provençale au XIXe siècle,* 33.

30. BDR 150 E 1N 2: "Roquefort-la-Bédoule: Pacage. Redevance sur les troupeaux paissant sur les terres gastes (1807–1882)"; BDR 7 M 286: "Bois Communaux: Gémenos."

31. Baudrillart, *Code forestier (du 21 mai 1827),* Article 90.

32. Allard, *Les forêts et le régime forestier en Provence,* 64.

33. BDR 150 E 1N 4: "Roquefort-la-Bédoule. Pacage. États des propriétaires dont les troupeaux paissent sur les terres gastes et les bois communaux, 1809–1884."

34. BDR 167 E 2N1 (Les Baux-de-Provence).

35. "Loi du 22 juin 1891 [*sic*] modifiant le titre II du code rural (vaine pâture)," *Les textes fondateurs du monde agricole,* accessed 11 April 2019, https://agriculture .gouv.fr/histoire/4_textes/IIIrepublique/22juin1891.htm.

36. BDR 134 J 106: "Procès entre la ville de Salon et Couderc et autres, au sujet de la vaine pâture" (1901); BDR 135 J 7: "Dépaissance et vaine pâture: Situation juridique de la Camargue (1909)"; Bloch, *French Rural History,* 135, 197–234.

37. Baudrillart, *Code forestier (du 21 mai 1827),* Article 90.

38. Kalaora and Savoie, "Aménagement et ménagement," 307–28, cited in Simon, Clément, and Pech, "Forestry Disputes in Provincial France," 346.

39. Orange and Amalbert, *Le mérinos d'Arles*, 25; Belleval, "Réflexions sur la transhumance des troupeaux en Provence," 23.

40. Orange and Amalbert, *Le mérinos d'Arles*, 9. Joseph-Étienne Michel provided a higher figure of three hundred thousand. See Michel, *Observations sur le commerce des bêtes à laine dans les départements des Bouches-du-Rhône*, 3.

41. Orange and Amalbert, *Le mérinos d'Arles*, 26, 36.

42. ANOM ALG GGA/P9a: "Forêts de l'Algérie, Statistique générale." See also Bourdieu and Wacquant, "On the Cunning of Imperialist Reason."

43. Ageron, *Modern Algeria*, 8.

44. See, for example, Combe, *Les forêts de l'Algérie*, 14, 16. See also Julien, *Histoire de l'Algérie contemporaine*, 404.

45. Mahé, *Histoire de la Grande Kabylie*, 8, 10.

46. Ford, "Reforestation, Landscape Conservation, and the Anxieties of Empire," 343–44, 357.

47. Fillias, *État actuel de l'Algérie*, 79.

48. Hardin's essay, "The Tragedy of the Commons," was published in 1968.

49. Fillias, *État actuel de l'Algérie*, 10–11.

50. ANOM ALG GGA/P13, Third File: "Forêts: Organisation: Constitution du Sol forestier: Commissions forestières (Constatation des droits de propriété): Exécution de la Loi du 16 juin 1851."

51. Quoted in Julien, *Histoire de l'Algérie contemporaine*, 405.

52. Ford, "Reforestation, Landscape Conservation, and the Anxieties of Empire," 344; Julien, *Histoire de l'Algérie contemporaine*, 404, 406–8.

53. ANOM ALG GGA/P13, Minute de la lettre écrite, Gouverneur Général de l'Algérie, Secrétariat Général, 2e Bureau, à M. le Ministre de l'Agriculture (Administration des forêts), Paris, 14 février 1882.

54. Julien, *Histoire de l'Algérie contemporaine*, 405; Davis, *Resurrecting the Granary of Rome*, 100.

55. Quoted in Julien, *Histoire de l'Algérie contemporaine*, 405.

56. ANOM FM F80 1679, Letter: Napoléon III to Jean-Jacques Pélissier, Gouverneur Général de l'Algérie (1 November 1861), quoted in Julien, *Histoire de l'Algérie contemporaine*, 425; Abi-Mershed, *Apostles of Modernity*, 167; Spillmann, *Napoléon III et le royaume arabe d'Algérie*, 26–27. See also Rey-Goldzeiguer, *Le royaume arabe*.

57. Julien, *Histoire de l'Algérie contemporaine*, 419; Ageron, *Modern Algeria*, 36–38; Lorcin, *Imperial Identities*, 77.

58. ANOM FM F80 1679, *Sénatus-consulte du 22 avril 1863*. See also Abi-Mershed, *Apostles of Modernity*, 167; and Davis, *Resurrecting the Granary of Rome*, 83–88.

59. *Projet de Sénatus-consulte relatif à la constitution de la propriété en Algérie*, Article 1, printed in *Statistique et documents relatifs au Sénatus-consulte sur la propriété arabe*, 45.

60. *Projet de Sénatus-consulte relatif à la constitution de la propriété en Algérie*, Articles 2, 3, and 7, printed in *Statistique et documents relatifs au Sénatus-consulte sur la propriété arabe*, 45. Article 7 nullified Article 14 of the law of 16 June 1851.

61. "Éditorial," *La Seybouse*, 12 July 1861, quoted in Julien, *Histoire de l'Algérie contemporaine*, 406.

62. Abi-Mershed, *Apostles of Modernity*, 168.

63. Julien, *Histoire de l'Algérie contemporaine*, 426; Yacono, *Les Bureaux Arabes et l'évolution des genres de vie indigènes dans l'Ouest du Tell algérois*, 160–71.

64. Fillias, *État actuel de l'Algérie*, 6–7.

65. Lorcin, *Imperial Identities*, 78.

66. Ageron, *Modern Algeria*, 52.

67. Ruedy, *Modern Algeria*, 93; Davis, *Resurrecting the Granary of Rome*, 92.

68. Ageron, *Modern Algeria*, 52. See also Davis, *Resurrecting the Granary of Rome*, 92.

69. Warnier, *L'Algérie devant l'Empereur*, 5. See also Warnier, *L'Algérie devant le Sénat*; Duval and Warnier, *Bureaux arabes et colons*; Warnier, *L'Algérie et les victimes de la guerre*; Davis, *Resurrecting the Granary of Rome*, 96; Lorcin, *Imperial Identities*, 129–30, 309.

70. Eyssautier, *Le statut réel français en Algérie*, 29–56.

71. Ageron, *Histoire de l'Algérie contemporaine*, 94.

72. Quoted in Ageron, *Histoire de l'Algérie contemporaine*, 95.

73. Ageron, *Histoire de l'Algérie contemporaine*, 94–95.

74. İslamoğlu-İnan, "Politics of Administering Property," 16–18.

75. Ágoston, "Flexible Empire," 19.

76. S. Shaw, *History of the Ottoman Empire and Modern Turkey*, 1:150–51; Gould, "Pashas and Brigands," 3.

77. Imber, *Ottoman Empire, 1300–1600*, 188–89; Ágoston, "Flexible Empire," 17–19; Khoury and Kostiner, *Tribes and State Formation in the Middle East*, 43.

78. İnalcık and Quataert, *Economic and Social History of the Ottoman Empire*, 71, 73.

79. Salzmann, "Ancien Regime Revisited," 399.

80. S. Shaw, *History of the Ottoman Empire and Modern Turkey*, 1:125–27; Imber, *Ottoman Empire, 1300–1600*, 190–91; Ágoston "Flexible Empire," 16–17.

81. Hanioğlu, *Brief History of the Late Ottoman Empire*, 21; Pamuk, *Monetary History of the Ottoman Empire*, 84–87, 189.

82. Salzmann, "Ancien Regime Revisited," 400–402; Imber, *Ottoman Empire, 1300–1600*, 211, 285; Hanioğlu, *Brief History of the Late Ottoman Empire*, 21; Genç,

Osmanlı İmparatorluğu'nda Devlet ve Ekonomi, 117; Genç, "Osmanlı Maliyesinde Malikâne Sistemi," 231–96.

83. İslamoğlu-İnan, "Politics of Administering Property," 19–20; Imber, *Brief History of the Late Ottoman Empire*, 211–12, 285.

84. İslamoğlu-İnan, "Politics of Administering Property," 29; Hanioğlu, *Brief History of the Late Ottoman Empire*, 89.

85. Karpat, *Ottoman Population, 1830–1914*, 49–50; Karal, *Osmanlı İmparatorluğu'nda İlk Nüfus Sayımı, 1831*, 189, cited in Dursun, "Population Policies of the Ottoman State in the Tanzimat Era," 14.

86. İslamoğlu-İnan, "Politics of Administering Property"; İslamoğlu, "Statistical Constitution of Property Rights on Land in the 19th Century Ottoman Empire," cited in Dursun, "Population Policies of the Ottoman State in the Tanzimat Era," 15; Kütükoğlu, "Osmanlı Sosyal ve İktisadi Kaynaklarından Temettü Defterleri," 395–418. See also Karpat, "Transformation of the Ottoman State," 258–59.

87. Aytekin, "Hukuk, Tarih ve Tarihyazımı"; İslamoğlu-İnan, "Politics of Administering Property"; Jorgens, "Comparative Examination of the Ottoman Land Code and Khedive Sa'id's Law of 1858"; Karpat, "Land Regime, Social Structure and Modernization in the Ottoman Empire"; Barkan, *Türk Toprak Hukuku Tarihinde Tanzimat*; Barkan, *Türkiye'de Toprak Meselesi*, 291–375.

88. İslamoğlu-İnan, "Politics of Administering Property," 9–11, 36–37; Dursun, "Forest and the State," 219; S. Shaw, *History of the Ottoman Empire and Modern Turkey*, 2:65; Karpat, "Transformation of the Ottoman State," 258–59.

89. Dursun, "Forest and the State," 15; İslamoğlu-İnan, "Politics of Administering Property," 40–41.

90. Ongley and Miller, *Ottoman Land Code*, 1; S. Shaw, *History of the Ottoman Empire and Modern Turkey*, 2:114; Batmaz and Koç, *Osmanlı Ormancılığı ile İlgili Belgeler*, 1:xvi.

91. *Ottoman Land Code of 1858*, Articles 25–28, printed in Ongley and Miller, *Ottoman Land Code*, 13–15. See also Dursun, "Forest and the State," 222; and Imber, "Law of the Land," 43.

92. Aytekin, "Hukuk, Tarih ve Tarihyazımı," 733–34; Jorgens, "Comparative Examination of the Ottoman Land Code and Khedive Sai'id's Law of 1858," 103; İslamoğlu-İnan, "Politics of Administering Property," 32–33.

93. İslamoğlu-İnan, "Politics of Administering Property," 27; S. Shaw, *History of the Ottoman Empire and Modern Turkey*, 2:114.

94. Sluglett and Farouk-Sluglett, "Application of the 1858 Land Code in Greater Syria," 413.

95. *Ottoman Land Code of 1858*, Article 84, printed in Ongley and Miller, *Ottoman Land Code*, 44.

96. Hütteroth, "Influence of Social Structure on Land Division and Settlement in Inner Anatolia," 23.

97. Batmaz and Koç, *Osmanlı Ormancılığı ile İlgili Belgeler*, 1:xvi; Dursun, "Forest and the State," 218–31.

98. S. Shaw, *History of the Ottoman Empire and Modern Turkey*, 2:235.

99. S. Shaw, *History of the Ottoman Empire and Modern Turkey*, 2:235; Dursun, "Forest and the State," 228.

100. *Ottoman Land Code of 1858*, Article 19, printed in Ongley and Miller, *Ottoman Land Code*, 11; Batmaz and Koç, *Osmanlı Ormancılığı ile İlgili Belgeler*, 1:xvi.

101. Kasaba, *Moveable Empire*, 7, 114; Deringil, *Well-Protected Domains*, 41; Deringil, "'They Live in a State of Nomadism and Savagery.'"

102. Gould, "Pashas and Brigands," xiii, 119–39; Toksöz, *Nomads, Migrants and Cotton*, 65–73. See also chapter 7.

103. Bates, *Nomads and Farmers*, 34; Gould, "Pashas and Brigands"; Johnson, *Nature of Nomadism*, 23–25; Kasaba, "Do States Always Favor Stasis?," 28; Kasaba, *Moveable Empire*; Köksal, "Coercion and Mediation," 470; Saydam, "XIX. Yüzyılın İlk Yarısında Aşiretlerin İskânına Dair Gözlemler," 229.

104. Kasaba, "Do States Always Favor Stasis?," 37; İnalcık and Quataert, *Economic and Social History of the Ottoman Empire*, 793–95.

105. Gould, "Pashas and Brigands," 64; A. Fisher, "Emigration of Muslims from the Russian Empire," 181, cited in Kasaba, *Moveable Empire*, 109.

106. Gould, "Pashas and Brigands," 64; İnalcık and Quataert, *Economic and Social History of the Ottoman Empire*, 861–62; Toksöz, *Nomads, Migrants and Cotton*, 73–81.

107. Kasaba, *Moveable Empire*, 57.

108. Hütteroth, "Influence of Social Structure on Land Division and Settlement in Inner Anatolia," 23; Gould, "Pashas and Brigands," 64; Toksöz, *Nomads, Migrants and Cotton*, 76–77.

109. TNA FO 78/2987, Letter: Wilson to Layard (Sivas, 14 October 1879).

110. Gould, "Pashas and Brigands," 2, 4; Toksöz, *Nomads, Migrants and Cotton*, 76–78; Yılmaz, "Policy of Immigrant Settlement of the Ottoman State"; İnalcık and Quataert, *Economic and Social History of the Ottoman Empire*, 787–88; Kasaba, *Moveable Empire*, 117–18.

111. Planhol, *De la plaine pamphylienne aux lacs pisidiens*, 115; Tabak, *Waning of the Mediterranean*, 229.

112. Cevdet, *Tarih-i Cevdet*, 10:148; Hanioğlu, *Brief History of the Late Ottoman Empire*, 61; TNA FO 222/2: Correspondence 1879–1880, Adalia (Antalya) Vilayet, 13 January 1880.

113. Gould, "Pashas and Brigands," 12; Planhol, *De la plaine pamphylienne aux lacs pisidiens*, 124–25; McNeill, *Mountains of the Mediterranean World*, 287; Tabak, *Waning of the Mediterranean*, 294.

114. Kasaba, *Moveable Empire*, 35–36; Griswold, *Great Anatolian Rebellion*; Gould, "Pashas and Brigands."

115. Barkan, "Osmanlı Imparatorluğunda Bir İskan ve Kolonizasyon Metodu Olarak Vakıflar ve Temlikler," cited in Kasaba, *Moveable Empire*, 36.

6. NATURE'S SCAPEGOATS

1. Prochaska, "Fire on the Mountain," 240; Marc, *Notes sur les forêts de l'Algérie*, 364–65, 435; Ford, "Reforestation, Landscape Conservation, and the Anxieties of Empire," 347; ANOM FM F80 1785, "Rapport à l'Empereur, Sujet: Incendies qui viennent de ravager les cantons forestiers de la Province de Constantine" (Paris, 22 September 1863).

2. ANOM FM F80 1785, Incendies: Responsabilite collective des Tribus, 1863; Besson-Lecousturier, *Note sur les forêts de chênes-liège*. The term "wildfire" can designate a fire of natural (nonhuman) origin or a fire occurring outside (in nature). I use the term in the latter sense.

3. ANOM FM F80 1785, Ministère de la Guerre: Cabinet du Ministre, "Application de la responsabilité collective aux tribus du territoire civil." For another early reference, see ANOM ALG GGA/P62, Letter: Gouverneur Général de l'Algérie to Ministère de la Guerre, "Au sujet de la responsabilité des tribus appliquée à la conservation des forêts" (Alger, 18 July 1854).

4. Campbell, *Botanist and the Vintner*.

5. Simon, Clément, and Pech, "Forestry Disputes in Provincial France," 347; Corvol, *L'homme aux bois*, 21, 26; Corvol, "Les partenaires de l'environnementalisme," 51–52; Whited, "Extinguishing Disaster in Alpine France," 263–70.

6. Rosenthal, *Fruits of Revolution*, 51–58.

7. BDR 7 M 135, First File: Agricultural Calamities, 1802–96.

8. BDR 7 M 135, Second File (Tarascon, 2 November 1840).

9. BDR 7 M 135, Second File (Tarascon, 2 November 1840).

10. Venture, "Arles et le Rhône," 3.

11. La Gorce, *Histoire du Second Empire*, 2:40–41, cited in Whited, *Forests and Peasant Politics*, 58.

12. Ford, *Natural Interests*, 67.

13. BDR 7 M 248: Herbages et pacages des troupeaux, titres de recettes des droits de pâturage, rôles de redevances (Arles, 7 December 1886).

14. "Discussion devant l'assemblée nationale du projet de transfert des forêts au Ministère de l'Agriculture," 196.

15. BDR 7 M 163, Circulaire: Ministère de l'Intérieur, Sciences et Beaux-Arts (Paris, 25 April 1821).

16. Surell, *Étude sur les torrents des Hautes-Alpes*, was first published in 1841. See also Whited, *Forests and Peasant Politics*, 56; and Simon, Clément, and Pech, "Forestry Disputes in Provincial France," 338.

17. Surell, *Étude sur les torrents des Hautes-Alpes*, 1:iv–v.

18. Surell, *Étude sur les torrents des Hautes-Alpes*, 1:272–73.

19. Blanqui, *Du déboisement des montanges*, 27.

20. Venture, "L'inondation de 1856." See also Bess, *Light-Green Society*, 57.

21. BDR 7 M 163, Circulaire: Ministère de l'Intérieur, Sciences et Beaux-Arts (Paris, 25 April 1821); BDR 7 M 163, "Reboisement, 1822, Première Arrondissement"; BDR 7 M 163, M. Bompar, "Mémoire sur le Reboisement des Terres Incultes des Montagnes, 1846."

22. Corvol, *L'homme aux bois*, 321, 330–31; Tassy, *La restauration des montagnes*, 1–2.

23. Ribbe, *La Provence au point de vue des bois*, 6.

24. Ribbe, *La Provence au point de vue des bois*, 9.

25. Whited, *Forests and Peasant Politics*, 59.

26. Ducrocq, *Cours de droit administratif*, 67; Demontzey, *Étude sur les travaux de reboisement et de gazonnement des montagnes*, iv.

27. *Loi relative au reboisement des montagnes du 28 juillet 1860*, Articles 1 and 2, printed in Deville and Rez, *Recueil des lois*, 170.

28. *Loi relative au reboisement des montagnes du 28 juillet 1860*, Article 4, printed in Deville and Rez, *Recueil des lois*, 171.

29. Demontzey, *Étude sur les travaux de reboisement et de gazonnement des montagnes*, iii.

30. Ford, *Natural Interests*, 60–63; Demontzey, *Étude sur les travaux de reboisement et de gazonnement des Montagnes*, vi; Tassy, *La restauration des montagnes*, 17.

31. Carrière, "Prosper Demontzey," 195.

32. Carrière, "Prosper Demontzey," 210.

33. *Gazon* means grass, turf, or lawn. *Gazonnement* refers to seeding, fertilizing, tending, and replanting grass. Thus, the title translates roughly to "Study on the work [being done] to reforest and replant the mountains."

34. Demontzey, *Étude sur les travaux de reboisement et de gazonnement des montagnes*, i.

35. Demontzey, *Étude sur les travaux de reboisement et de gazonnement des montagnes*, vi.

36. Carrière, "Prosper Demontzey," 199.

37. Demontzey published numerous other works on this theme, including, *Traité pratique du reboisement et du gazonnement des montagnes* (1882), *La restauration des terrains en montagne au pavillon des forêts* (1889), *Le reboisement des montagnes et l'extinction des torrents* (1891), *L'extinction des torrents en France par le reboisement* (1894), and *Les retenues d'eau et le reboisement dans le bassin de la Durance* (1896). In addition, a German edition of his *Étude sur les travaux de reboisement et de gazonnement des montagnes* was published in 1880.

38. Demontzey, *Étude sur les travaux de reboisement et de gazonnement des montagnes*, 253.

39. See, for example, Tassy, *La restauration des montagnes*, 14, 16, 33, 69. Tassy's other publications on the same theme include *Réorganisation du Service Forestier* and *État des forêts en France*, as well as two more editions of his *Études sur l'aménagement des forêts*, published in 1872 and 1887, respectively.

40. Printed in Bourdeaux, *Code forestier*, 131–39.

41. *Loi relative à la restauration et conservation des terrains en montagne*, Article 2, printed in Bourdeaux, *Code forestier*, 131.

42. *Loi relative à la restauration et conservation des terrains en montagne*, Article 1, printed in Bourdeaux, *Code forestier*, 131.

43. *Loi relative à la restauration et conservation des terrains en montagne*, Article 2, printed in Bourdeaux, *Code forestier*, 131.

44. Corvol, *L'homme aux bois*, 343; Larrère and Nougarede, *L'homme et la forêt*, 86.

45. Tétreau, *Commentaire de la loi du 4 avril 1882 sur la restauration et la conservation des terrains en montagne*, 7.

46. Métailie, "Lutter contre l'érosion," 105.

47. Tassy, *La restauration des montagnes*. See also Corvol, *L'homme aux bois*, 379.

48. Tassy, *La restauration des montagnes*, 111.

49. Corvol, "Les partenaires de l'environnementalisme," 44.

50. BDR 7 M 192, "Direction générale des forêts: Activités (1871–1909), Marseille"; BDR 7 M 192, "Inspection d'Aix, cantonnement d'Aubagne, 1871–1874."

51. BDR 112E 2N 5 (Gémenos), Third File: Commune de Pierrefeu, Toulon, Var; Letter: Maire (V. Maurel) to Conseiller Général (2 April 1886).

52. Corvol, "Région Rhône, Alpes, Provence et Côte d'Azur," 321.

53. Planhol, *Historical Geography of France*, 360; Simon, Clément, and Pech, "Forestry Disputes in Provincial France," 348–49; Fabre, *L'exode montagneux en France*; Fabre, *L'exode du montagnard et la transhumance du mouton en France*.

54. Métailie, "Lutter contre l'érosion," 101–3; Larrère and Nougarede, *L'homme et la forêt*, 86.

55. *Bulletin de la Ligue du Reboisement de l'Algérie* 25 (1884): 526, 527.

56. Trolard, "Appel aux Algériens," 13.

57. Ligue du reboisement de l'Algérie, *De la promulgation en Algérie de la loi du 4 avril 1882*, 510.

58. ANOM FM F80 1787: Forêts (Algérie), 1845–98.

59. Julien, *Histoire de l'Algérie contemporaine*, 409.

60. See various reports in ANOM FM F80 731: Agriculture et élevage (Algérie), 1849–58.

61. Nouschi, *Enquête sur le niveau de vie des populations rurales constantinoises de la conquête jusqu'en 1919*, 345; "Faits Divers," *Le Mobacher*, 20 June 1867.

62. ANOM FM F80 1785, M. Gustave Denis, Sénateur, "Rapport fait au nom de la Commission des finances chargée d'examiner le projet de loi, adopte par la chambre des députes, ayant pour objet le règlement des indemnités dues aux communes et aux particuliers victimes des incendies de forêts survenus au mois d'août 1881 dans le département de Constantine" (Sénat, session de 1897), 3; Lorcin, *Imperial Identities*, 78.

63. ANOM ALG GGA/P128, "Note sur les incendies en forêt."

64. Nouschi, *Enquête sur le niveau de vie des populations rurales constantinoises de la conquête jusqu'en 1919*, 338–74.

65. Nouschi, *Enquête sur le niveau de vie des populations rurales constantinoises de la conquête jusqu'en 1919*, 369.

66. Prochaska, "Fire on the Mountain," 230.

67. A. Grove and Rackham, *Nature of Mediterranean Europe*, 228–29; J. Hughes, *Mediterranean*, 93.

68. Allard, *Les forêts et le régime forestier en Provence*, 13.

69. Ribbe, *Des incendies de forêts dans la région des Maures et de l'Estérel (Provence) leurs causes, leur histoire, moyens d'y remédier*, 137.

70. MM. L'Hôpital et Faré, Conseillers d'État, Commissaires du gouvernement, "Loi sur les incendies dans la région des Maures et de l'Esterel," 348.

71. "Chronique forestière" (1864), 325.

72. Ribbe, *Des incendies de forêts dans la région des Maures et de l'Estérel (Provence) leurs causes, leur histoire, moyens d'y remédier*, 45.

73. Morel, "L'incendie de forêts dans le sud-est de la France," 30.

74. ANOM ALG GGA/P128, "Note sur les incendies en forêt," printed in Ford, "Reforestation, Landscape Conservation, and the Anxieties of Empire," 347.

75. Trottier, *Boisement et colonisation*, 22.

76. Quoted in Bernard and Lacroix, *L'évolution du nomadisme en Algérie*, 47.

77. Allard, *Les forêts et le régime forestier en Provence*, 132; Ribbe, *Des incendies de forêts dans la région des Maures et de l'Estérel (Provence) leurs causes, leur histoire, moyens d'y remédier*, 35.

78. *Revue des Eaux et Forêts* 6 (1867): 267.

79. Morel, "L'incendie de forêts dans le sud-est de la France," chap. 2.

80. Ribbe, *Des incendies de forêts dans la région des Maures et de l'Estérel (Provence) leurs causes, leur histoire, moyens d'y remédier*, 67.

81. Ribbe, *Des incendies de forêts dans la région des Maures et de l'Estérel (Provence) leurs causes, leur histoire, moyens d'y remédier*, 31, 35–36.

82. Morel, "L'incendie de forêts dans le sud-est de la France," chap. 2.

83. ANOM FM F80 981, Letter: Inspecteur Général Directeur des Finances to Gouverneur Général de l'Algérie (Alger, 18 June 1840).

84. Specific cases are detailed in ANOM ALG GGA/P1, FM F80 1785, ALG GGA/P62, and FM F80 986.

85. ANOM ALG GGA/P9a, "Statistique générale," 14 March 1845.

86. ANOM ALG GGA/P62, E. Vidal, "Rapport à la commission des indemnités" (20 July 1871).

87. ANOM MI MIOM/78, Letter, État-major de la Division, Territoires militaires, Section des affaires indigenes to Gouverneur-General Civil de l'Algérie (Constantine, 12 August 1871).

88. ANOM 22K/10, "Division de Constantine, subdivision de Batna, Cercle de Tébessa, Statistiques et Renseignements, Forêts."

89. See ANOM FM F80 987. European settlers were indicted for (very) approximately 1 percent or less of reported fires. One example is in ANOM FM F80 1785, Letter: Sous-Gouverneur to Ministre de la Guerre (23 August 1866).

90. ANOM ALG GGA/P62, Letter: Préfet d'Oran to Gouverneur Général de l'Algérie (Oran, 12 August 1887). For another, similar critique, see Trolard, *Incendies forestiers en Algérie*, 31.

91. ANOM 9 X 121, Rapport de la Commission d'Enquête, *Incendies de forêts en 1902 dans la région de Bône* (Algiers: Franceschi, 1903), 103, quoted in and translated by Prochaska, "Fire on the Mountain," 239.

92. Notices of the burn bans are in ANOM 22K/10, "Division de Constantine, subdivision de Batna, Cercle de Tébessa, Statistiques et Renseignements, Forêts."

93. Ageron, *Les Algériens musulmans et la France*.

94. ANOM ALG GGA/P62, E. Vidal, "Rapport à la commission des indemnités" (20 July 1871).

95. ANOM FM F80 1785, Ministère de la Guerre: Cabinet du Ministre, "Application de la responsabilité collective aux tribus du territoire civil." See also ANOM ALG GGA/P62 for another early reference (1854).

96. Besson-Lecousturier, *Note sur les forêts de chênes-liège*, 7.

97. *Loi relative aux mesures à prendre en vue de prévenir les incendies dans les régions boisées de l'Algérie du 17 juillet 1874*, Articles 5 and 6, printed in Estoublon and

Lefébure, *Code de l'Algérie annoté*, 435–36. See also Boudy, *Économie forestière Nord-Africaine*, 247.

98. Prochaska, "Fire on the Mountain," 241; ANOM ALG GGA/P62, "11e Section. Incendies. Responsabilité Collective. Procès-Verbaux de notification de propositions d'amendes collectives" (1877).

99. Ligue du reboisement de l'Algérie, *De la promulgation en Algérie de la loi du 4 avril 1882*; Davis, *Resurrecting the Granary of Rome*, 81–82; Boudy, *Économie forestière Nord-Africaine*, 246–47.

100. *Loi relative aux mesures à prendre en vue de prévenir les incendies dans les régions boisées de l'Algérie du 17 juillet 1874*, Article 7, printed in Estoublon and Lefébure, *Code de l'Algérie annoté*, 436.

101. Marc, *Notes sur les forêts de l'Algérie*, cited in Prochaska, "Fire on the Mountain," 242.

102. Poujoulat, *Voyage dans l'Asie Mineure*, 396.

103. 1858 Ceza Kanunnâme-i Hümayunu [Ottoman penal code], madde [item] 164, cited in Koç "1870 Orman Nizamnâmesi'nin Osmanlı Ormancılığına Katkısı Üzerine Bazı Notlar," 245 (see also 241). Koç cites as examples BOA A.MKT.UM no. 125/16; and BOA A.MKT.UM no. 122/44. For other examples of Ottoman antifire initiatives, see the *Orman Nizamnamesi* (1870), printed in Koç, "1870 Orman Nizamnâmesi'nin Osmanlı Ormancılığına Katkısı Üzerine Bazı Notlar"; Young, *Corps de droit Ottoman*; and Dursun, "Forest and the State," 332. For a case involving a fire violation, see BOA A.MKT.UM no. 502/66 (1909), cited in Batmaz and Koç, *Osmanlı Ormancılığı ile İlgili Belgeler*, 135.

104. Yiğitoğlu, *Türkiye'de Ormancılığın Temelleri*, 1–4. See also Dursun, "Forest and the State," 8–9.

105. Consular Reports, no. 589, Foreign Office, May 1903, 24, quoted in Young, *Corps de droit ottoman*, 2; Dursun, "Forest and the State," 332.

106. Ayalon, *Natural Disasters in the Ottoman Empire*, 183–92.

107. Gökmen, "Batı Anadolu'da Çekirge Felâketi," 128; Corancez, *Itinéraire d'une partie peu connue de l'Asie Mineure*, 238; Jennings, "Locust Problem in Cyprus"; Veinstein, "Sur les sauterelles à Chypre, en Thrace et en Macédoine à époque ottomane."

108. Gökmen, "Batı Anadolu'da Çekirge Felâketi," 145; Veinstein, "Sur les sauterelles à Chypre, en Thrace et en Macédoine à époque ottomane," 215.

109. TNA FO 222/2, Letter from Captain Stewart, Adalia (Antalya) Vilayet, 13 January 1880; TNA FO 222/2, Letter: Satheral to Layard (Konya, 16 April 1880).

110. Gül, "Osmanlı Devletinde Kuraklık ve Kıtlık," 146; Erler, "Ankara ve Konya Vilayetlerinde Kuraklık ve Kıtlık," 82–89; Akdemir, Köse, Aras, and Nüzhet Dalfes, "Anadolu'nun 350 Yılında Yaşanan Önemli Kurak ve Yağışlı Yıllar," 129–35; Erler, "XIX. Yüzyıldaki Bazı Doğal Afetler ve Osmanlı Yönetimi"; Ayalon,

Natural Disasters in the Ottoman Empire; Erler, *Osmanlı Devleti'nde Kuraklık*; Kılıç, "Osmanlı Devleti'nde Meydana Gelen Kıtlıklar"; and Özdeğer, "XIX. Yüzyıl Sonlarında Meydana Gelen Bir Kuraklık ve Kıtlık Hadisesi ile Bunun Sosyo-Ekonomik Sonuçları."

111. Burnaby, *On Horseback through Asia Minor*, 83. See also *Famine in Asia Minor*; "Informations," 120; S. Shaw, *History of the Ottoman Empire and Modern Turkey*, 2:156.

112. Burnaby, *On Horseback through Asia Minor*, 66.

113. TNA FO 222/2, Dispatch nos. 1–11, 1880: J. D. H. Stewart, "Report on Konieh Province."

114. Gould, "Pashas and Brigands," 85–118. See also chapter 7.

115. Ayalon, *Natural Disasters in the Ottoman Empire*, 104–5, 110–33, 182.

116. Gül, "Osmanlı Devletinde Kuraklık ve Kıtlık," 147–48; Burnaby, *On Horseback through Asia Minor*, 83.

117. TNA FO 195/55, Letter book, consulate general, 1857; Perrot, *Souvenirs d'un voyage en Asie Mineure*, 67; Vogt, "Sismicité historique du domaine ottoman"; Gül, "Osmanlı Devletinde Kuraklık ve Kıtlık," 144.

118. TNA FO 78/2987, Letter: Col. Wilson to Sir Layard (Sivas, 28 October 1879, copy no. 33); [D. A.] J., "Obituary: Major-General Sir Charles William Wilson."

119. Toksöz, *Nomads, Migrants and Cotton*, 72–73.

120. Toksöz, *Nomads, Migrants and Cotton*, 79; Gould, "Pashas and Brigands," 151–56.

121. TNA FO 78/2987, Letter: Col. Wilson to Herbert Chromeside (10 August 1879).

122. TNA FO 78/2987, Letter: Col. Wilson to Sir Layard (Sivas, 28 October 1879, copy no. 33).

123. Burnaby, *On Horseback through Asia Minor*, 66.

124. Poujoulat, *Voyage dans l'Asie Mineure*, 203.

125. Perrot, *Souvenirs d'un voyage en Asie Mineure*, 105.

126. Ayalon, *Natural Disasters in the Ottoman Empire*, 153, 171–72.

127. See, for example, Walker, *Toxic Archipelago*, 42.

128. Simon, Clément, and Pech, "Forestry Disputes in Provincial France," 347.

7. SHEEP TO THE SLAUGHTER

For the epigraph Braudel cites Planhol, *De la plaine pamphylienne aux lacs pisidiens*, 194.

1. Larrère and Nougarede, *L'homme et la forêt*, 89.

2. See Kalaora and Savoye, "Aménagement et ménagement"; and Le Play, "Des forêts considérées dans leur rapport avec la constitution physique du globe et l'économie des sociétés."

3. Ford, *Natural Interests*, 61–62; Kalaora and Savoye, "Aménagement et ménagement," 31.

4. Briot, *Les Alpes françaises*, iv.

5. Briot, *Les Alpes françaises*, vi.

6. Puyo, "La science forestière vue par les géographes français," 627. See also CARAN F10 1855: Fabre (Lucien-Albert).

7. Fabre, "L'état et la dépopulation montagneuse en France," 9, quoted in Puyo, "La science forestière vue par les géographes français," 627.

8. Puyo, "La science forestière vue par les géographes français," 628; CARAN F10 1855: Fabre (Lucien-Albert).

9. Among the more notable were Pierre Buffault, Auguste Calvet, and Emile Trutat. Puyo, "La science forestière vue par les géographes français," 627.

10. Simon, Clément, and Pech, "Forestry Disputes in Provincial France," 343; Dumoulin, "Communes et pâturages forestiers en Provence au XIXe siècle," 88; Lhomme, "La crise agricole à la fin du XIXe siècle en France," 533. For the expansion of the Australian market, see Guthrie, *World History of Sheep and Wool*, 75–78; Ryder, *Sheep and Men*, 610–32; McIvor, *History and Development of Sheep Farming*; Carter, *Sheep and Wool Correspondence of Sir Joseph Banks*; and Lee, *Wanganella and the Merino Aristocrats*. For Spain, see Phillips and Phillips, *Spain's Golden Fleece*.

11. Orange and Amalbert, *Le mérinos d'Arles*, 36.

12. "Société d'économie politique (réunion de septembre)," *Journal des Économistes* (Paris, 1869): 459; Daumas, "L'industrie lainière en France," 14–20; Orange and Amalbert, *Le mérinos d'Arles*, 35–36.

13. Arbos, *La vie pastorale dans les Alpes françaises*; Fabre, *L'exode du montagnard et la transhumance du mouton en France*; Fabre, *La fuite des populations pastorales françaises*; Fabre, *L'échec du coton à la laine au début du XXe siècle et à propos de la désertion des montagnes françaises*.

14. Orange and Amalbert, *Le mérinos d'Arles*, 36–38; *Le mouton en Provence*, 96; Clout, "La transhumance," 227; Shippers, "Le cycle annuel d'un berger transhumant," 64.

15. "Les forêts algériennes."

16. ANOM ALG GGA/P6, "Reconnaissance définitive et délimitation des forêts de l'Inspection de Sétif [Constantine]" (1895).

17. Trabut and Mathieu, *Les Hauts-Plateaux oranais*. See also Davis, *Resurrecting the Granary of Rome*, 124–35.

18. Trabut and Mathieu, *Les Hauts-Plateaux oranais*, 40.

19. Trabut and Mathieu, *Les Hauts-Plateaux oranais*, 42.

20. ANOM ALG GGA/P1, Jules Ferry, "Organisation et attributions du gouvernement général de l'Algérie," *Supplément du Journal des Débats*, 29 October 1892. See also Ageron, *Les Algériens musulmans et la France*, 1:447–58; Davis, *Resurrecting the Granary of Rome*, 94–95; Ford, *Natural Interests*, 357–58. Ford also cites ANOM ALG GGA/P89, Letters: 11 January 1893, 13 June 1893, 26 February 1895; and ANOM ALG GGA/P91, Letter: 2 September 1908.

21. Quoted in Ageron, *Les Algériens musulmans et la France*, 1:454.

22. Davis, *Resurrecting the Granary of Rome*, 99, 118.

23. Trolard, *La question forestière algérienne devant le Senat*, 57, quoted in Ford, *Natural Interests*, 358. See also "Les forêts algériennes."

24. ANOM ALG GGA/P1, Jules Ferry, "Organisation et attributions du gouvernement général de l'Algérie," *Supplément du Journal des Débats*, 29 October 1892.

25. Quoted in Ageron, *Les Algériens musulmans et la France*, 1:460.

26. Ageron, *Les Algériens musulmans et la France*, 1:460.

27. ANOM ALG GGA/P1, "Règlements forestiers Algériens & de la promulgation du code forestier."

28. *Rapport fait au nom de la commission par M. Ernest Picard, le 8 juillet 1874*, printed in Estoublon and Lefébure, *Code de l'Algérie annoté*, 434.

29. ANOM ALG GGA/P8, Gouverneur Général de l'Algérie, "Minute de la lettre écrite" (27 April 1889).

30. Woolsey, *French Forests and Forestry*, 161–208. See also Estoublon and Lefébure, *Code de l'Algérie annoté*.

31. *Algerian Forest Code of 1903*, Sec. VI, Articles 53 and 63, printed in Woolsey, *French Forests and Forestry*, 161–208.

32. Davis, *Resurrecting the Granary of Rome*, 120.

33. *Algerian Forest Code of 1903*, Articles 63 and 66, printed in Woolsey, *French Forests and Forestry*, 161–208.

34. *Algerian Forest Code of 1903*, Articles 53, 60–63, 89, and 91, printed in Woolsey, *French Forests and Forestry*, 161–208.

35. *Algerian Forest Code of 1903*, Articles 76 and 106–8, printed in Woolsey, *French Forests and Forestry*, 161–208.

36. *Algerian Forest Code of 1903*, Articles 104, 177, and 187, printed in Woolsey, *French Forests and Forestry*, 161–208.

37. Ageron, *Les Algériens musulmans et la France*, 1:563.

38. F. White, *Vegetation of Africa*, 225–31, cited in Davis, *Resurrecting the Granary of Rome*, 100.

39. Boukhobza, "Nomadisme et colonisation," 157–60.

40. Carayol, *La législation forestière de l'Algérie*, 153, quoted in Davis, *Resurrecting the Granary of Rome*, 120.

41. Boukhobza, "Nomadisme et colonisation," 145.
42. Quoted in Montagne, *La civilisation du désert*, 47.
43. Boukhobza, "Nomadisme et colonisation," 43, 143, 146; Woolsey, *French Forests and Forestry*, 88.
44. Boukhobza, "Nomadisme et colonisation," 299.
45. Bates, *Nomads and Farmers*, 34; Gould, "Pashas and Brigands"; Johnson, *Nature of Nomadism*, 23–25; Kasaba, "Do States Always Favor Stasis?," 28; Kasaba, *Moveable Empire*, 54–58; Köksal, "Coercion and Mediation," 470.
46. Halaçoğlu, "Fırka-i İslâhiye ve Yapmış Olduğu İskân," 13; Orhonlu, *Osmanlı İmparatorluğu'nda Aşiretlerin İskâni*, 116.
47. For a description of the environmental calamities of Anatolia in the 1870s, see chapter 6.
48. Köksal, "Coercion and Mediation," 473.
49. Gould, "Pashas and Brigands," 27, 208–9; Karpat, "Ottoman Population Records," 274; Kasaba, *Moveable Empire*, 116. See also Halaçoğlu, "Fırka-i İslâhiye ve Yapmış Olduğu İskân," 12–14; and Köksal, "Coercion and Mediation," 474.
50. The German traveler Friedrich Sarre encountered such tribes during his travels in 1895. See Sarre, *Reise in Kleinasien*, 116.
51. See, for example, the works of Arnold van Gennep, Xavier de Planhol, Jean-Paul Roux, John Kolars, Wolf-Dieter Hütteroth, Daniel Bates, and Douglas Johnson.
52. Hütteroth, "Influence of Social Structure on Land Division," 24.
53. Quataert, *Ottoman Empire, 1700–1922*, 117–24; Bates, *Nomads and Farmers*, 10–11.
54. Hütteroth, "Influence of Social Structure on Land Division," 22.
55. Planhol, *De la plaine pamphylienne aux lacs pisidiens*, 125; Hütteroth, "Influence of Social Structure on Land Division," 22.
56. Dursun, "Forest and the State," 175.
57. Dewdney, *Turkey*, 25, 118.
58. Halaçoğlu, "Fırka-i İslâhiye ve Yapmış Olduğu İskân," 13; Orhonlu, *Osmanlı İmparatorluğu'nda Aşiretlerin İskâni*, 116.
59. Kasaba, *Moveable Empire*, 109–10; Özbay, "Tanzimat Sonrasında Akdağ Kazası Civarına Afşar Türkmenlerinin İskâni," 454.
60. Keskin, "Osmanlı Ormancılığı'nın Gelişiminde Fransız Uzmanların Rolü," 141; Kutluk, "Türkiye'de Yabancı Ormancılar," 185–86.
61. Yund, *Türkiye Orman Umum Müdürleri Albümü*, 8.
62. Bricogne, "La mission forestière en Turquie," 224; Dursun, "Forest and the State," 216–17. See also "Orman Mektebi Nizamnamesi," *Düstûr* 1. Tertib, Vol. II (11 L 1286/13 January 1870), cited in Dursun, "Forest and the State," 182.

63. *Orman Nizamnamesi* (1870), Articles 1, 19–20, and 27, printed in Koç, "1870 Orman Nizamnamesi'nin Osmanlı Ormancılığına Katkısı Üzerine Bazı Notlar," 251, 253–54. See also Thirgood, *Cyprus*, 85; and Dursun, "Forest and the State," 265–69.

64. *Orman Nizamnamesi* (1870), Article 5, printed in Koç, "1870 Orman Nizamnamesi'nin Osmanlı Ormancılığına Katkısı Üzerine Bazı Notlar," 251. See also Dursun, "Forest and the State," 216–17; and S. Harris, "Colonial Forestry and Environmental History," 181–82.

65. Koç, "1870 Orman Nizamnamesi'nin Osmanlı Ormancılığına Katkısı Üzerine Bazı Notlar," 231, 243; Keskin, "Osmanlı Ormancılığı'nın Gelişiminde Fransız Uzmanların Rolü," 142.

66. *Orman Nizamnamesi* (1870), Articles 5, 13, 14, 15, 16, 30, 43, 44, 45, and 50, printed in Koç, "1870 Orman Nizamnamesi'nin Osmanlı Ormancılığına Katkısı Üzerine Bazı Notlar," 245, 251–57.

67. Dursun, "Forest and the State," 282.

68. A new, more extensive forest law replaced the 1870 one in 1917, though the latter was based on this precedent.

69. "Bi'l-Cümle Orman Memurlarının Suret-i Tertib ve Veza'ifine Da'ir 51 Maddelik Talimat," *Düstûr* 2, Tertib, Vol. IV (7 Mart 1292/19 March 1876), cited in Dursun, "Forest and the State," 205; *Takvim-i Vakayi*, no. 1843, 14 L 1293/2 November 1876, cited in Dursun, "Forest and the State," 217.

70. Dursun, "Forest and the State," 205; BOA Y.A.RES no. 36/14 (1886), printed in Batmaz and Koç, *Osmanlı Ormancılığı ile İlgili Belgeler*, 3:41.

71. Bricogne, "La mission forestière en Turquie," 224.

72. Koç, "1870 Orman Nizamnamesi'nin Osmanlı Ormancılığına Katkısı Üzerine Bazı Notlar," 231–57; Dursun, "Forest and the State," 217–18.

73. Dursun, "Forest and the State," 205.

74. Y.PRK.OMZ 1/33 (29 B 1308/10 March 1891); and Güran, *Osmanlı Devleti'nin İlk İstatistik Yıllığı 1897*, 181, both cited in Dursun, "Forest and the State," 207.

75. ANOM ALG GGA/P5, "Annuaire Militaire."

76. Dursun, "Forest and the State," 204.

77. Montrichard, "Une excursion en Asie Mineure," 85.

78. Dursun, "Forest and the State," 217; Montrichard, "L'Ile de Chypre," *Revue des Eaux et Forêts* 13 (1874): 39, cited in S. Harris, "Colonial Forestry and Environmental History," 183.

79. Rıza, *Orman ve Mer'a Kanununun Esbab-ı Mucibe Layıhası*, 230, quoted in Dursun, "Forest and the State," 216.

80. Dursun, "Forest and the State," 287–89.

81. BOA ŞD. no. 530/20 (1901); and BOA ŞD. no. 520/17 (1312/1894), printed in Batmaz and Koç, *Osmanlı Ormancılığı ile İlgili Belgeler*, 3:79, 83.

82. Dursun, "Forest and the State," 356–57; BOA I.OM Genel no. 830 (156), Hususi no. 1 (1896), printed in Batmaz and Koç, *Osmanlı Ormancılığı ile İlgili Belgeler*, 1:199; BOA ŞD. no. 528/5 (1317/1900), printed in Batmaz and Koç, *Osmanlı Ormancılığı ile İlgili Belgeler*, 2:77–78.

83. T.C. Orman Genel Müdürlüğü (General Directorate of Turkish Forestry), accessed 10 October 2018, https://ogm.gov.tr/SitePages/OGM/OGMDefault.aspx.

84. Food and Agriculture Organization of the United Nations (FAO), "Forest Area Statistics," accessed 10 October 2018, http://www.fao.org/forestry/. The contrast is largely due to the fact that the majority of Algeria's surface area is covered by the Sahara Desert.

85. In Turkish it is referred to as Tassy Mektebi. Akgür, "Yurdumuzda Ormancılık Öğretiminin Başlangıç Tarihi Hakkında Bir Takvim Araştırması," 123.

86. "Notice sur Louis Tassy," 13.

87. Planhol, *De la plaine pamphylienne aux lacs pisidiens*, 48.

88. Boukhobza, "Nomadisme et colonisation," 8.

89. M. Trouette, inspecteur du service de l'élevage, quoted in Lehuraux, *Le nomadisme et la colonisation dans les Hauts Plateaux de l'Algérie*, 227.

90. Planhol, *De la plaine pamphylienne aux lacs pisidiens*, 125.

91. Dominian, "Peoples of Northern and Central Asiatic Turkey," 847.

92. Prochaska, "Fire on the Mountain," 229–30.

93. Clout, "La transhumance," 227.

94. Tolley, "Qui sont les bergers?," 76.

CONCLUSION

1. For a review of these arguments, see Thornes, "Land Degradation," 563–66; Allen, "Vegetation and Ecosystem Dynamics," 204; and Geeson, Brandt, and Thornes, *Mediterranean Desertification*.

2. A. Grove and Rackham, *Nature of Mediterranean Europe*; Davis, *Resurrecting the Granary of Rome*; Woodward, *Physical Geography of the Mediterranean*; Enne et al., "Agro-pastoral Activities and Land Degradation in Mediterranean Areas." For a different perspective, see J. Hughes, *Mediterranean*, 119.

3. Allen, "Vegetation and Ecosystem Dynamics," 203; Wainwright, "Weathering, Soils, and Slope Processes," 173; Davis, *Resurrecting the Granary of Rome*, 184.

4. Enne et al., "Agro-pastoral Activities and Land Degradation in Mediterranean Areas," 75.

5. Behnke, Scoones, and Kerven, *Range Ecology at Disequilibrium*; Scoones, "Range Management Science & Policy," 48; Niamir-Fuller and Turner, "Review of Recent Literature on Pastoralism and Transhumance in Africa," 30–34.

6. Niamir-Fuller and Turner, "Review of Recent Literature on Pastoralism and Transhumance in Africa," 22–23.

7. Niamir-Fuller and Turner, "Review of Recent Literature on Pastoralism and Transhumance in Africa," 33; Leach and Mearns, "Environmental Change and Policy," 12–13.

8. Niamir-Fuller and Turner, "Review of Recent Literature on Pastoralism and Transhumance in Africa," 22; Horden and Purcell, *Corrupting Sea*, chap. 3; A. Grove and Rackham, *Nature of Mediterranean Europe*; Davis, *Resurrecting the Granary of Rome*.

9. For other versions of this argument, see especially Mahé, *Histoire de la Grande Kabylie*; Boukhobza, "Nomadisme et colonisation"; Ford, *Natural Interests*; Julien, *Histoire de l'Algérie contemporaine*; Cutler, "Evoking the State"; and Sivak, "Law, Territory, and the Legal Geography of French Rule in Algeria." See also the works of Robert Ageron and of Diana K. Davis.

10. Ryder, *Sheep and Men*, 249.

11. Study of M. Trouette, inspecteur du service de l'élevage, quoted in Lehuraux, *Le nomadisme et la colonisation dans les Hauts Plateaux de l'Algérie*, 227. See also Comité des Travaux Historiques et Scientifiques, *Bulletin de la section des sciences économiques et sociales*, 42, 51.

12. Lists of current members, including their full name and occupation, were printed in the *Bulletin de la Ligue du Reboisement*.

13. Bernard and Lacroix, *L'évolution du nomadisme en Algérie*, i.

BIBLIOGRAPHY

ARCHIVAL SOURCES

Archives Départementales des Bouches-du-Rhône (BDR), Aix-en-Provence and Marseille

    Série C: Correspondance

    Série E

    Série J

    Série M: Agriculture, Eaux et Forêts

    Série O

    Série U: Cours d'appel

    Ancien Série W (in process of reclassification)

    Série PER (periodicals)

Archives Nationales d'Outre-Mer (ANOM), Aix-en-Provence

    Gouverneur Général de l'Algérie (ALG GGA)

        Séries I & K: Administration des Musulmans

        Série P: Forêts

    Fonds Ministériels (FM)

    MI 18 MIOM (Microform): Correspondance politique

Başbakanlık Osmanlı Arşivi (BOA), Istanbul

    Bab-ı Ali Evrak Odası, Sadaret Evrakı, Mektubi

        Kalemi-Umum Vilayat (A.MKT.UM)

    Cevdet İktisat (C.IKT)

    Dahiliye (DH) and Cevdet Dahiliye (C.DH)

    Hariciye Nezareti (HR)

    İrade Dahiliye (I.DH)

    İrade Meclis-i Vala (I.MV)

    İrade Orman ve Ma'adin (I.OM)

    Sadaret Evrakı Mektubi Mühimme-Meclis-i Vala (A.MKT.MVL)

    Yıldız Sadaret Resmi Maruzat Evrakı (Y.A.RES)

Centre d'Accueil et de Recherche des Archives Nationales (CARAN), Paris
SÉRIE F10: AGRICULTURE
Centre des Archives Diplomatiques, Nantes (CADN)
Constantinople: Ambassade
2 Mi: Correspondance politique arrivée et départ
avec le Ministère des Affaires étrangères
166PO: Dossiers thématiques
The National Archives of the United Kingdom (TNA), Kew, London
Foreign Office Records (FO): General Correspondence, Ottoman Empire

PUBLISHED SOURCES

Abbé, André, Henri Bresc, and Jean-Paul Ollivier. *Bergers de Provence et du pays niçois.* 2nd ed. Nice: Serre Editeur, 1996.

Abi-Mershed, Osama. *Apostles of Modernity: Saint-Simonians and the Civilizing Mission in Algeria.* Stanford: Stanford University Press, 2010.

Acte du Colloque international l'élevage et la vie pastorale dans les montagnes de l'Europe au Moyen âge et à l'époque moderne. Clermont-Ferrand: Inst. d'Études du Massif Central, 1984.

Acun, Nyazi. *Ormanlarımız ve Cumhuriyet Hükümeti'nin Orman Dâvası.* Ankara: Recep Ulusoğlu Basımevi, 1945.

Africa: Bulletin de la Société de géographie d'Alger. Alger: Société de géographie d'Alger, 1880.

Ageron, Charles-Robert. *Histoire de l'Algérie contemporaine.* Volume 2, *De l'insurrection de 1871 au déclenchement de la guerre de libération (1954).* Paris: Presses Universitaires de France, 1979.

———. *Les Algériens musulmans et la France (1871–1919).* 2 vols. Paris: Presses Universitaires de France, 1968.

———. *Modern Algeria: A History from 1830 to the Present.* Translated by Michael Brett. Trenton NJ: Africa World Press, 1991.

Ágoston, Gábor. "A Flexible Empire: Authority and Its Limits on the Ottoman Frontiers." *International Journal of Turkish Studies* 9, no. 1–2 (2003): 15–31.

Agulhon, Maurice. *Histoire de la Provence.* Paris: Presses Universitaires de France, 1987.

———. *La vie sociale en Provence intérieure au lendemain de la Révolution.* Paris: Société des études robespierristes, 1970.

———. *1848 ou l'apprentissage de la république (1848–1852).* Paris: Seuil, 1973.

Ahrweiler, Hélène. *Géographie historique du monde méditerranéen.* Paris: Fondation européenne de la science, 1988.

Ainsworth, W. F. *Travels and Researches in Asia Minor, Mesopotamia, Chaldaea and Armenia.* London: J. W. Parker, 1842.

Akdağ, Mustafa. "Celâli İsyanlarından Büyük Kaçgunluk." *Tarih Araştırmaları Dergisi* 2 (1964): 1–49.

——— . *Türk Halkının Dirlik ve Düzenlik Kavgası*. Ankara: Bilgi Yayınevi, 1975.

Akdemir, Ünal, Nesibe Köse, Aliye Aras, and H. Nüzhet Dalfes. "Anadolu'nun 350 Yılında Yaşanan Önemli Kurak ve Yağışlı Yıllar." *Türkiye Kuvarterner Sempozyumu, TORQUA-V, ITÜ, Avrasya Yer Bilimleri Enstitüsü*, 2–5 June 2005, 129–35.

Akgür, Necati. "Yurdumuzda Ormancılık Öğretiminin Başlangıç Tarihi Hakkında Bir Takvim Araştırması." *İstanbul Üniversitesi Orman Fakültesi Dergisi, Seri B* 24, no. 1 (1974): 121–35.

Akpınar, Alişan, and Eugene L. Rogan. *Aşiret, Mektep, Devlet: Osmanlı Devleti'nde Aşiret Mektebi*. İstanbul: Aram, 2001.

Aksan, Virginia. *Ottomans and Europeans: Contacts and Conflicts*. Istanbul: Isis Press, 2004.

Aksoy, Erdal. "Konar-Göçer Yörük Alt-Kültüründe Kadın Kimliği." In *Osmanlı'dan Cumhuriyet'e Yörükler ve Türkmenler*, edited by Hayati Beşirli and Ibrahim Erdal, 15–24. Ankara: Phoenix, 2008.

Albion, Robert Greenhalgh. *Forests and Sea Power: The Timber Problem of the Royal Navy, 1652–1862*. Cambridge MA: Harvard University Press, 1926.

Algérie. Commission d'études forestières. *Compte-rendu des séances et rapport de la commission*. Alger: J. Torrent, 1904.

——— . *État actuel de l'Algérie: Publié d'après les documents officiels*. Alger: Imprimerie de l'Association ouvrière, 1861.

Allard, Ferdinand. *Les forêts et le régime forestier en Provence: Thèse pour le doctorat ès sciences politiques et économiques, Université d'Aix-Marseille*. Paris: A. Rousseau, 1901.

Allen, Harriet. "Vegetation and Ecosystem Dynamics." In *The Physical Geography of the Mediterranean*, edited by Jamie C. Woodward, 203–28. Oxford: Oxford University Press, 2009.

Almond, Gabriel A., Marvin Chodorow, and Roy Harvey Pearce. "Progress and Its Discontents." *Bulletin of the American Academy of Arts and Sciences* 35, no. 3 (1981): 4–23.

Almond, Gabriel, Marvin Chodorow, and Roy Harvey Pearce, eds. *Progress and Its Discontents*. Berkeley: University of California Press, 1982.

Amalbert, Maurice. *L'élevage du mouton dans les plaines du Bas-Rhône: Le mérinos d'Arles*. Tain: Union Rhodaniens, 1931.

Ames, Glenn J., and Ronald S. Love, eds. *Distant Lands and Diverse Cultures: The French Experience in Asia, 1600–1700*. Westport CT: Praeger, 2003.

Amitai, Reuven. *Mongols, Turks, and Others: Eurasian Nomads and the Sedentary World*. Leiden: Brill, 2005.

Ammer, Christine. *The American Heritage Dictionary of Idioms*. Boston: Houghton Mifflin Harcourt, 1997.

Annequin, Colette. *Aux origines de la transhumance: Les Alpes et la vie pastorale d'hier à aujourd'hui*. Paris: Picard, 2006.

Antoine, Annie. *Le paysage de l'historien: Archéologie des bocages de l'Ouest de la France à l'époque moderne*. Rennes: Presses Universitaires de Rennes, 2000.

Appleby, Andrew. "Epidemics and Famine in the Little Ice Age." *Journal of Interdisciplinary History* 10, no. 4 (1980): 643–63.

Arbos, Philippe. *Évolution économique et démographique des Alpes françaises du Sud*. Paris: Imprimerie Nationale, 1913.

———. "The Geography of Pastoral Life: Illustrated with European Examples." *Geographical Review* 13, no. 4 (1923): 559.

———. *L'économie pastorale dans quelques vallées savoyardes*. Paris, 1912.

———. *La vie pastorale dans les Alpes françaises: Étude de géographie humaine*. Paris: A. Colin, 1922.

———. *Questionnaires pour l'étude de la vie pastorale en montagne*. N.p., 1919.

Archiloque, Alain. "D'hier à aujourd'hui, la passion d'être berger." In *Bergers* (Revue éditée par Pays et gens du Verdon). Turin: Imprimerie Mariogros, 1999.

Archives départementales des Bouches-du-Rhône. *Forêts perdues, forêts retrouvées: Exposition, Marseille, 14 juin–31 octobre 1997*. Marseille: Archives départementales des Bouches-du-Rhône, 1997.

Armağan, Abdüllatif. "Yüzyıllarda Teke Sancağı'nda Konar-Göçerler: Sosyo-Ekonomik ve Demografik Durumları." In *Osmanlı'dan Cumhuriyet'e Yörükler ve Türkmenler*, edited by Hayati Beşirli and Ibrahim Erdal, 71–133. Ankara: Phoenix, 2008.

Arnould, P. "Les nouvelles forêts françaises." *Information Géographique* 60, no. 4 (1996): 141–56.

Arslan, Hüseyin. *Osmanlı'da Nüfus Hareketleri (XVI. Yüzyıl): Yönetim, Nüfus, Göçler, İskânlar, Sürgünler*. Üsküdar, İstanbul: Kaknüs Yayınları, 2001.

Asher, Michael. *Death in the Sahara: The Lords of the Desert and the Timbuktu Railway Expedition Massacre*. New York: Skyhorse, 2008.

Ayalon, Yaron. *Natural Disasters in the Ottoman Empire: Plague, Famine, and Other Misfortunes*. Cambridge: Cambridge University Press, 2015.

Aydın, Mehmet Akif. "Türk Hukuk Tarihçiliği." *Türkiye Araştırmaları Literatür Dergisi* (TALID) 3, no. 5 (2005): 9–25.

Aydın, Yusuf Alperen. *Sultanın Kalyonları: Osmanlı Donanmasının Yelkenli Savaş Gemileri 1701–1770*. İstanbul: Küre Yayınları, 2011.

Ayoubi, Fadl. *Contribution à l'étude de la désertification des hautes plaines algériennes, l'exemple: Le bassin versant du Hodna.* Paris, 1990.

Aytekin, E. Attila. "Hukuk, Tarih ve Tarihyazımı: 1858 Osmanlı Arazi Kanunnâme-si'ne Yönelik Yaklaşımlar" [Law, history, and historiography: Approaches to the Ottoman Land Code of 1858]. *TALID* 3, no. 5 (2005): 723–44.

Bacqué-Grammont, Jean-Louis. *Économie et sociétés dans l'Empire ottoman: Fin du XVIII au début du XX siècle.* Paris: Éditions du CNRS, 1983.

———. *Représentants permanents de la France en Turquie (1536–1991) et de la Turquie en France (1797–1991).* Istanbul and Paris: ISIS, 1991.

Badré, Louis. *Histoire de la forêt française.* Paris: Arthaud, 1983.

Baehrel, René, and Maurice Aymard. *Une croissance: La Basse-Provence rurale de la fin du seizième siècle à 1789.* Paris: SEVPEN, 1988.

Balla, E. "Fiscal Crisis and Institutional Change in the Ottoman Empire and France." *Journal of Economic History* 69, no. 3 (2009): 809–45.

Bamford, Paul. *Forests and French Sea Power, 1660–1789.* Toronto: University of Toronto Press, 1956.

———. "French Forest Legislation and Administration, 1660–1789." *Agricultural History* 29, no. 3 (1955): 97–107.

Banse, Ewald. *Die Türkei: Eine moderne Geographie.* Braunschweig, 1916.

Barbier, J. V. *A travers le Sahara: Les missions de Colonel Flatters d'après des documents absolument inédits.* Paris, 1884.

Barfield, Thomas J. *The Nomadic Alternative.* Englewood Cliffs NJ: Prentice Hall, 1993.

Barkan, Ömer Lutfi. "Essai sur les données statistiques des registres de recensement dans l'Empire ottoman aux XVe et XVIe siècles." *Journal of the Economic and Social History of the Orient* 1, no. 1 (1957): 9–36.

———. "Osmanlı Imparatorluğunda Bir İskan ve Kolonizasyon Metodu Olarak Vakıflar ve Temlikler" [Grants of *vakıf* and freehold property rights as a method of colonization in the Ottoman Empire]. *Vakiflar Dergisi* 26 (1942): 279–386.

———. *Türkiye'de Toprak Meselesi.* İstanbul: Gözlem Yayınları, 1980.

———. *Türk Toprak Hukuku Tarihinde Tanzimat ve 1274 (1858) Tarihli Arazi Kanun-namesi.* İstanbul: Maarif Matbaası, 1960.

Barker, William Burckhardt. *Lares and Penates, or, Cilicia and Its Governors: Being a short historical account of that province from the earliest times to the present day, together with a description of some household gods.* Edited by William Francis Ainsworth. London: Ingram, Cooke, 1853.

Barkey, Karen. *Bandits and Bureaucrats: The Ottoman Route to State Centralization.* Ithaca NY: Cornell University Press, 1994.

———. "Rebellious Alliances: The State and Peasant Unrest in Early Seventeenth-Century France and the Ottoman Empire." *American Sociological Review* 56, no. 6 (1991): 699–715.

Baron, François Louis Jerome. *Projet de l'organisation de l'administration des Eaux et Forêts*. Paris: Imprimerie Nationale, 1790.

Barth, Fredrik. *Nomads of South Persia: The Basseri Tribe of the Khamseh Confederacy*. Boston: Little Brown, 1961.

Barton, Greg. *Empire Forestry and the Origins of Environmentalism*. Cambridge: Cambridge University Press, 2002.

Bar-Yosef, Ofer. *Pastoralism in the Levant: Archaeological Materials in Anthropological Perspectives*. Madison WI: Prehistory Press, 1992.

Bates, Daniel G. "Differential Access to Pasture in a Nomadic Society: Yörük of S. E. Turkey." *Journal of Asian and African Studies* 8 (1972): 48–59.

———. *Nomads and Farmers: A Study of the Yörük of Southeastern Turkey*. Ann Arbor: University of Michigan, 1973.

———. "Shepherd becomes Farmer: A Study of Sedentarization and Social Change in Southeastern Turkey." In *Turkey: Geography and Social Perspectives*, edited by Peter Benedict. Leiden: Brill, 1974.

———. "Yörük Settlement in Southeast Turkey." In *When Nomads Settle: Processes of Sedentarization as Adaptation and Response*. New York: Praeger, 1980.

Batmaz, Eftal, and Bekir Koç, eds. *Osmanlı Ormancılığı ile İlgili Belgeler* [Documents on Ottoman forestry]. 3 vols. Ankara: T. C. Orman Bakanlığı (Republic of Turkey Ministry of Forestry), 1999.

Battistini, Eugène. *Les forêts de chêne-liège de l'Algérie*. Alger: Victor Heintz, 1937.

Batu, Hâmit. *L'Empire ottoman, la République de Turquie et la France*. Istanbul and Paris: Éditions Isis, 1986.

Baudrillart, Jacques-Joseph. *Code forestier (du 21 mai 1827), suivi de l'ordonnance réglementaire*. Paris: A. Bertrand, 1827.

———. *Instruction sur la culture du bois, à l'usage des forestiers: Ouvrage traduit de l'allemand de G. I. Hartig, Mattre des Forêts de la principauté de Solms, et Membre honoraire de la Société de physique de Berlin*. Paris: Levrault, 1805.

———. *Nouveau manuel forestier, à l'usage des agens forestiers de tous grades, des arpenteurs, des Gardes des Bois Impériaux et communaux, des Préposés de la Marine pour la recherche des bois propres aux constructions navales: Traduit sur la 4e édition de l'ouvrage allemande de M. de Burgsdorf, Grand maitre des Forêts de la Prusse et adapte à notre système d'administration d'après l'ordre du Gouvernement*. 2 vols. Paris: Arthus-Bertrand, 1808.

———. *Traité général des eaux et forêts, chasses et pêches*. Paris: Huzard, 1821–45.

Baytop, Asuman. *Türkiye'de Botanik Tarihi Araştirmaları*. Ankara: Türkiye Bilimsel ve Teknik Araştirma Kurumu, 2003.

Bazire, Pierre, and Jean Gadant. *La forêt en France*. Paris: Documentation française, 1991.

Becquerel, Alfred. *Traité élémentaire d'hygiène publique et privée*. 3rd ed. Paris, 1864.

Behnke, Roy H., Ian Scoones, and Carol Kerven. *Range Ecology at Disequilibrium: New Models of Natural Variability and Pastoral Adaptation in African Savannas*. London: Overseas Development Institute, 1993.

Beinart, William, and Lotte Hughes. *Environment and Empire*. Oxford: Oxford University Press, 2007.

Belleval, Amphoux de. "Réflexions sur la transhumance des troupeaux en Provence." In *Annales Provençales d'Agriculture*. Marseille: Typographie Barlatier-Feissat et Demonchy, 1846.

Benchetrit, Maurice. "Les modalités de la dégradation des forêts dans le Tell oranais." *Revue de géographie de Lyon* 41, no. 4 (1966): 303–38.

Benedict, Peter, Erol Tümertekin, and Fatma Mansur, eds. *Turkey: Geographic and Social Perspectives*. Leiden: Brill, 1974.

Bent, J. Theodore. "Exploration in Cilicia Tracheia." *Proceedings of the Royal Geographical Society and Monthly Record of Geography*, new monthly series 12, no. 8 (1890): 445–63.

Bergerie, Rougier de la. *Mémoire et observations sur les abus des défrichements et la destruction des bois et forêts, avec un projet d'organisation forestière*. Auxerre: Imprimerie Laurent Fournier, An IX [1801].

Bergers (Revue éditée par Pays et gens du Verdon). Turin: Imprimerie Mariogros, 1999.

Berkel, Adnan. *Ormancılık İş Bilgisi*. İstanbul: Kutulmus Matbaasi, 1976.

Berktay, Halil. *New Approaches to State and Peasant in Ottoman History*. London: Frank Cass, 1992.

Bernard, Augustin, and N. Lacroix. *L'évolution du nomadisme en Algérie*. Alger: Adolphe Jourdan, 1906.

Bernardeau, F. *Atlas forestier de la France par département*. [Paris]: Administration des Forêts, 1889.

Bernhard, Robert. *Türkiye Ormancılığının Mevzuatı, Tarihi ve Vazifeleri*. Translated by N. B. Somel. Ankara: Yüksek Ziraat Enstitüsü, 1935. Also published in German as *Grundlagen, Geschichte und Aufgaben der Forstwirtschaft in der Türkei*.

Berque, Jacques. *L'intérieur du Maghreb*. Paris: Gallimard, 1978.

——— . *Maghreb: Histoire et sociétés*. Alger: SNED, 1974.

Beşirli, Hayati, and Ibrahim Erdal, eds. *Osmanlı'dan Cumhuriyet'e Yörükler ve Türkmenler*. Ankara: Phoenix, 2008.

Bess, Michael. *The Light-Green Society*. Chicago: University of Chicago, 2003.

Besson-Lecousturier. *Note sur les forêts de chênes-liège: Le rôle qu'elles ont joué dans la colonisation; L'indemnité qui doit leur être allouée pour les dommages causés par l'insurrection.* Paris: Imprimerie Central des Chemins de Fer, A. Chaix et Cie, 1872.

Blache, Jules. *L'homme et la montagne.* Paris: Librairie Gallimard, 1933.

Blackbourn, David. *The Conquest of Nature: Water, Landscape, and the Making of Modern Germany.* New York: Norton, 2006.

Black-Michaud, Jacob. *Sheep and Land: The Economics of Power in a Tribal Society.* Cambridge: Cambridge University Press, 1986.

Blais, R. "Contribution à une histoire des gardes forestiers au XVIII siècle." *Revue Forestière Française* 38, no. 1 (1986): 17–26.

Blanchard, Raoul. "The Natural Regions of the French Alps." *Geographical Review* 11, no. 1 (1921): 31–49.

Blanchard, W. O. "The Landes: Reclaimed Waste Lands of France." *Economic Geography* 2, no. 2 (1926): 249–255.

Blanqui, Adolphe-Jérôme. *Du déboisement des montagnes.* Paris: Chez Reynard, 1846.

Blau, Gerda. "Wool in the World Economy." *Journal of the Royal Statistical Society* 109, no. 3 (1946): 179–242.

Bloch, Marc. *French Rural History: An Essay on Its Basic Characteristics.* Berkeley: University of California Press, 1966.

———. *Les caractères originaux de l'histoire rurale française.* Oslo: H. Aschehoug, 1931.

Blondel, Jacques. "The 'Design' of Mediterranean Landscapes: A Millennial Story of Humans and Ecological Systems during the Historic Period." *Human Ecology* 34, no. 5 (2006): 713–29.

———. "The Nature and Origin of the Vertebrate Fauna." In *The Physical Geography of the Mediterranean*, edited by Jamie C. Woodward, 139–64. Oxford: Oxford University Press, 2009.

Blum, Jerome. "Agricultural History and Nineteenth-Century European Ideologies." *Agricultural History* 56, no. 4 (1982): 621–31.

Bode, Christoph. *West Meets East: Klassiker der britischen Orient-Reiseliteratur.* Heidelberg: C. Winter, 1997.

Bonnet, P. "Aspects de la transhumance ovine provençale." *Ethnozootechnie* 2 (1975): 11–18.

———. *Grande transhumance ovine et briançonnais.* Grenoble: CTGREF Groupement de Grenoble, Première et Deuxième Parties, 1978.

Bonnin, Bernard. "L'élevage dans les hautes terres dauphinoises aux XVIIe et XVIIIe siècles." In *L'élevage et la vie pastorale dans les montagnes de l'Europe au Moyen âge et à l'époque moderne.* Clermont-Ferrand: Institut d'études du Massif central, 1984.

Borchhardt, Jutta. *Von Nomaden zu Gemüsebauern: Auf der Suche nach yoruk-Identitat bei den Sacikarali in der Sudwest-Turkei*. Münster: Lit, 2001.

Bordessoule, Éric. *Les "montagnes" du Massif central: Espaces pastoraux et transformations du milieu rural dans les monts d'Auvergne*. Clermont-Ferrand: Presses Universitaires Blaise Pascal, 2001.

Bordier, A. *La colonisation scientifique et les colonies françaises*. Paris: Reinwald, 1884.

Bory de Saint-Vincent, Jean-Baptiste-Geneviève-Marcellin. *Note sur la commission exploratrice et scientifique d'Algérie: Présentée à S. Exc. le ministre de la Guerre*. Paris: Imprimerie de Cosson, 1838.

Bostan, İdris. *Osmanlı Bahriye Teskilâtı: XVII. Yüzyılda Tersâne-i Âmire*. Ankara: Türk Tarih Kurumu Basimevi, 1992.

Bouche, Charles-François. *Essai sur l'histoire de Provence*. Marseille: Jean Mossy, 1775.

Boudy, Paul. *Économie forestière Nord-Africaine*. Volume 1, *Milieu physique et milieu humain*. Paris: Larose, 1948.

Boukhobza, Mohammed. "Nomadisme et colonisation: Analyse des mécanismes de déstructuration et de disparition de la société pastorale traditionnelle en Algérie." Thèse du troisième cycle. Paris, 1976.

Bourdeaux, Henry. *Code forestier; Suivi des lois sur la pêche et la chasse; et Code rural*. Paris: Jurisprudence générale Dalloz, 1931.

Bourdieu, Pierre, and Loïc Wacquant. "On the Cunning of Imperialist Reason." *Theory, Culture, & Society* 16, no. 1 (1999): 41–58.

Boutilly, V. *Recueil de législation forestière algérienne: Lois, décrets et règlements divers*. Paris and Nancy: Berger-Levrault, 1905.

Boye, Édouard. *Le Var, considérations au point de vue forestier, pastoral, agricole*. Lille: Imprimerie de L. Danel, 1888.

Boyer de Fonscolombe, M. E. H. "Mémoire sur la destruction et le rétablissement des bois dans les départemens qui composoient la Provence." In *Recueil des mémoires de la Société des amis des sciences, des lettres, de l'agriculture et des arts, établie à Aix, département des Bouches-du-Rhône*. Aix: Chez Augustin Pontier, Imprimeur du Roi, rue du Pont-Moreau, 1819.

Bradburd, Daniel. "Nomads and Their Trade Partners: Historical Context and Trade Relations in Southwest Iran, 1840–1975." *American Ethnologist* 24, no. 4 (1997): 895–909.

Braudel, Fernand. *Capitalism and Material Life*. London: Weidenfeld and Nicolson, 1973.

——— . *The Mediterranean and the Mediterranean World in the Age of Philip II*. Translated by Siân Reynolds. 2 vols. 1972. Berkeley: University of California Press, 1995.

Brauer, Ralph W. "Boundaries and Frontiers in Medieval Muslim Geography." *Transactions of the American Philosophical Society* 85, no. 6 (1995): 1–73.

Brett, Michael. "Ibn Khaldun and the Arabisation of North Africa." *Maghreb Review* 4, no. 1 (1979).

———. "Legislating for Inequality in Algeria: The Sénatus-consulte of 14 July 1865." *Bulletin of the School of Oriental and African Studies, University of London* 51, no. 3 (1988): 440–61.

Brewer, Mitra Jonah. "Gold, Frankincense, and Myrrh: French Consuls and Commercial Diplomacy in the Ottoman Levant, 1660–1699." PhD diss., Georgetown University, 2003.

Brice, William. *The Environmental History of the Near and Middle East since the Last Ice Age.* London: Academic Press, 1978.

Bricogne, Louis A. "Les fôrets de l'Empire ottoman." *Revue des Eaux et Fôrets* 16 (1877).

———. "La mission forestière en Turquie." *Revue des Eaux et Forêts* 9 (1870).

———. "La mission forestière en Turquie." *Revue des Eaux et Forêts* 15 (1876).

Briot, Félix. *Les Alpes françaises: Nouvelles études sur l'économie alpestre.* Paris: Berger-Levrault, 1896.

Brisebarre, Anne-Marie. *Bergers des Cévennes: Histoire et ethnographie du monde pastoral et de la transhumance en Cévennes.* Paris: Berger-Levrault, 1978.

Broadbridge, Anne F. "Royal Authority, Justice, and Order in Society: The Influence of Ibn Khaldun on the Writings of al-Maqrizi and Ibn Taghribirdi." *Mamluk Studies Review* 7, no. 2 (2003): 231–45.

Broc, Numa. "Les grandes missions scientifiques françaises au XIXe siècle (Morée, Algérie, Mexique) et leurs travaux géographiques." *Revue d'Histoire des Sciences* 34, no. 3 (1981): 319–58.

Brosselin, Arlette. "Pour une histoire de la forêt française." *Revue d'Histoire, Économie et Société* 55, nos. 1–2 (1977): 92–111.

Brossier, J. "Évolution des idées en matière forestière et conséquences sur la gestion des forêts alpines." *Revue Forestière Française.* Spécial: Éléments d'histoire forestière. *ENGREF* (1977): 153–62.

Brown, John Croumbie. *French Forest Ordinance of 1669, with Historical Sketch of Previous Treatment of Forests in France.* Edinburgh: Oliver and Boyd, 1883.

Brown, Karen. "'Trees, Forests and Communities': Some Historiographical Approaches to Environmental History on Africa." *Area* 35, no. 4 (2003): 343–56.

Brun, Jean-Pierre, Gaétan Congés, and Otello Badan. "Les bergeries romaines de la Crau d'Arles: Les origines de la transhumance en Provence." *Gallia* 52, no. 1 (1995): 263–310.

Buffault, Pierre. *Le briançonnais forestier et pastoral: Essai de monographie.* Paris, 1913.

———. "Le déboisement: Ses causes, ses conséquences." *Revue des Eaux et Forêts* (1937): 432–41, 506–18.

Buffon, Georges-Louis Leclerc de. *Les époques de la nature*. Paris: Imprimerie Royale, 1790.

———. *Histoire naturelle, générale et particulière, avec la description du cabinet du roi*. 44 vols. Paris: Imprimerie royale, 1749–1804.

———. *Mémoire sur la conservation et le rétablissement des forêts*. Paris: Académie Royale des Sciences, 1739.

Bulliet, Richard. *The Camel and the Wheel*. Cambridge MA: Harvard University Press, 1975.

Burke, Edmund. "The Sociology of Islam." In *Genealogies of Orientalism: History, Theory, Politics*, edited by Edmund Burke and David Prochaska, 154–73. Lincoln: University of Nebraska Press, 2008.

Burnaby, Frederick. *On Horseback through Asia Minor*. 1877. New York: Hippocrene, 1985.

Buttoud, Gérard. *Les conservateurs des eaux et forêts sous la Troisième République (1870–1940): Matériaux biographiques pour une sociologie historique de la haute administration forestière française*. Nancy: Laboratoire d'économie forestière de l'école nationale du génie rural des eaux et des forêts, 1981.

———. *L'état forestier: Politique et administration des forêts dans l'histoire française contemporaine*. Nancy: Université de Nancy, 1983.

———. *Les politiques forestières*. Paris: Presses Universitaires de France, 1998.

———. *Les propriétaires forestiers privés en France: Anatomie d'un groupe de pression*. Nancy: ENGREF, 1979.

———. "Prix et marché du bois à la fin du XIXe siècle." *Revue Forestière Française*. Spécial: Éléments d'histoire forestière. *ENGREF* (1977): 153–62.

Butzer, Karl W. "Environmental History in the Mediterranean World: Cross-Disciplinary Investigation of Cause-and-Effect for Degradation and Soil Erosion." *Journal of Archaeological Science* 32, no. 12 (2005): 1773–1800.

Çağlar, Yücel. *Türkiye Ormanları ve Ormancılık*. İstanbul: İletişim Yayınları, 1992.

Cameron, Iain. "The Policing of Forests in Eighteenth-Century France." In *Police and Policing: Past and Present Society Colloquium*, 1–27. Oxford: Oxford University Press, 1983.

Campbell, Christy. *The Botanist and the Vintner: How Wine Was Saved for the World*. Chapel Hill NC: Algonquin Books, 2006.

Carayol, Angel-Paul. *La législation forestière de l'Algérie*. Paris: A. Rousseau, 1906.

Carbonnieres, Ramond de. "De l'économie pastorale dans les Hautes-Pyrénées, de ses vices et des moyens d'y porter remède." *Bulletin de la Société Ramond* An II (1794).

Carette, Antoine-Ernest-Hippolyte. *Études sur la Kabilie*. Volume 4 in the subseries Sciences historiques et géographiques of the series Exploration scientifique de l'Algérie. Paris: Imprimerie Nationale, 1848.

————. *Recherches sur l'origine et les migrations des principales tribus de l'Afrique septentrionale et particulièrement de l'Algérie*. Paris: Imprimerie Impériale, 1853.

Carrière, P. "Prosper Demontzey." *Revue des Eaux et Forêts* 38 (1898): 193–222.

Carter, Harold B. *The Sheep and Wool Correspondence of Sir Joseph Banks, 1781–1820*. Norwich: Fletcher and Son, 1979.

Carteron, Charles. *Voyages en Algérie: Tous les usages des Arabes, leur vie intime et extérieure, ainsi que celle des Européens dans la colonie*. Paris: J. Hetzel, 1866.

Castagne, Louis. *Observations sur le reboisement des montagnes et des terres vagues, dans le département des Bouches-du-Rhône*. S.I., 1845.

Cavaillès, Henri. *La transhumance pyrénéenne et la circulation des troupeaux dans les plaines de Gascogne*. Paris: Armand Colin, 1931.

————. *La vie pastorale et agricole dans les Pyrénées des Gaves, de l'Adour et des Nestes: Étude de géographie humaine*. Paris: Armand Colin, 1931.

Çelebi, Evliya. *Narrative of Travels in Europe, Asia, and Africa in the Seventeenth Century*. Edited and translated by Joseph von Hammer-Purgstall. [London]: Oriental Translation Fund, 1834.

————. *Seyahatname*. Edited by Y. Dağlı et al. 8 vols. İstanbul: Yapı Kredi Yayınları, 2001–2005.

Çelik, Şenol. "Yüzyılda İçel Yörükleri Hakkında Bazı Değerlendirmeler." In *Anadolu'da ve Rumeli'de Yörükler ve Türkmenler: Sempozyumu Bildirileri, Tarsus, 14 Mayis 2000*, edited by Tufan Gündüz, 83–101. Ankara: Yörtürk, 2000.

Çelik, Zeynep. *Urban Forms and Colonial Confrontations: Algiers under French Rule*. Berkeley: University of California Press, 1997.

Cevdet, Ahmed. *Tarih-i Cevdet*. İstanbul: Matbaa-i Osmaniye, 1891.

Cezar, Mustafa. *Osmanlı Tarihinde Levendler*. İstanbul: Çelikcilt Matbaası, 1965.

Chang, Claudia, and Harold A. Koster. "Beyond Bones: Toward an Archaeology of Pastoralism." *Advances in Archaeological Method and Theory* 9 (1986): 97–148.

Chastagnaret, Gérard, ed. *Les sociétés méditerranéennes face au risque: Disciplines, temps, espaces*. Cairo: Institut Français d'Archéologie Orientale, 2008.

Chateaubriand, François-René, vicomte de. *Itinéraire de Paris à Jérusalem*. In *Œuvres complètes*, volume 5. Paris: Garnier, 1872.

Chauvet, Pierre, and Paul Pons. *Les Hautes-Alpes hier, aujourd'hui, demain: L'économie, le cadre de vie, les régions*. Gap: Société d'études des Hautes-Alpes, 1975.

Chevallier, P., and M. J. Couailhac. "Sauvegarde des forêts de montagne en France au XIXe siècle (l'exemple du Dauphine)." Paper presented at La Forêt: Actes du 113e congrès national des sociétés savantes, Strasbourg, 1988.

Christensen, Peter. *The Decline of Iranshahr: Irrigation and Environments in the History of the Middle East, 500 B.C. to A.D. 1500.* Copenhagen: Museum Tusculanum Press and University of Copenhagen, 1993.

Christian, P. *L'Afrique française, l'empire du Maroc et les déserts du Sahara.* Paris: A. Barbier, 1846.

Christiansen, Keith, and Pierre Rosenberg, eds. *Poussin and Nature: Arcadian Visions.* New Haven: Yale University Press, 2008.

"Chronique forestière." *Revue des Eaux et Forêts* 3 (1864).

"Chronique forestière." *Revue des Eaux et Forêts* 4 (1865).

"Chronique forestière." *Revue des Eaux et Forêts* 6 (1867).

"Chronique forestière." *Revue des Eaux et Forêts* 9 (1870).

"Chronique forestière." *Revue des Eaux et Forêts*, série 2, 29, vol. 4 (1890).

"Chronologie forestière (de Louis XVIII au IIe Empire)." Spécial: Éléments d'histoire forestière, *Revue Forestière Française* (1977): 116–18.

Çiçek, Kemal, ed. *The Great Ottoman-Turkish Civilisation.* Ankara: Yeni Türkiye, 2000.

Clancy-Smith, Julia Ann. *Mediterraneans: North Africa and Europe in an Age of Migration, c. 1800–1900.* Berkeley: University of California Press, 2011.

——— . *Rebel and Saint: Muslim Notables, Populist Protest, Colonial Encounters (Algeria and Tunisia, 1800–1904).* Berkeley: University of California Press, 1994.

Clark, Edward C. "The Ottoman Industrial Revolution." *International Journal of Middle East Studies* 5, no. 1 (1974): 65–76.

Clark, E. "Le pastoralisme dans les Alpes: Histoire parallèle/Alpine grazing, a tale of two states." *Revue de Géographie Alpine* 80, no. 2 (1992): 128–55.

Clave, J. *Études sur l'économie forestière.* Paris, 1862.

Cleary, M. C. "Contemporary Transhumance in Languedoc and Provence." *Geografiska Annaler: Series B, Human Geography* 69, no. 2 (1987): 107–13.

Clout, Hugh. "La transhumance: Passé, présent, avenir?" *Modern and Contemporary France* 13, no. 2 (2005): 225–28.

Clutton-Brock, Juliet. *A Natural History of Domesticated Mammals.* Cambridge: Cambridge University Press, 1999.

Cointat, Michel. "La dégradation des forêts dans le Département du Gard." *Revue Forestière Française* 2 (1954).

Collantes, Fernando. "The Demise of European Mountain Pastoralism: Spain 1500–2000." *Nomadic Peoples* 13, no. 2 (2009): 124–45.

——— . "Rural Europe Reshaped: The Economic Transformation of Upland Regions, 1850–2000." *Economic History Review*, n.s. 62, no. 2 (2009): 306–23.

Collins, James B. *The State in Early Modern France.* Cambridge: Cambridge University Press, 1995.

Collomp, Alain. "Les draps de laine, leur fabrication et leur transport en Haute-Provence du XVIIe au XIXe siècle." *Mélanges de l'École française de Rome, Moyen-Âge, Temps modernes* 99, no. 2 (1987): 1085–96.

Combe, A. D. *Les forêts de l'Algérie*. Alger: Imprimerie de Giralt, 1889.

Combes, F. "Restauration des terrains en montagne." *Revue Forestière Française* 41, no. 2 (1989): 91–106.

Comité des Travaux Historiques et Scientifiques. *Bulletin de la section des sciences économiques et sociales*. Paris: Imprimerie Nationale, 1931.

Commission des délégués des concessionnaires des chênes-liège de l'Algérie: Observations sur le Rapport de la Commission instituée à Constantine . . . pour l'enquête sur les incendies de forêts en Algérie (1863–1865). Paris: Imprimerie de Chaix, 1866.

Conklin, Alice. *In the Museum of Man: Race, Anthropology, and Empire in France, 1850–1950*. Ithaca NY: Cornell University Press, 2013.

Cook, Michael. *Population Pressure in Rural Anatolia, 1450–1600*. London: Oxford University Press, 1972.

Corancez, Louis-Alexandre-Olivier de. *Itinéraire d'une partie peu connue de l'Asie Mineure*. Paris: Chez J.-M. Eberhard, 1816.

Corbier, Mireille. "La transhumance dans les pays de la Méditerranée antique." In *Transhumance et estivage en Occident des origines aux enjeux actuels*, edited by Pierre-Yves Laffont. Toulouse: Presses Universitaires du Mirail, 2006.

Corvol, Andrée. *Forêt et eau: XIIIe–XXIe siècle*. Paris: L'Harmattan, 2007.

———, ed. *Forêts d'Occident du Moyen âge à nos jours: Actes des XXIVes Journées internationales d'histoire de l'Abbaye de Flaran, 6–8 septembre 2002*. Toulouse: Presses Universitaires du Mirail, 2004.

———, ed. *La nature en révolution, 1760–1800*. Paris: L'Harmattan, 1993.

———. "La privatisation des forêts nationales aux XVIIIe et XIXe siècles." *Histoire économique et financière de la France: Études et documents* 2 (1990): 211–22.

———. "Les partenaires de l'environnementalisme." In *Les sources de l'histoire de l'environnement: Le XIXe siècle*, edited by Andrée Corvol. Paris: L'Harmattan, 1999.

———, ed. *Les sources de l'histoire de l'environnement: Le XIXe siècle*. Paris: L'Harmattan, 1999.

———. *L'homme aux bois: Histoire des relations de l'homme et de la forêt, XVIIe–XXe siècle*. Paris: Fayard, 1987.

———. "Région Rhône, Alpes, Provence et Côte d'Azur." In *Les sources de l'histoire de l'environnement : Le XIXe siècle*, edited by Andrée Corvol, 305–22. Paris: L'Harmattan, 1999.

———. *Tempêtes sur la forêt française: XVIe–XXe siècle*. Paris: L'Harmattan, 2005.

Corvol, Andrée, Paul Arnould, and Micheline Hotyat, eds. *La forêt: Perceptions et représentations*. Paris, L'Harmattan, 1997.

Coste, Pierre. "La vie pastorale en Provence au milieu du XIVe siècle." *Études Rurales* 46 (1972): 61–75.

Costes, A. *La statistique agricole de 1814.* Paris: CTHS, 1914.

Coulston Gillispie, Charles. "Scientific Aspects of the French Egyptian Expedition 1798–1801." *Proceedings of the American Philosophical Society* 133, no. 4 (1989): 447–74.

Coutin, Pierre. "La politique agricole de la Ve République." *Revue Économique* 10, no. 5 (1959): 784–792.

Crane, Howard. "Evliya Çelebi's Journey through the Pamphylian Plain in 1671–72." *Muqarnas* 10 (1993): 157–68.

Cribb, Roger. *Nomads in Archaeology.* Cambridge: Cambridge University Press, 1991.

Crone, Patricia. "The Tribe and the State." In *States in History,* edited by J. Hall, 48–77. Oxford: Blackwell, 1986.

Cronon, William. *Changes in the Land: Indians, Colonists, and the Ecology of New England.* New York: Hill and Wang, 1983.

Cuinet, Vital. *La Turquie d'Asie: Géographie administrative, statistique descriptive et raisonnée de chaque province de l'Asie-mineure.* Volume 1. Paris: Ernest Leroux, 1890.

Cummins, Bryan D. *Bear Country: Predation, Politics, and the Changing Face of Pyrenean Pastoralism.* Durham NC: Carolina Academic Press, 2008.

Cutler, Brock William. "Evoking the State: Environmental Disaster and Colonial Policy in Algeria, 1840–1870." PhD diss., University of California, Irvine, 2011.

Darling, Linda T. *Revenue-Raising and Legitimacy: Tax Collection and Finance Administration in the Ottoman Empire, 1560–1660.* Leiden: Brill, 1996.

Daubrée, L. *Statistique et atlas des forêts de France.* Paris, 1912.

Daum, Andreas W. "Science, Politics, and Religion: Humboldtian Thinking and the Transformations of Civil Society in Germany, 1830–1870." *Osiris* 17 (2002): 107–40.

Daumas, Jean-Claude. "L'industrie lainière en France: Un siècle de mutations (1870–1973)," *Matériaux pour l'histoire de notre temps* 47, no. 1 (1997): 14–20.

Davis, Diana K. "Desert 'Wastes' of the Maghreb: Desertification Narratives in French Colonial Environmental History of North Africa." *Cultural Geographies* 11, no. 4 (2004): 359–387.

——— . "Neoliberalism, Environmentalism, and Agricultural Restructuring in Morocco." *Geographical Journal* 172, no. 2 (2006): 88–105.

——— . "Potential Forests: Degradation Narratives, Science, and Environmental Policy in Protectorate Morocco, 1912–1956." *Environmental History* 10, no. 2 (2005): 211–38.

——— . *Resurrecting the Granary of Rome: Environmental History and French Colonial Expansion in North Africa.* Athens: Ohio University Press, 2007.

Davis, Diana K., and Edmund Burke III, eds. *Environmental Imaginaries of the Middle East and North Africa*. Athens: Ohio University Press, 2013.

Davison, Roderic. *Reform in the Ottoman Empire, 1856–1876*. Princeton: Princeton University Press, 1963.

Deacon, Robert T. "Deforestation and Ownership: Evidence from Historical Accounts and Contemporary Data." *Land Economics* 75, no. 3 (1999): 341–359.

Dean, Warren. *With Broadax and Firebrand: The Destruction of the Brazilian Atlantic Forest*. Berkeley: University of California Press, 1995.

Debussche, Max, Jacques Lepart, and Alain Dervieux. "Mediterranean Landscape Changes: Evidence from Old Postcards." *Global Ecology and Biogeography* 8, no. 1 (1999): 3–15.

Delamare, Adolphe Hedwige Alphonse. *Exploration scientifique de l'Algérie pendant les années 1840–1845*. Paris: Imprimerie Nationale, 1850.

Delamarre, Mariel Jean-Brunhes. "Bergers de France: Exposition ayant eu lieu au Musée des arts et traditions populaires du 26 juillet au 19 novembre 1962." *Arts et Traditions Populaires* 10 (1962): 7–327.

Delort, Robert, François Walter, and Jacques Le Goff. *Histoire de l'environnement européen*. Paris: Presses Universitaires de France (PUF), 2001.

Delsalle, Paul, and Laurence Delobette. *La Franche-Comté à la charnière du Moyen âge et de la Renaissance, 1450–1550: Actes du colloque de Besançon, 10–11 octobre 2002*. Besançon: Presses Univ. Franche-Comté, 2003.

Demircioğlu, Yusuf Ziya. *Yürükler ve Köylülerde Hikâyeler, Masallar*. İstanbul: Güneş Matbaası, 1934.

Demontzey, Prosper. *Étude sur les travaux de reboisement et de gazonnement des montagnes*. Paris: Imprimerie nationale, 1878.

——— . *Traité pratique du reboisement et du gazonnement des montagnes*. Paris: J. Rothschild, 1882.

DeNovo, John A. "A Railroad for Turkey: The Chester Project, 1908–1913." *Business History Review* 33, no. 3 (1959): 300–329.

Deprest, Florence. "Using the Concept of Genre de Vie: French Geographers and Colonial Algeria, c. 1880–1949." *Journal of Historical Geography* 37, no. 2 (2011): 158–66.

Deringil, Selim. "'There Is No Compulsion in Religion': On Conversion and Apostasy in the Late Ottoman Empire: 1839–1856." *Comparative Studies in Society and History* 4, no. 3 (2000): 547–75.

——— . *The Well-Protected Domains: Ideology and the Legitimation of Power in the Ottoman Empire, 1876–1909*. London: I. B. Tauris, 2011.

———. "'They Live in a State of Nomadism and Savagery': The Late Ottoman Empire and the Post-Colonial Debate." *Comparative Studies in Society and History* 45, no. 2 (2003): 311–42.

Devèze, Michel. "La crise forestière en France dans la première moitié du XVIIIe siècle, et les suggestions de Vauban, Réaumur, Buffon." *Actes du 88e Congrès National des Sociétés Savantes, Clérmont-Ferrand* (1963): 595–616.

———. *La vie de la forêt française au XVIe siècle.* 2 vols. Paris: Éditions Jean Touzot, 1961.

Deville, A., and L. Rez. *Recueil des lois, ordonnances, décrets et règlements relatifs aux alignements, à l'expropriation pour cause d'utilité.* Paris: Imprimerie Nouvelle, 1886.

Dewdney, John. *Turkey: An Introductory Geography.* New York: Praeger, 1971.

Dietrich, Bruno F. A. "European Forests and Their Utilization." *Economic Geography* 4, no. 2 (1928): 140–58.

Diker, Mazhar. *Türkiye'de Ormancılık: Dün-Bugün-Yarın.* Ankara: T.C. Tarım Bakanlığı Orman Genel Müdürlüğü, 1947.

Dion, Emmanuel, and Sébastien Jahan. *Le peuple de la forêt: Nomadisme ouvrier et identités dans la France du centre-ouest aux temps modernes.* Rennes: Presses Universitaires de Rennes, 2002.

"Discussion devant l'assemblée nationale du projet de transfert des forêts au Ministère de l'Agriculture: Extrait du compte rendu des séances des 19 et 20 février 1873." *Revue des Eaux et Forêts* 12 (1873): 196.

Dombrowski, Raoul von. *Allgemeine Encyklopädie der gesammten Forst-und Jagdwissenschaften.* Wien: Berlag von Moriss Perles, 1894.

Dominian, Leon. "The Peoples of Northern and Central Asiatic Turkey." *Bulletin of the American Geographical Society* 47, no. 11 (1915): 832–71.

Drayton, Richard Harry. *Nature's Government: Science, Imperial Britain, and the "Improvement" of the World.* New Haven: Yale University Press, 2000.

Duby, Georges, and Armand Wallon, eds. *Histoire de la France rurale.* Paris: Seuil, 1975.

Duclos, Jean-Claude. *L'homme et le mouton dans l'espace de la transhumance.* Grenoble: Glénat, 1994.

Ducrocq, Théophile. *Cours de droit administratif.* Volume 2. 5th ed. Paris: E. Thorin, 1877.

Ducrot, L. *La reforme forestière et la propriété privée, étude historique, droite comparé.* Lyon, 1910.

Duffy, Andrea. "Civilizing through Cork: Conservationism and la Mission Civilisatrice in French Colonial Algeria." *Environmental History* 23, no. 2 (2018).

————. "Fighting Fire with Fire: Mobile Pastoralists and French Discourse on Wildfires in Nineteenth-Century Algeria." *Resilience: A Journal of the Environmental Humanities* 3 (2016).

Duguid, S. "The Politics of Unity: Hamidian Policy in Eastern Anatolia." *Middle Eastern Studies* 9 (1973): 139–55.

Duhamel du Monceau, M. [Henri-Louis]. *De l'exploitation des bois, ou, Moyens de tirer un parti avantageux des taillis, demi-futaies et hautes-futaies, et d'en faire une juste estimation avec la description des arts qui se pratiquent dans les forêts: Faisant partie du traité complet des bois & des forests.* 2 vols. Paris: Chez H. L. Guerin & L. F. Delatour, 1764.

————. *Du transport, de la conservation et de la force des bois.* Paris: Chez L. F. Delatour, 1767.

Duma, Jean. *Les Bourbon-Penthièvre (1678–1793): Une nébuleuse aristocratique au XVIIIe siècle.* Paris: Publications de la Sorbonne, 1995.

Dumoulin, Jacqueline. "Communes et pâturages forestiers en Provence au XIXe siècle: Le témoignage des comptabilités communales." *Provence Historique* 183 (1996): 57–96.

————. *La forêt provençale au XIXe siècle: Histoire des communaux boisés soumis au régime forestier.* Salon de Provence: Ixalog, 2002.

Dursun, Selçuk. "Forest and the State: History of Forestry and Forest Administration in the Ottoman Empire." PhD diss., Sabanci University, 2007.

————. "Population Policies of the Ottoman State in the Tanzimat Era: 1840–1870." Master's thesis, Sabanci University, 2001.

Duval, Jules, and Auguste-Hubert Warnier. *Bureaux arabes et colons.* Paris: Challamel Ainé, 1869.

Duvergier, J. B. *Collection complète des lois, décrets, ordonnances, règlemens, avis du Conseil-d'État.* Paris: Directeur de l'Administration, 1859.

Duveyrier, Henri. *Exploration du Sahara: Les Touareg du nord.* Paris, 1864.

Duvigneau, Guy. "L'évolution de la sédentarisation dans les hautes plaines de l'ouest algérien." *Revue de l'Occident musulman et de la Méditerranée* 45, no. 1 (1987): 80–93.

Dwyer, Cathy M. *The Welfare of Sheep.* Dordrecht: Springer, 2008.

Dyson-Hudson, Rada, and Neville Dyson-Hudson. "Nomadic Pastoralism." *Annual Review of Anthropology* 9 (1980): 15–61.

Earle, Edward. *Turkey, the Great Powers, and the Bagdad Railway: A Study in Imperialism.* New York: Macmillan, 1923.

Elwitt, Sanford H. "French Imperialism and Social Policy: The Case of Tunisia." *Science and Society* 31, no. 2 (1967): 129–48.

Emmanuelli, François-Xavier, ed. *L'intendance de Provence à la fin du XVIIe siècle: Édition critique des mémoires "pour l'instruction du duc de Bourgogne."* Paris: Bibliothèque Nationale, 1980.

Enfantin, Prosper. *Colonisation de l'Algérie.* Paris: P. Bertrand, 1843.

Enne, G., G. Pulina, M. D'Angelo, F. Previtali, S. Madrau, S. Caredda, and A. H. D. Francesconi. "Agro-pastoral Activities and Land Degradation in Mediterranean Areas: Case Study of Sardinia." In *Mediterranean Desertification: A Mosaic of Processes and Responses,* edited by N. A. Geeson, C. M. Brandt, and J. B. Thornes, 71–81. Chichester: John Wiley & Sons, 2002.

Equipe écologie et anthropologie des sociétés pastorales. *Pastoral Production and Society/Production pastorale et société: Proceedings of the International Meeting on Nomadic Pastoralism, Paris 1–3 Dec. 1976.* Cambridge: Cambridge University Press, 1979.

Eraslan, İsmail. *Türkiye'de Ormancılık Öğretim ve Eğitim Kurumlarının Gelişimi.* İstanbul: Ormancılık Eğitim ve Kültür Vakfı, 1989.

Erel, T. Levent, and Fatih Adatepe. "Tarihsel Depremlerin Akdeniz Bölgesi Antik Kent Yaşamındaki İzleri" [Traces of historic earthquakes in the ancient city life of the Mediterranean region]. *Black Sea/Mediterranean Environment* 13 (2007): 244–45.

Erinç, Sirri, and Necdet Tunçdilek. "The Agricultural Regions of Turkey." *Geographical Review* 42, no. 2 (1952): 179–203.

Erler, Mehmet Yavuz. "Ankara ve Konya Vilayetlerinde Kuraklık ve Kıtlık (1845 ve 1874 Yılları)." Ondokuz Mayıs University Institute of Social Sciences, 1997.

———. "XIX. Yüzyıldaki Bazı Doğal Afetler ve Osmanlı Yönetimi." In *Türkler,* edited by Hasan Celal Güzel, Kemal Çiçek, Salim Koca, and Murat Ocak, 763–68. Ankara: Yeni Türkiye, 2002.

———. *Osmanlı Devleti'nde Kuraklık (1800–1880).* İstanbul: Libra Kitap, 2010.

Eroz, Mehmet. *Yörükler.* İstanbul: Türk Dünyası Araştırmaları Vakfı, 1991.

Estoublon, Robert, and Adolphe Lefébure. *Code de l'Algérie annoté: Recueil chronologique des lois, ordonnances, décrets, arrêtés, circulaires, etc., formant la législation algérienne actuellement en vigueur, avec les travaux préparatoires et l'indication de la jurisprudence, suivi d'une table alphabétique de concordance.* Alger: A. Jourdan, 1907.

"Excursion de MM Tassy et Sthème en Asie Mineure." *Annales Forestières* 16 (1857): 335–37.

Eyssautier, L.-A. *Le statut réel français en Algérie, ou législation et jurisprudence sur la propriété, depuis 1830 jusqu'à la loi du 28 avril 1887.* Alger: A. Jourdan, 1887.

Fabre, J. *Des habitants des montagnes considérés dans leurs rapports avec le régime forestier.* Marseille: Imprimerie de Senès, 1849.

Fabre, Lucien. *Hommes de la Crau: Des coussouls aux alpages.* Le Coudray-Macouard: Éditions Cheminements, 2000.

Fabre, Lucien-Albert. *Boisements, irrigations, barrages-réservoirs.* Rome: Imprimerie de l'Unione cooperativa editrice, 1904.

———. *Études économiques et sociologiques dans les hautes montagnes françaises.* N.p., 1902.

———. *La fuite des populations pastorales françaises.* Paris: Au Secrétariat de la Société d'Économie Sociale, 1909.

———. *L'échec du coton à la laine au début du XXe siècle et à propos de la désertion des montagnes françaises.* Besançon: Imprimerie de Jacquin, 1910.

———. "L'état et la dépopulation montagneuse en France." *Revue Internationale de Sociologie.* Paris: V. Giard et E. Brière libraires-éditeurs, 1909.

———. *L'exode du montagnard et la transhumance du mouton en France.* Lyon: A. Rey, 1909.

———. *L'exode montagneux en France.* Besançon: Imprimerie de Jacquin, 1908.

———. *Les territoires sylvo-pastoraux du département de la Côte-d'Or.* Dijon: Imprimerie d'E. Jacquot, 1911.

Faiseau-Lavanne, J.-B.-F. *Recherches statistiques sur les forêts de la France tendant à signaler le danger qu'il y aurait pour elles d'ouvrir nos frontières aux fers étrangers.* Paris: A. J. Kilian, 1829.

Falsan, Albert. *Les Alpes françaises, la flore et la faune, le rôle de l'homme dans les alpes, la transhumance.* N.p.: Nabu Press, 2010.

The Famine in Asia Minor: Its History, Compiled from the Pages of the "Levant Herald." Istanbul: Isis, 1989.

Faré, H. *Rapports sur les forêts de l'État.* Paris, 1868.

Faroqhi, Suraiya. "Camels, Wagons, and the Ottoman State in the Sixteenth and Seventeenth Centuries." *International Journal of Middle East Studies* 14, no. 4 (1982): 523–39.

———. "A Natural Disaster as an Indicator of Agricultural Change: Flooding in the Edirne Area, 1100/1688–89." In *Natural Disasters in the Ottoman Empire: Halcyon Days in Crete III: A Symposium Held in Rethymnon 10–12 January 1997,* edited by Elisavet Zachariadou. Rethymnon: Crete University Press, 1999.

———. *The Ottoman Empire and the World around It.* London: I. B. Tauris, 2004.

———. "Towns, Agriculture and the State in Sixteenth-Century Ottoman Anatolia." *Journal of the Economic and Social History of the Orient* 33, no. 2 (1990): 125–56.

Fassin, Paul. "Le droit d'esplèche dans la Crau d'Arles." PhD diss., Université d'Aix-Marseille, 1898.

Faure, Louis Etienne. *Le berger des Alpes, ou, Mémoire sur la manière d'élever, de propager les bêtes á laine d'Espagne mérinos, et la race indigène dans le département des Hautes-Alpes.* Paris: Fantin, 1807.

Fernow, B. *Economics of Forestry: A Reference Book for Students of Political Economy and Professional and Lay Students of Forestry.* New York: T. Y. Crowell, 1902.

Ferrand, Joseph. *De la propriété communale en France et de sa mise en valeur, étude historique et administrative.* Paris: P. Dupont, 1859.

Fesler, James W. "French Field Administration: The Beginnings." *Comparative Studies in Society and History* 5, no. 1 (1962): 76–111.

Fesquet, F. "La défense de la forêt méditerranéenne au 19ème siècle: Un conflit forestier-paysan pour le contrôle des espaces communaux." Communication au séminaire Civilisation et patrimoine: Problématique de la forêt méditerranéenne. Maison Méditerranéenne des Sciences de l'Homme, Aix-en-Provence, 8 March 2001.

———. "La lutte contre les inondations au XIXème siècle: Aménagement des cours d'eau ou reboisement des montagnes; Entre complémentarité et opposition des démarches." Communication au colloque international: La rivière aménagée entre héritage et modernité; Formes, techniques et mise en oeuvre, Orléans, 15–16 October 2004. Printed in *AESTUARIA: Cultures et Développement Durable* 7 (2005): 299–314.

———. "Restauration des terrains en montagne et dynamiques démographiques en France au XIXème siècle: La gestion du risque a-t-elle désertifié la montagne?" Communication au colloque international de l'EHESS: Terrains communs, regards croisés; Intégrer le social et l'environnemental en histoire, Paris, 11–13 September 2008.

———. "Un corps quasi-militaire dans l'aménagement du territoire: Le corps forestier et le reboisement des montagnes méditerranéennes en France et en Italie aux XIX et XXèmes siècles." Thèse de doctorat, Université Paul Valéry, November 1997.

Festy, Octave. *Les animaux ruraux en l'an III: Dossier de l'enquête de la Commission d'agriculture et des arts.* Paris, 1941–46.

Feynman, Richard. *Ecology and Empire: Nomads in the Cultural Evolution of the Old World.* Los Angeles: Ethnographics, 1989.

Fillias, Achille. *État actuel de l'Algérie: Géographie physique et politique, description, population, mœurs et coutumes, commerce et industrie, administration, dictionnaire de toutes les localités.* Alger: Tissier, 1862.

Findley, Carter Vaughn. "Economic Bases of Revolution and Repression in the Late Ottoman Empire." *Comparative Studies in Society and History* 28, no. 1 (1986): 81–106.

———. *Turkey, Islam, Nationalism, and Modernity: A History, 1789–2007*. New Haven: Yale University Press, 2010.

Finkel, Caroline. *Osman's Dream: The History of the Ottoman Empire*. New York: Basic Books, 2005.

Finkel, Caroline, and N. N. Amraseys. *The Seismicity of Turkey and Adjacent Areas: A Historical Review, 1500–1800*. Istanbul: Eren, 1995.

Firat, Fehim. *Ormancılık İşletme İktisadı*. İstanbul: Kutulmuş Matbaası, 1971.

Fischlin, Andreas, and Dimitrios Gyalistras. "Assessing Impacts of Climatic Change on Forests in the Alps." *Global Ecology and Biogeography Letters* 6, no. 1 (1997): 19–37.

Fisher, Alan. "Emigration of Muslims from the Russian Empire in the Years after the Crimean War." In *A Precarious Balance: Conflict, Trade and Diplomacy on the Russian-Ottoman Frontier*. Istanbul: Isis, 1999.

Fisher, Stanley. *Ottoman Land Laws: Containing the Ottoman Land Code and Later Legislation Affecting Land with Notes and an Appendix of Cyprus Laws and Rules Relating to Land*. Oxford: Oxford University Press, 1919.

Fleischer, Cornell. *Bureaucrat and Intellectual in the Ottoman Empire: The Historian Mustafa Ali (1541–1600)*. Princeton: Princeton University Press, 1986.

———. "Royal Authority, Dynastic Cyclism and 'Ibn Khaldunism' in Sixteenth-Century Letters." *Journal of Asian and African Studies* 18 (1983): 198–219.

Ford, Caroline. *Natural Interests: The Contest over Environment in Modern France*. Cambridge MA: Harvard University Press, 2016.

———. "Reforestation, Landscape Conservation, and the Anxieties of Empire in French Colonial Algeria." *American Historical Review* 113, no. 2 (2008): 341–62.

"Forest Battalions for Service in France." *Science* 46, no. 1187 (1917): 306–7.

"Forêts françaises et forêts allemandes: Étude historique compare." *Revue Historique* (1967): 47–68.

Forêts perdues, forêts retrouvées. N.p.: Archives départementales des Bouches-du-Rhône, 1997.

Foucault, Michel. *Discipline and Punish: The Birth of the Prison*. New York: Pantheon, 1977.

———. "Governmentality." In *The Foucault Effect: Studies in Governmentality*, edited by Graham Burchell, Colin Gordon, and Peter Miller, 87–104. Chicago: University of Chicago Press, 1991.

———. "Omnes et Singulatim: Towards a Criticism of 'Political Reason.'" In *The Tanner Lectures on Human Values*, edited by S. McMurrin. Salt Lake City: University of Utah Press, 1981.

Fourchy, Pierre. "Remarques sur la question du déboisement des Alpes." *Revue de Géographie Alpine* 32, no. 1 (1944): 113–28.

Fournier, Joseph. *Les chemins de transhumance en Provence et en Dauphiné, d'après les journaux de route des conducteurs de troupeaux, au XVIIIe siècle.* Paris: Imprimerie Nationale, 1901.

———. *Histoire politique du département des Bouches-du-Rhône (1789-1914).* Marseille: Société Anonyme du Sémaphore, 1928.

Frängsmyr, Tore, J. L. Heilborn, and Robin E. Rider, eds. *The Quantifying Spirit in the Eighteenth Century.* Berkeley: University of California Press, 1990.

Freeman, John F. "Forest Conservancy in the Alps of Dauphiné, 1287–1870." *Forest and Conservation History* 38, no. 4 (1994): 171–80.

Frödin, John. "Les formes de la vie pastorale en Turquie." *Geografiska Annaler* 26 (1944): 219–72.

———. "Quelques traits de la végétation et de l'habitat pastoral de la Turquie du Nord." *Geografiska Annaler* 14 (1932): 209–43.

Fromentin, Eugène. *Between Sea and Sahara.* New York: I. B. Taurus, 2004.

Furetière, Antoine. *Dictionnaire universel, contenant généralement tous les mots français, tant vieux que modernes et les termes des sciences des arts.* 3 vols. 1690. La Haye: P. Husson, 1727.

Gadant, Jean. *L'atlas des forêts de France.* Paris: J.-P. de Monza, 1991.

Galaty, John. *The World of Pastoralism: Herding Systems in Comparative Perspective.* New York: Guilford Press, 1990.

Galloy, Pierre. *Nomades et paysans d'Afrique Noire Occidentale: Études de géographie soudanaise.* Nancy: Berger-Levrault, 1963.

Garcin, E. *Dictionnaire historique et topographie de la Provence ancienne et moderne.* Draguignan, 1835.

Gasquet, F.-H. de. *Observations sur le pâturage des moutons dans les forêts royales du Midi de la France.* Draguignan, 1830.

Gaussen, Henri. "A View from Canigou: Nature and Man in the Eastern Pyrenees." *Geographical Review* 26, no. 2 (1936): 190–204.

Gautier, [E.-F.], et al. *Histoire et historiens de l'Algérie.* Paris: Presses Universitaires de France, 1932.

Gay, Peter. *The Age of Enlightenment.* New York: Time, 1966.

Geeson, N. A., C. M Brandt, and J. B. Thornes, eds. *Mediterranean Desertification: A Mosaic of Processes and Responses.* Chichester: John Wiley & Sons, 2002.

Genç, Mehmet. *Osmanlı İmparatorluğu'nda Devlet ve Ekonomi.* İstanbul: Ötüken, 2000.

———. "Osmanlı Maliyesinde Malikâne Sistemi." In *Türkiye İktisat Tarihi Semineri: Metinler, Tartışmalar, 8–10 Haziran, 1973,* edited by Osman Okyar and Ünal Nabantoğlu, 231–96. Ankara: Hacettepe Üniversitesi Yayınları, 1975.

———. "A Study on the Feasibility of Using Eighteenth-Century Ottoman Financial Records as an Indicator of Economic Activity." In *The Ottoman Empire in the*

World Economy, edited by Huri Islamoğlu-Inan, 345–73. Cambridge: Cambridge University Press, 1987.

Geny-Mothe, Muriel. "Aménagement direct des forêts, propriété des ecclésiastiques (Le gruerie de Fleurance, 1669–1789)." In *Forêts d'Occident du Moyen âge à nos jours: Actes des XXIVes Journées internationales d'histoire de l'Abbaye de Flaran, 6–8 septembre 2002.* Toulouse: Presses Universitaires du Mirail, 2004.

George, P. "Anciennes et nouvelles forêts en région méditerranéenne." *Les Études Rhodaniennes* 9, no. 2 (1933): 85–120.

Gibelin, M. "Observations sur les chèvres." *Recueil des mémoires de la Société des amis des sciences, des lettres, de l'agriculture et des arts, établie à Aix, département des Bouches-du-Rhône.* Aix-en-Provence: Chez Augustin Pontier, Imprimeur du Roi, rue du Pont-Moreau, 1819.

Ginat, J. *Changing Nomads in a Changing World.* Brighton: Sussex Academic Press, 1998.

Girel, Pautou. "Interventions humaines et changements de la végétation alluviale dans la vallée de l'Isère (de Montmélian au Port de St-Gervais)." *Revue de Géographie Alpine* 82, no. 2 (1994): 127–46.

Göçek, Fatma Müge. *East Encounters West: France and the Ottoman Empire in the Eighteenth Century.* New York: Oxford University Press, 1987.

Godechot, Jacques Léon. *Les institutions de la France sous la Révolution et l'Empire.* Paris: Presses Universitaires de France, 1985.

Gökbilgin, M. Tayyib. *Rumeli'de Yürükler, Tatarlar, ve Evlad-ı Fatihan.* İstanbul: Osman Yalçın Matbaası, 1957.

Gökmen, Ertan. "Batı Anadolu'da Çekirge Felâketi (1850–1915)." *Belleten* 74, no. 269 (2010): 127–81.

Gomez-Ibanez, Daniel A. "Energy, Economics, and the Decline of Transhumance." *Geographical Review* 67, no. 3 (1977): 284–98.

Göney, S. "The Nomads and Reason of Their Sedentarization in Western Anatolia, Especially in the Buyuk Menderes Region of Turkey." *Review of the Geographical Institute of the University of Istanbul* 16:87–98.

Gonzalez, George A. "The Conservation Policy Network, 1890–1910: The Development and Implementation of 'Practical' Forestry." *Polity* 31, no. 2 (1998): 269–99.

Gould, Andrew G. "The Burning of the Tents: The Forcible Settlement of Nomads in Southern Anatolia." In *Humanist and Scholar: Essays in Honor of Andreas Tietze,* edited by Heath W. Lowry and Donald Quataert, 71–85. Istanbul: Isis, 1993.

———. "Pashas and Brigands: Ottoman Provincial Reform and Its Impact on the Nomadic Tribes of Southern Anatolia, 1840–1885." PhD diss., University of California, Los Angeles, 1973.

Graham, Hamish. "Policing the Forests of Pre-Industrial France: Round Up the Usual Suspects." *European History Quarterly* 33, no. 2 (2003): 157–82.

———. "Profits and Privileges: Forest and Commercial Interests in Ancien Régime France." *French History* 16, no. 4 (2002): 381–401.

Grant, Jonathan. "Rethinking the Ottoman 'Decline': Military Technology Diffusion in the Ottoman Empire, Fifteenth to Eighteenth Centuries." *Journal of World History* 10, no. 1 (1999): 179–201.

Grantham, George. "Agricultural Supply during the Industrial Revolution: French Evidence and European Implications." *Journal of Economic History* 49, no. 1 (1989): 43–72.

Gravius, Georges. *Les incendies de forêts en Algérie, leurs causes vraies et leurs remèdes: Quelques considérations générales sur la colonie.* Constantine: L. Marle, 1866.

Greeley, W. B. "Economic Aspects of Forestry." *Journal of Land and Public Utility Economics* 1, no. 2 (1925): 129–37.

Grellier, Bernard. "A Transhumant Shepherd on Mount Aigoual: Sheep Transhumance and the Shepherd's Knowledge." *International Social Science Journal* 58, no. 187 (2006): 161–64.

Grévoz, Daniel. *Sahara 1830–1881: Les mirages français et la tragédie Flatters.* Paris: L'Harmattan, 2000.

Griffiths, Tom. *Ecology and Empire: Environmental History of Settler Societies.* Seattle: University of Washington Press, 1997.

Griswold, William J. "Climatic Change: A Possible Factor in the Social Unrest of Seventeenth Century Anatolia." In *Humanist and Scholar: Essays in Honor of Andreas Tietze,* edited by Heath W. Lowry and Donald Quataert. Istanbul: Isis, 1993.

———. *The Great Anatolian Rebellion 1000–1020 / 1591–1611.* Berlin: K. Schwarz, 1983.

Grove, A. T., and Oliver Rackham. *The Nature of Mediterranean Europe: An Ecological History.* New Haven: Yale University Press, 2001.

Grove, Jean. "The Little Ice Age in the Massif of Mont Blanc." *Transactions of the Institute of British Geographers* 40 (1966): 129–43.

Grove, Richard. "Colonial Conservation, Ecological Hegemony and Popular Resistance: Towards a Global Synthesis." In *Imperialism and the Natural World,* edited by J. M. MacKenzie. Manchester: Manchester University Press, 1990.

———. *Green Imperialism: Colonial Expansion, Tropical Island Edens, and the Origins of Environmentalism, 1600–1860.* Cambridge: Cambridge University Press, 1995.

Grove, Richard H., Vinita Damodaran, and Satpal Sangwan, eds. *Nature and the Orient: The Environmental History of South and Southeast Asia.* Delhi: Oxford University Press, 1998.

Guénot, Stanislas. "Le déboisement des Pyrénées: De la gravite et de l'opportunité du sujet." *Bulletin de la Société de Géographie de Toulouse*, no. 1 (1900): 62–81.

Guiot, Léonide. *Les forêts et les pâturages du comté de Nice*. Paris, 1875.

Gül, Abdülkadir. "Osmanlı Devletinde Kuraklık ve Kıtlık (Erzurum Vilayeti Örneği: 1892–1893 ve 1906–1908 Yılları)." *Uluslar Arası Sosyal Araştırmalar Dergisi* (2009): 144–58.

Güldali, Nuri. *Geomorphologie der Türkei*. Wiesbaden: Reichert, 1979.

Güler, İbrahim. "XVIII. Yüzyılda Osmanlı Devleti'nde Nüfus Hareketleri Olarak 'İç Göçler.'" *İstanbul Üniversitesi Edebiyat Fakültesi Tarih Dergisi* 36 (2000): 155–211.

Gündüz, Tufan, ed. *Anadolu'da ve Rumeli'de Yörükler ve Türkmenler: Sempozyumu Bildirileri, Tarsus, 14 Mayıs 2000*. Ankara: Yörtürk, 2000.

Güran, Tevfik, ed. *Osmanlı Devleti'nin İlk İstatistik Yıllığı, 1897* [First statistical yearbook of the Ottoman Empire]. Ankara: Devlet İstatistik Enstitüsü, 1997.

Guthrie, J. F. *A World History of Sheep and Wool*. [Melbourne]: McCarron Bird, 1957.

Halaçoğlu, Yusuf. *XVIII. Yüzyılda Osmanlı İmparatorluğu'nun İskân Siyaseti ve Aşiretlerin Yerleştirilmesi*. Ankara: TTK Yayınları, 1988.

———. "Fırka-i İslâhiye ve Yapmış Olduğu İskân." *Tarih Dergisi* 27 (1973): 1–20.

Hamilton, William John. "Extracts from Notes Made on a Journey in Asia Minor in 1836." *Journal of the Royal Geographical Society of London* 7 (1837): 34–61.

———. *Researches in Asia Minor, Pontus and Armenia: With Some Account of Their Antiquities and Geology*. 2 vols. London: J. Murray, 1842.

Hanioğlu, M. Şükrü. *A Brief History of the Late Ottoman Empire*. Princeton: Princeton University Press, 2008.

Hannoum, Abdelmajid. *Violent Modernity: France in Algeria*. Cambridge MA: Harvard University Press, 2010.

Hardin, Garrett. "Extensions of 'The Tragedy of the Commons.'" *Science* 280, no. 5364 (1998): 682–83.

———. "The Tragedy of the Commons." *Science* 162, no. 3859 (1968): 1243–48.

Harding, Andrew, Jean Palutikof, and Tom Holt. "The Climate System." In *The Physical Geography of the Mediterranean*, edited by Jamie C. Woodward, 69–88. Oxford: Oxford University Press, 2009.

Harris, Cole. "How Did Colonialism Dispossess? Comments from an Edge of Empire." *Annals of the Association of American Geographers* 94, no. 1 (2004): 165–82.

Harris, Sarah Elizabeth. "Colonial Forestry and Environmental History: British Policies in Cyprus, 1878–1960." PhD diss., University of Texas at Austin, 2007.

Hart, David. *Banditry in Islam: Case Studies from Morocco, Algeria and the Pakistan NW Frontier*. Wisbech: MENAS, 1987.

Hartig, Georg Ludwig. *Grundsätze der Forst-Direction*. Hadamar: Neue Gelehrten-Buchhandlung, 1803.

Hasluck, F. W. "Heterodox Tribes of Asia Minor." *Journal of the Royal Anthropological Institute of Great Britain and Ireland* 51 (1921): 310–42.

Hattendorf, John. *Naval Policy and Strategy in the Mediterranean Sea: Past, Present and Future.* Hoboken NJ: Taylor & Francis, 2000.

Hayhoe, Jeremy. *Enlightened Feudalism: Seigneurial Justice and Village Society in Eighteenth-Century Northern Burgundy.* Rochester NY: University of Rochester Press, 2008.

Heffernan, Michael J. "A State Scholarship: The Political Geography of French International Science during the Nineteenth Century." *Transactions of the Institute of British Geographers* 19, no. 1 (1994): 21–45.

Heffernan, Michael J., and Keith Sutton. "The Landscape of Colonialism: The Impact of French Colonial Rule on the Algerian Rural Settlement Pattern, 1830–1987." In *Colonialism and Development in the Contemporary World*, edited by Christopher Dixon and Michael Heffernan, 121–52. London: Mansell, 1991.

Henderson, W. O. "The Cotton Famine on the Continent, 1861–5." *Economic History Review* 4, no. 2 (1933): 195–207.

———. "German Economic Penetration in the Middle East, 1870–1914." *Economic History Review* 18, no. 1–2 (1948): 54–64.

Henry, Frédéric. "The Age and History of the French Mediterranean Steppe Revisited by Soil Wood Charcoal Analysis." *The Holocene* 20, no. 1 (2010): 25–34.

Heper, Metin. "Center and Periphery in the Ottoman Empire: With Special Reference to the Nineteenth Century." *International Political Science Review/Revue Internationale de Science Politique* 1, no. 1 (1980): 81–105.

Herzfeld, Michael. *Anthropology through the Looking-Glass: Critical Ethnography on the Margins of Europe.* Cambridge: Cambridge University Press, 1987.

Heske, Franz. *Türkiye'de Orman ve Ormancılık.* İstanbul: Hüsnü Tabiat Basimevi, 1952.

Hirtz, Georges. *L'Algérie nomade et ksourienne: 1830–1954.* Marseille: Editions P. Tacussel, 1989.

Hogarth, D. *The Nearer East.* New York: Appleton, 1902.

Holmes, G. D. "History of Forestry and Forest Management." *Philosophical Transactions of the Royal Society of London. Series B, Biological Sciences* 271, no. 911 (1975): 69–80.

Homewood, K. M., and W. A. Rodgers. "Pastoralism and Conservation." *Human Ecology* 12, no. 4 (1984): 431–41.

Horden, Peregrine, and Nicholas Purcell. *The Corrupting Sea: A Study of Mediterranean History.* Oxford: Wiley-Blackwell, 2000.

Horn, Jeff. *The Path Not Taken: French Industrialization in the Age of Revolution, 1750–1830.* Cambridge MA: MIT Press, 2006.

Hourani, Albert. "Ottoman Reform and the Politics of Notables." In *The Modern Middle East*, edited by A. Hourani, P. S. Khoury, and M. C. Wilson, 83–110. Berkeley: University of California Press, 1993.

House, J. W. "The Franco-Italian Boundary in the Alpes Maritimes." *Transactions and Papers* (Institute of British Geographers) 26 (1959): 107–31.

Huffel, Gustave. *Économie forestière*. Paris, 1904–1907.

———. *Histoire des forêts françaises, de l'origine jusqu'à la suppression des maîtrises des eaux et forêts*. Nancy: École nationale des eaux et forêts, 1955.

———. *Influence des forêts sur le climat*. Besançon: Imprimerie de P. Jacquin, 1895.

Hughes, J. Donald. *The Mediterranean: An Environmental History*. Santa Barbara CA: ABC-CLIO, 2005.

Hughes, Philip, and Jamie Woodward. "Glacial and Periglacial Environments." In *The Physical Geography of the Mediterranean*, edited by Jamie C. Woodward, 353–84. Oxford: Oxford University Press, 2009.

Hundeshagen, J. Ch. *Encyclopädie der Forstwissenschaft, systematisch abgefasst*. 3 vols. 4th ed. Tübingen: Laupp, 1842.

Hurewitz, J. *Diplomacy in the Near and Middle East: A Documentary Record*. Princeton: Van Nostrand, 1956.

Husson, Jean-Pierre. *Les forêts françaises*. Nancy: Presses Universitaires de Nancy, 1995.

Hütteroth, Wolf-Dieter. *Die Siedlungen im Südlichen Inneranatolien um 1870–1900*. Göttingen: Göttinger geographische Abhandlungen, 1968.

———. "The Influence of Social Structure on Land Division and Settlement in Inner Anatolia." In *Turkey: Geographic and Social Perspectives*, edited by Peter Benedict, Erol Tümertekin, and Fatma Mansur, 19–47. Leiden: Brill, 1974.

———. *Landliche Siedlungen im Südlichen Inneranatolien*. Göttingen: Universitat Göttingen, 1968.

Ibn Battuta, and Sir Hamilton Alexander Rosskeen Gibb. *The Travels of Ibn Battūta: A.D. 1325–1354, Volume 3*. Translated and edited by H. A. R. Gibb. Cambridge: published for the Hakluyt Society by Cambridge University Press, 1971.

Ibn Khaldun. *The Muqaddimah: An Introduction to History*. Translated by Franz Rosenthal. Edited and abridged by N. J. Dawood. 2nd ed. Princeton: Princeton University Press, 1967.

İhsanoğlu, Ekmeleddin, ed. *Osmanlı Coğrafya Literatürü Tarihi / History of Geographical Literature during the Ottoman Period*. İstanbul: IRCICA, 2000.

Imber, Colin. "The Law of the Land." In *The Ottoman World*, edited by Cristine Woodhead. Milton Park: Routledge, 2012.

———. "The Navy of Suleyman the Magnificent." *Archivum Ottomanicum* 6 (1980): 211–82.

————. *The Ottoman Empire, 1300–1600*. New York: Palgrave Macmillan, 2002.

İnal, Selâhattin, and Yılmaz Bozkurt. *Vorträge über die Türkische Forstwirtschaft, gehalten in Deutschland*. İstanbul, 1962.

İnalcık, Halil. *The Middle East and the Balkans under the Ottoman Empire: Essays on Economy and Society*. Bloomington: Indiana University Turkish Studies Department, 1993.

————. "The Yuruks: Their Origins, Expansion and Economic Role." In *Oriental Carpet and Textile Studies II: Carpets of the Mediterranean Countries, 1400–1600*, edited by Robert Pinner and Walter B. Denny, 39–65. London: Hali Magazine, 1986.

İnalcık, Halil, and Donald Quataert. *An Economic and Social History of the Ottoman Empire, 1300–1914*. Cambridge: Cambridge University Press, 1994.

"Informations." *L'explorateur: Journal Géographique et Commercial* [edited by Charles Hertz and Adolphe Puissant] 1 (1875): 120.

İpek, Nedim. *Rumeli'den Anadolu'ya Türk Göçleri (1877–1890)*. Ankara: TTK Basımevi, 1994.

Irons, William, ed. *Perspectives on Nomadism*. Leiden: E. J. Brill, 1972.

Irwin, Robert. "Toynbee and Ibn Khaldun." *Middle Eastern Studies* 33, no. 3 (1997): 461–79.

İslamoğlu-İnan, Huri. *The Ottoman Empire and the World Economy*. Cambridge: Cambridge University Press; Paris: Éditions de la Maison des sciences de l'homme, 1987.

————. "Politics of Administering Property: Law and Statistics in the Nineteenth-Century Ottoman Empire." In *Constituting Modernity: Private Property in the East and West*, edited by Huri İslamoğlu, 276–320. London: I. B. Tauris, 2004.

————. *State and Peasant in the Ottoman Empire: Agrarian Power Relations and Regional Economic Development in Ottoman Anatolia during the Sixteenth Century*. Leiden: Brill, 1994.

————. "Statistical Constitution of Property Rights on Land in the 19th Century Ottoman Empire: An Evaluation of *Temmetuat* Registers." Paper delivered at the Conference on Land Issues in the Middle East, Harvard University, March 1996.

Isom-Verhaaren, Christine. "Shifting Identities: Foreign State Servants in France and the Ottoman Empire." *Journal of Early Modern History* 8, no. 1–2 (April 2004): 109–34.

Issawi, Charles. *The Economic History of Turkey, 1800–1914*. Chicago: University of Chicago Press, 1980.

J., D. A. "Obituary: Major-General Sir Charles William Wilson." *Geographical Journal* 26, no. 6 (1905): 682–84.

Jay, J.-L., ed. *Bulletin des lois des justices de paix, recueil chronologique des édits, décrets, arrêtés, lois, ordonnances et circulaires ministérielles, depuis 1563 jusqu'en 1852.* Paris: Chez Durand, 1852.

Jennings, R. "The Locust Problem in Cyprus." *Bulletin of the School of Oriental and African Studies* 51 (1988): 281–313.

Johansen, Ulla. "Die Nomadenzelte Südostanatoliens." *Zeitschrift fur Kultur, Politik und Wirtschaft der Islamischen Landern* 7 (1965): 33–37.

Johnson, Douglas. "The Human Dimensions of Desertification." *Economic Geography* 53, no. 4 (1977): 317–21.

——— . *The Nature of Nomadism: A Comparative Study of Pastoral Migrations in Southwestern Asia and Northern Africa.* Chicago: University of Chicago, Department of Geography, 1969.

——— . "Nomadism and Desertification in Africa and the Middle East." *GeoJournal* 31, no. 1 (1993): 51–66.

Jones, Dallas L. "A Geographic Survey of the French Economy." *American Journal of Economics and Sociology* 7, no. 1 (1947): 33–51.

Jorda, Maurice, and Jean-Christophe Roditis. "Les épisodes de gel du Rhône depuis l'an mil: Périodisation, fréquence, interprétation paléoclimatique." *Méditerranée* 78, no. 3 (1993): 19–30.

Jorgens, Denise. "A Comparative Examination of the Provisions of the Ottoman Land Code and Khedive Sa'id's Law of 1858." In *New Perspectives on Property and Land in the Middle East,* edited by Roger Owen, 93–120. Cambridge MA: Harvard University Press, 2000.

Jourdan, A.-J.-L., et al., eds. *Recueil des anciennes lois françaises, depuis l'an 420 jusqu'à la révolution de 1789: Contenant la notice des principaux monumens des Merovingiens, des Carlovingiens et des Capétiens, et le texte des ordonnances, édits, déclarations, lettres-patentes, règlements, arrêts du Conseil, etc., de la troisième race, qui ne sont pas abroges, ou qui peuvent servir, soit à l'interprétation, soit à l'histoire du droit public et prive, avec notes de concordance, table chronologique et table générale analytique et alphabétique des matières.* 29 vols. 1821. Ridgewood NJ: Gregg, 1964–66.

Jousse, Daniel. *Commentaire sur l'Ordonnance des eaux et forêts, du mois d'août 1669.* Paris: Chez Debure, père, 1772.

Julien, Charles-André. *Histoire de l'Algérie contemporaine.* Volume 1, *La conquête et les débuts de la colonisation (1827–1871).* Paris: Presses Universitaires de France, 1964.

Kalaora, Bernard. *La forêt pacifiée: Les forestiers de l'école de Le Play, experts des sociétés pastorales.* Paris: L'Harmattan, 1986.

Kalaora, Bernard, and A. Savoie. "Aménagement et ménagement: Le cas de la politique forestière au XIXe siècle." In *La forêt,* compiled and edited by Andrée

Corvol, 307–28. Paris: Editions du Comité des Travaux Historiques et Scientifiques (CTHS), 1991.

Karal, Enver Ziya. *Osmanlı İmparatorluğu'nda İlk Nüfus Sayımı, 1831.* Ankara: T.C. Başvekâlet İstatistik Müdürlüğü, 1943.

Karkar, Yaqub. *Railway Development in the Ottoman Empire, 1856–1914.* New York: Vantage Press, 1972.

Karpat, Kemal H. "The Land Regime, Social Structure and Modernization in the Ottoman Empire." In *Beginnings of Modernization in the Middle East: The Nineteenth Century,* edited by William R. Polk. Chicago: University of Chicago Press, 1968.

——. *Ottoman Population, 1830–1914: Demographic and Social Characteristics.* Madison: University of Wisconsin Press, 1985.

——. "Ottoman Population Records and the Census of 1881/82–1893." *International Journal of Middle East Studies* 9, no. 3 (1978): 237–74.

——. "The Transformation of the Ottoman State, 1789–1908." *International Journal of Middle East Studies* 3, no. 3 (1972): 243–81.

Karpat, Kemal H., and Robert W. Zens, eds. *Ottoman Borderlands: Issues, Personalities, and Political Changes.* Madison: University of Wisconsin, 2003.

Kasaba, Reşat. "Do States Always Favor Stasis?" In *Boundaries and Belonging: States and Societies in the Struggle to Shape Identities and Local Practices,* edited by Joel S. Midgal, 27–48. Cambridge: Cambridge University Press, 2004.

——. *A Moveable Empire: Ottoman Nomads, Migrants, and Refugees.* Seattle: University of Washington Press, 2009.

——. *The Ottoman Empire and the World Economy: The Nineteenth Century.* Albany: State University of New York Press, 1988.

Keskin, Özkan. "Osmanlı Ormancılığı'nın Gelişiminde Fransız Uzmanların Rolü." *Tarih Dergisi* 44 (2006): 123–42.

Keyder, Çağlar. *Landholding and Commercial Agriculture in the Middle East.* Albany: State University of New York Press, 1991.

Khalidi, Tarif, ed. *Land Tenure and Social Transformation in the Middle East.* Beirut: American University, 1984.

Khazanov, Anatoly. *Nomads in the Sedentary World.* Richmond, Surrey: Curzon, 2001.

——. *Nomads and the Outside World.* Cambridge: Cambridge University Press, 1984.

Khoury, Philip, and J. Kostiner, eds. *Tribes and State Formation in the Middle East.* Berkeley: University of California Press, 1990.

Kılıç, Orhan. "Osmanlı Devleti'nde Meydana Gelen Kıtlıklar." In *Türkler,* edited by Hasan Celal Güzel, Kemal Çiçek, Salim Koca, and Murat Ocak, 718–30. Ankara: Yeni Türkiye, 2002.

Kiray, Mübeccel, and Mediterranean Social Sciences Research Council. *Social Strat-ification and Development in the Mediterranean Basin*. The Hague: Mouton Institute, 1973.

Klein, Julius. *The Mesta: A Study in Spanish Economic History, 1273–1836*. Cambridge MA: Harvard University Press, 1963.

Klooster, Daniel James. "Toward Adaptive Community Forest Management: Inte-grating Local Forest Knowledge with Scientific Forestry." *Economic Geography* 78, no. 1 (2002): 43–70.

Koç, Bekir. "1870 Orman Nizamnâmesi'nin Osmanlı Ormancılığına Katkısı Üzerine Bazı Notlar." *Ankara Üniversitesi Dil ve Tarih-Coğrafya Fakültesi Tarih Bölümü Tarih Araştırmaları Dergisi* 24, no. 37 (2005): 231–57.

——— . "Osmanlı Devleti'ndeki Orman ve Korularin Tasarruf Yöntemleri ve İdarel-erine İlişkin Bir Araştırma." *OTAM (Ankara Üniversitesi Osmanlı Tarihi Araştırma ve Uygulama Merkezi Dergisi)* 10 (1999): 139–58.

——— . "Tanzimat Sonrası Hukuk Metinlerinde Çevre Bilincinin Arka-planı Olarak 'Av Yasak ve Sınırlılıkları' Üzerine Bazı Düşünceler." *OTAM (Ankara Üniversitesi Osmanlı Tarihi Araştırma ve Uygulama Merkezi Dergisi)* 19 (2006): 271–81.

——— . "Tanzimat Sonrası Hukuk Metinlerinde Av Yasakları." *Acta Turcica* 1, no. 1 (2009): 153–63.

Koerner, W., J. L. Dupouey, E. Dambrine, and M. Benoit. "Influence of Past Land Use on the Vegetation and Soils of Present Day Forest in the Vosges Mountains, France." *Journal of Ecology* 85, no. 3 (1997): 351–58.

Köksal, Yonca. "Coercion and Mediation: Centralization and Sedentarization of Tribes in the Ottoman Empire." *Middle Eastern Studies* 42 (2006): 469–91.

Kolars, John. "Locational Aspects of Cultural Ecology: The Case of the Goat in Non-Western Agriculture." *Geographical Review* 56, no. 4 (1966): 577–84.

——— . *Tradition, Season, and Change in a Turkish Village*. Chicago: University of Chicago, Department of Geography, 1963.

Kolars, John, and Henry J. Malin. "Population and Accessibility: An Analysis of Turkish Railroads." *Geographical Review* 60, no. 2 (1970): 229–46.

Konukçu, Mustafa. *Ormanlar ve Ormancılığımız: Faydaları, İstatistiki Gerçekler Anayasa, Kalkınma Planları, Hükümet Programları ve Yıllık Programlar'da Ormancılık / Forests and Turkish Forestry*. Ankara: Devlet Planlama Teşkilatı, 2001.

——— . *Statistical Profile of Turkish Forestry*. Ankara: Devlet Panlama Teşkilatı Yayın, 1998.

Köprülü, Mehmed Fuad. *Les origines de l'Empire ottoman*. Paris: E. De Boccard, 1935.

Kotschy, T. *Reise in den Kilikischen Taurus*. Gotha, 1858.

Krader, Lawrence. "The Ecology of Nomadic Pastoralism." *International Social Science Journal* 11 (1959): 499–510.

———. *Social Organization of the Mongol-Turkic Pastoral Nomads*. The Hague: Mouton, 1963.

Kutluk, Halil. "Türkiye'de Yabancı Ormancılar." In *Türk Ormancılığı Yüzüncü Tedris Yılına Girerken, 1857–1957*. Ankara: Doğuş Matbaası, 1957.

———. *Türkiye Ormancılığı ile İlgili Tarihi Vesikalar, 893–1339 (1487–1923)*. İstanbul: Osmanbey Matbaası, 1948.

Kütükoğlu, Mübahat. "Osmanlı Sosyal ve İktisadi Kaynaklarından Temettü Defterleri." *Belleten* 59, no. 225 (1995): 395–418.

Labrouche, Laurence. *Ariane Mnouchkine: Un parcours théâtral: Le terrassier, l'enfant et le voyageur*. Paris: L'Harmattan, 1999.

Lachiver, Marcel. *Les années de misère: La famine au temps du Grand Roi, 1680–1720*. Paris: Fayard, 1991.

Lacoste, Yves. *Ibn Khaldun: The Birth of History and the Past of the Third World*. London: Verso, 1984.

Ladoucette, Jean-Charles-François. *Histoire, antiquités, usages, dialectes des Hautes-Alpes*. Paris: Imprimerie de Madame Hérissant le Doux, 1820.

———. *Histoire, topographie, antiquités, usages, dialectes des Hautes-Alpes: Avec un atlas et des notes*. 3rd ed. Paris: Gide, 1848.

Laffont, Pierre-Yves, ed. *Transhumance et estivage en Occident des origines aux enjeux actuels*. Toulouse: Presses Universitaires du Mirail, 2006.

La Forest, Antoine. *Correspondance du comte de la Forest, ambassadeur de France en Espagne 1808–1813*. Paris: A. Picard et fils, 1905.

La Gorce, Pierre de. *Histoire du Second Empire*. 7 vols. New York: AMS Press, 1969.

Lamazou, Etienne. *L'ours et les brebis: Mémoires d'un berger transhumant des Pyrénées à la Gironde (Mémoire Vive)*. Paris: Seghers, 1988.

Lantz, François. *Chemins de fer et perception de l'espace dans les provinces arabes de l'Empire ottoman, 1890–1914*. Paris: L'Harmattan, 2005.

Lapidus, Ira M. "Tribes and State Formation in Islamic History." In *Tribes and State Formation in the Middle East*, edited by P. Khoury and J. Kostiner, 25–47. Berkeley: University of California Press, 1990.

Larrère, Raphaël. "Rauch ou Rougier de la Bergerie: Utopie ou réforme?" In *Revolution et espaces forestiers*, edited by Denis Woronoff, 232–46. Paris: L'Harmattan, 1988.

Larrère, Raphaël, and Olivier Nougarede. *L'homme et la forêt*. Paris: Gallimard, 1993.

Leach, Melissa, and Robert Mearns. "Environmental Change and Policy." In *The Lie of the Land: Challenging Received Wisdom on the African Environment*, edited by Melissa Leach and Robin Mearns. Oxford: International African Institute in association with James Currey; Portsmouth NH: Heinemann, 1996.

———, eds. *The Lie of the Land: Challenging Received Wisdom on the African Environment*. Oxford: International African Institute in association with James Currey; Portsmouth NH: Heinemann, 1996.

Leake, William Martin. *Journal of a Tour in Asia Minor: With Comparative Remarks on the Ancient and Modern Geography of That Country*. Hildesheim NY: Olms, 1976.

Lecugy, Jacques. "A mort les chèvres!" In *Bergers* (Revue éditée par Pays et gens du Verdon). Turin: Imprimerie Mariogros, 1999.

Lee, Timothy. *Wanganella and the Merino Aristocrats*. Richmond, Victoria: Hardie Grant Books, 2011.

Leeuwen, Carel van, ed. *Nomads in Central Asia: Animal Husbandry and Culture in Transition (19th–20th Century)*. Amsterdam: Royal Tropical Institute, 1994.

Lefebure, Claude. "Pastoral Production and Society." *Current Anthropology* 19, no. 1 (1978): 131–32.

Lefebvre, Theodore. "La transhumance dans les Basses-Pyrénées." *Annales de Géographie* 37, no. 205 (1928): 35–60.

Le Gall, Michel, and Kenneth J. Perkins. *The Maghrib in Question: Essays in History and Historiography*. Austin: University of Texas Press, 1997.

Le Guat, François. *The Voyage of François Leguat of Bresse, to Rodriguez, Mauritius, Java, and the Cape of Good Hope*. London: printed for the Hakluyt Society, 1891.

Lehning, James R. *Peasant and French: Cultural Contact in Rural France during the Nineteenth Century*. Cambridge University Press, 1995.

Lehuraux, Léon. *Le nomadisme et la colonisation dans les Hauts Plateaux de l'Algérie*. Paris: Editions du Comité de l'Afrique Française, 1931.

Le nouveau code rural, ou Le jurisconsulte campagnard: Contenant le code rural de 1791, le code forestier, la loi sur la pêche fluviale, le texte des dispositions du code civil. Avignon: P. Chaillot jeune, 1838.

Le Play, Frédéric. "Des forêts considérées dans leur rapport avec la constitution physique du globe et l'économie des sociétés." Undated manuscript placed in the Bibliothèque de l'Institut catholique, Paris.

Le Roy Ladurie, Emmanuel. *Histoire humaine et comparée du climat*. Volume 1, *Canicules et glaciers (XIIIe–XVIIIe siècles)*. Paris: Fayard, 2004.

———. *Histoire humaine et comparée du climat*. Volume 2, *Disettes et révolutions, 1740–1860*. Paris: Fayard, 2006.

———. *Histoire humaine et comparée du climat*. Volume 3, *Le réchauffement de 1860 à nos jours*. Paris: Fayard, 2009.

———. *Times of Feast, Times of Famine: A History of Climate since the Year 1000*. Garden City NY: Doubleday, 1971.

Les eaux et forêts du XIIe au XXe siècle. Paris: CNRS Éditions, 1987.

"Les forêts algériennes: Réponse au rapport de M. Jules Ferry." *Revue des Eaux et Forêts* 31 (1892).

"Les forêts françaises à la veille de la révolution." *Revue Historique* (1966): 347–80.

Leveau, Philippe. "Entre le delta du Rhône, la Crau et les Alpes, les sequenciations du temps pastoral et les mouvements des troupeaux à l'époque Romaine." In *Transhumance et estivage en Occident des origines aux enjeux actuels*, edited by Pierre-Yves Laffont. Toulouse: Presses Universitaires du Mirail, 2006.

Lewis, Norman N. *Nomads and Settlers in Syria and Jordan, 1800–1980*. Cambridge: Cambridge University Press, 1987.

Leynaud, E., and M. Georges. "Aspects géographiques de l'élevage dans la zone de montagne du département des Hautes-Alpes." *Études Rurales* 18 (1965): 5–36.

Lhomme, Jean. "La crise agricole à la fin du XIXe siècle en France: Essai d'interprétation économique et sociale." *Revue Économique* 21, no. 4 (1970): 521–53.

Ligue du reboisement de l'Algérie. *De la promulgation en Algérie de la loi du 4 avril 1882*. Alger: Imprimerie Casabianca, 1883.

Lindner, Rudi Paul. "Nomadism, Horses and Huns." *Past and Present* 92 (1981): 3–19.

———. *Nomads and Ottomans in Medieval Anatolia*. Bloomington: Research Institute for Inner Asian Studies, Indiana University, 1983.

———. "What Was a Nomadic Tribe?" *Comparative Studies in Society and History* 24, no. 4 (1982): 689–711.

Livet, Roger. *Habitat rural et structures agraires en Basse-Provence*. Gap: L. Jean, 1962.

Livingstone, Ian. "Livestock Management and 'Overgrazing' Among Pastoralists." *Ambio* 20, no. 2 (1991): 80–85.

Llasat, María del Carmen. "Storms and Floods." In *The Physical Geography of the Mediterranean*, edited by Jamie C. Woodward, 513–40. Oxford: Oxford University Press, 2009.

Lloret, Francisco, Josep Piñol, and Marc Castellnou. "Wildfires." In *The Physical Geography of the Mediterranean*, edited by Jamie C. Woodward, 541–58. Oxford: Oxford University Press, 2009.

Locher, A. *With Star and Crescent*. Philadelphia: Ætna, 1890.

Lockman, Zachary. *Contending Visions of the Middle East: The History and Politics of Orientalism*. Cambridge: Cambridge University Press, 2004.

Lombard, Maurice. "Un problème cartographié: Le bois dans la Méditerranée musulmane (VIIe–XIe Siècles)." *Annales: Économies, Sociétés, Civilisations* 14, no. 2 (1959): 234–54.

Lorcin, Patricia M. E. *Imperial Identities: Stereotyping, Prejudice and Race in Colonial Algeria*. London: I. B. Tauris; New York: St. Martin's Press, 1995.

Lowenthal, David. *George Perkins Marsh: Prophet of Conservation*. Seattle: University of Washington Press, 2003.

Lowood, Henry E. "The Calculating Forester: Quantification, Cameral Science, and the Emergence of Scientific Forestry Management in Germany." In *The Quantifying Spirit in the Eighteenth Century*, edited by Tore Frängsmyr, J. L. Heilborn, and Robin E. Rider, 315–42. Berkeley: University of California Press, 1990.

——— . *Patriotism, Profit, and the Promotion of Science in the German Enlightenment: The Economic and Scientific Societies, 1760–1815*. New York: Garland, 1991.

Lucas, Hippolyte. *Exploration scientifique de l'Algérie pendant les années 1840, 1841, 1842, publiée par ordre du gouvernement: Sciences physiques, zoologie, histoire naturelle des animaux articulés*. Paris: Imprimerie nationale, 1849.

Luther, Usha. *Historical Route Network of Anatolia (Istanbul-Izmir-Konya) 1550s to 1850s: A Methodological Study*. Ankara: Turkish Historical Society, 1989.

M., L. C. "Des forêts de la Turquie." *Annales Forestières*, Troisième Série, 13 (November 1854) 405–7.

——— . "Mission forestière de M M Tassy et Sthème en Turqui." *Annales Forestières* 16 (1857): 113–15.

——— . "Recherches sur le mouvement d'importation et d'exportation des bois, des combustibles minéraux, de la fonte et des fers pendant la période quinquennale de 1842 à 1847." *Annales Forestières*, Deuxième Série, 7, no. 2 (1848): 28–33.

MacKenzie, John M., ed. *Imperialism and the Natural World*. Manchester: Manchester University Press, 1990.

Macklin, Mark, and Jamie Woodward. "Rivers and Environmental Change." In *The Physical Geography of the Mediterranean*, edited by Jamie C. Woodward, 319–52. Oxford: Oxford University Press, 2009.

Madeline, Philippe, and Jean-Marc Moriceau, eds. *Acteurs et espaces de l'élevage (XVIIe–XXIe siècle): Évolution, structuration, spécialisation*. Rennes: Association d'Histoire des Sociétés Rurales, 2006.

Mahé, Alain. *Histoire de la Grande Kabylie, XIXe–XXe siècles: Anthropologie historique du lien social dans les communautés villageoises*. Saint-Denis: Bouchene, 2001.

Mancall, Peter, ed. *Bringing the World to Early Modern Europe: Travel Accounts and Their Audiences*. Leiden: Brill, 2007.

Marc, H. *Notes sur les forêts de l'Algérie*. Paris: Larose, 1930.

Marcelin, Paul. "Contribution à l'étude géographique de la garrigue nîmoise." *Les Études Rhodaniennes* 2, no. 1 (1926): 35–180.

Mardin, Şerif. *Religion and Social Change in Modern Turkey: The Case of Bediüzzaman Said Nursi*. Albany: State University of New York Press, 1989.

Marie, F. *Moyens de prévenir les inondations et d'accroitre les bois et les pâturages dans la Haute et Basse-Provence*. Marseille, 1862.

Marino, John A. *Pastoral Economics in the Kingdom of Naples.* Baltimore: Johns Hopkins University Press, 1988.

Marx, Emanuel. "The Tribe as a Unit of Subsistence: Nomadic Pastoralism in the Middle East." *American Anthropologist,* n.s. 79, no. 2 (1977): 343–63.

Masson, Paul. *Les Bouches-du-Rhône: Encyclopédie départementale.* Marseille: Archives départmentales des Bouches-du-Rhône, 1937.

Massy, P. H. H. "Exploration in Asiatic Turkey." *Geographical Journal* 26, no. 3 (1905): 272–303.

Matteson, Carol Kieko. *Forests in Revolutionary France: Conservation, Community, and Conflict, 1669–1848.* Cambridge: Cambridge University Press, 2015.

———. "Masters of Their Woods: Conservation, Community, and Conflict in Revolutionary France, 1669–1848." PhD diss., Yale University, 2008.

Maulde La Clavière, René de. *Procédures politiques du règne de Louis XII.* Paris: Imprimerie Nationale, 1885.

Mauron, Marie. *La transhumance du pays d'Arles aux Grandes Alpes.* Paris: Amiot-Dumont, 1952.

Mavidal, J., and E. Laurent, eds. *Archives parlementaires de 1787 à 1860: Recueil complet des débats législatifs et politiques des chambres françaises; Deuxième série (1800 à 1860).* 127 vols. Paris, 1862–1913.

McCarthy, Justin. *Death and Exile: The Ethnic Cleansing of Ottoman Muslims, 1821–1922.* Princeton: Darwin Press, 1996.

McIvor, Clarence. *The History and Development of Sheep Farming from Antiquity to Modern Times.* Sydney: Tilghman & Barnett, 1893.

McMurray, Jonathan. *Distant Ties: Germany, the Ottoman Empire, and the Construction of the Baghdad Railway.* Westport CT: Praeger, 2001.

McNeill, John R. *The Mountains of the Mediterranean World.* Cambridge: Cambridge University Press, 1992.

———. "Tragedies of Privatization: Land, Liberty, and Environmental Change in Spain and Italy, 1800–1910." In *Land, Property, and the Environment,* edited by John F. Richards. Oakland CA: ICS, 2002.

McPhee, Peter. *Revolution and Environment in Southern France, 1780–1830: Peasants, Lords, and Murder in the Corbières.* Oxford: Oxford University Press, 1999.

———. *A Social History of France: 1780–1880.* London: Taylor & Francis, 1992.

Mediterranean Forests and Maquis: Ecology, Conservation, and Management. Paris: UNESCO, 1977.

Mediterranean Landscapes. Berlin: Gesellschaft für Erdkunde zu Berlin, 2007.

Meiggs, Russell. *Trees and Timber in the Ancient Mediterranean World.* Oxford: Clarendon Press, 1982.

Mentzel, Peter. *Transportation Technology and Imperialism in the Ottoman Empire, 1800–1923*. Washington DC: American Historical Association, 2006.

Merchant, Carolyn. *The Death of Nature: Women, Ecology, and the Scientific Revolution*. San Francisco: Harper & Row, 1980.

———. *Reinventing Eden: The Fate of Nature in Western Culture*. New York: Routledge, 2013.

Merriam, Gordon P. "The Regional Geography of Anatolia." *Economic Geography* 2, no. 1 (1926): 86–107.

Merriam, John. "Turkish Agriculture and Ottoman Fiscal Policy, 1855–1870." Master's thesis, University of California, Berkeley, 1954.

Merriman, John M. *Police Stories: Building the French State, 1815–1851*. Oxford: Oxford University Press, 2006.

Métailie, Jean-Paul. "Lutter contre l'érosion: Le reboisement des montagnes." In *Les sources de l'histoire de l'environnement: le XIXe siècle*, edited by Andrée Corvol. Paris: L'Harmattan, 1999.

Meyer, Eve. "*Turquerie* and Eighteenth-Century Music." *Eighteenth-Century Studies* 7, no. 4 (1974): 474–88.

Michel, J. *Déboisement et reboisement en Provence*. Aix-en-Provence: Pourcel, 1907.

Michel, Joseph-Étienne. *Observations sur le commerce des bêtes à laine dans les départements des Bouches-du-Rhône, des Basses-Alpes et du Var*. Aix: P.-J. Calmen, 1790.

———. "Rapport instructif sur l'amélioration des bêtes a laine françaises dites transhumants." Eyguières, 1799.

Michelet, Jules. *Histoire de la Révolution française*. Paris: Gallimard, 1952.

———. *Tableau de la France*. Paris: Société Les Belles Lettres, 1934.

Mikesell, Marvin W. "The Deforestation of Mount Lebanon." *Geographical Review* 59, no. 1 (1969): 1–28.

Mikhail, Alan. *Nature and Empire in Ottoman Egypt: An Environmental History*. Cambridge: Cambridge University Press, 2011.

———. *Under Osman's Tree: The Ottoman Empire, Egypt, and Environmental History* Chicago: University of Chicago Press, 2017.

———, ed. *Water on Sand: Environmental Histories of the Middle East and North Africa*. Oxford: Oxford University Press, 2012.

Miller, Char. *Gifford Pinchot and the Making of Modern Environmentalism*. Washington DC: Island Press, 2001.

Millin, A.-L. *Voyage dans les départements du Midi de la France*. Paris, 1807–11.

Milward, Alan S., and S. B. Saul. *The Development of the Economies of Continental Europe, 1850–1914*. Cambridge MA: Harvard University Press, 1977.

Mistral, Frédéric. *Mirèio: A Provençal Poem*. Translated by Harriet Waters Preston. London: T. Fisher Unwin, 1890. Project Gutenberg online ed.

Mitchell, Brian R. *International Historical Statistics: Africa, Asia and Oceania, 1750–1993*. London: Macmillan Reference; New York: Stockton Press, 1998.

———. *International Historical Statistics: Europe, 1750–1993*. London: Macmillan Reference; New York: Stockton Press, 1998.

MM. L'Hôpital et Faré, Conseillers d'État, Commissaires du gouvernement. "Loi sur les incendies dans la région des Maures et de l'Esterel." *Revue des Eaux et Forêts* 9 (1870): 348.

Moltke, H. von. *Briefe über Zustände und Begebenheiten in der Turkei aus den Jahren 1835–1839*. Berlin, 1841.

Moncel, Delisle de. *Mémoire sur le repeuplement*. Nancy: Chez H. Haener, Imprimeur du Roi, 1791.

Montagne, R. *La civilisation du désert: Nomads d'Orient et d'Afrique*. Paris: Hachette, 1947.

Montagnon, Pierre. *Histoire de l'Algérie*. Paris: Pygmalion, 1998.

Montigny, Lucas de. *Causeries sur mes plantations*. Aix-en-Provence: Typographie Remondet-Aubin, 1862.

Montrichard. "L'Ile de Chypre." *Revue des Eaux et Forêts* 13 (1874): 39.

———. "Une excursion en Asie Mineure." *Revue des Eaux et Forêts* 12 (1873): 85–95.

Morel, Emmanuel. "L'incendie de forêts dans le sud-est de la France, étude de politique criminelle et de protection sociale." Thèse pour le Doctorat, Université d'Aix-en-Provence-Marseille, Faculté de Droit d'Aix, 1935.

Moriceau, Jean-Marc. *Histoire du méchant loup: 3000 attaques sur l'homme en France (XVe–XXe siècle)*. Paris: Fayard, 2007.

———. *Histoire et géographie de l'élevage français: XVe–XVIIIe siècles*. Paris: Fayard, 2005.

———. "Une question en renouvellement: L'histoire de l'élevage en France." *Annales de Bretagne et des pays de l'Ouest* 106, no. 1 (1999): 17–40.

Le mouton en Provence: 6000 ans d'histoire. Grans: Association pour la Sauvegarde de la Crau, 2007.

Moyal, Maurice. *Transhumance: Sur la route des alpages; 1951*. Marseille: Images en Manoeuvres Éd., 2002.

Mukerji, Chandra. "The Great Forestry Survey of 1669–1671: The Use of Archives for Political Reform." *Social Studies of Science* 37, no. 2 (2007): 227–53.

Mumcu, Uğur. *Hukuk, Devlet, Aşiret*. Ankara: Umag, 1997.

Murphey, Rhoads. "Some Features of Nomadism in the Ottoman Empire." *Journal of Turkish Studies* 8 (1984): 189–97.

Musset, Danielle, ed. *Histoire et actualité de la transhumance en Provence*. Mane: Les Alpes de Lumière, 1986.

Myres, John L. "Nomadism." *Journal of the Royal Anthropological Institute of Great Britain and Ireland* 71, no. 1–2 (1941): 19–42.

"Nécrologie, Prosper Demontzey." *Revue des Eaux et Forêts* 38 (1898): 193–222.

Nedonsel, Yves. "Contribution à l'étude de l'élevage ovin transhumant dans les Bouches-du-Rhône." Thèse, Université de Provence, 1976.

Nelson, Cynthia. *The Desert and the Sown: Nomads in the Wider Society*. Berkeley: Institute of International Studies, University of California, 1973.

Netting, Robert. *Balancing on an Alp: Ecological Change and Continuity in a Swiss Mountain Community*. Cambridge: Cambridge University Press, 1981.

Niamir-Fuller, Maryam, ed. *Managing Mobility in African Rangelands: The Legitimization of Transhumance*. London: Intermediate Technology Publications, 1999.

Niamir-Fuller, Maryam, and Matthew D. Turner. "A Review of Recent Literature on Pastoralism and Transhumance in Africa." In *Managing Mobility in African Rangelands: The Legitimization of Transhumance*, edited by Maryam Niamir-Fuller, 18–46. London: Intermediate Technology Publications, 1999.

Nicholson, Sharon E. *Dryland Climatology*. Cambridge: Cambridge University Press, 2011.

———. "The Historical Climatology of Africa." In *Climate and History*, edited by T. M. L. Wigley, M. J. Ingram, and G. Farmer, 249–70. Cambridge: Cambridge University Press, 1981.

"Notice sur Louis Tassy." *Revue des Eaux et Forêts* 35 (1896): 1–14.

Nougarede, O. "La transmission de la forêt paysanne-les bois dans la vie de la famille agricole." In *Bois et forêts des agriculteurs, Clermont-Ferrand, 20–21 octobre 1999*. Antony: Cemagref Editions, 1999.

———. "Processus historique de dissociation de l'agriculture et de la forêt." Paper presented at the conference Agriculteurs, agricultures et forêts, Paris, 12–13 November 1994. N.p.: Cemagref Editions, 1994.

Nouschi, André. *Enquête sur le niveau de vie des populations rurales constantinoises de la conquête jusqu'en 1919*. Paris: Presses Universitaires de France, 1961.

Nugent, Jeffrey B., and Nicolas Sanchez. "Tribes, Chiefs, and Transhumance: A Comparative Institutional Analysis." *Economic Development and Cultural Change* 42, no. 1 (1993): 87–113.

Observations sur l'aménagement des forêts, et particulièrement des forêts nationales. Edited by Société royale d'agriculture de Paris. Paris: Imprimerie de la Feuille du Cultivateur, 1791.

Oğuzoğlu, Yusuf. *Fluctuations in the Ottoman Social Order: According to Archival Sources: Reactions to Changes in the Ottoman Social Structure*. Cambridge MA: Department of Near Eastern Languages and Literatures, Harvard University, 2006.

Ongley, F., and Horace E. Miller, eds. *The Ottoman Land Code*. London: William Clowes and Sons, 1892.

Orange, A., and M. Amalbert. *Le mérinos d'Arles*. Antibes: F. Genre, 1924.

Ordonnance de Louis XIV: Sur le fait des eaux et forêts, du mois d'août 1669; Régistrée au Parlement de Besançon, le 27 avril, 1694. Augmentée, etc. Besançon, 1750.

Orhonlu, Cengiz. *Osmanlı İmparatorluğu'nda Aşiretleri İskân Teşebbüsü: 1691–1696.* İstanbul: İstanbul Üniversitesi Edebiyat Fakültesi, 1963.

———. *Osmanlı İmparatorluğu'nda Aşiretlerin İskâni.* İstanbul: Eren, 1987.

———. *Osmanlı İmparatorluğu'nda Derbend Teskilâtı.* İstanbul: Eren, 1990.

———. *Osmanlı İmparatorluğu'nda Şehircilik ve Ulaşım: Üzerine Araştırmalar.* İzmir: Ticaret Matbaacılık, 1984.

Osborne, Michael A. "Science and the French Empire." *Isis* 96, no. 1 (2005): 80–87.

Ostrom, Elinor. *Governing the Commons: The Evolution of Institutions for Collective Action.* Cambridge: Cambridge University Press, 1990.

Owen, Roger. *The Middle East in the World Economy 1800–1914.* London: I. B. Tauris, 2005.

———, ed. *New Perspectives on Property and Land in the Middle East.* Cambridge MA: Center for Middle Eastern Studies of Harvard University, distributed by Harvard University Press, 2000.

Özbay, Rahmi Deniz. "Tanzimat Sonrasında Akdağ Kazası Civarına Afşar Türkmenlerinin İskâni." In *XIV. Türk Tarih Kongresi, Ankara: 9–13 Eylül 2002*, 453–81. Ankara: Türk Tarih Kurumu, 2005.

Özbayri, Kemal. *Tahtacılar ve Yörükler/Tahtadjis et Yöruks (matériaux pour l'étude des nomades du Taurus).* Paris: A. Maisonneuve, 1972.

Özdeğer, Yunus. "XIX. Yüzyıl Sonlarında Meydana Gelen Bir Kuraklık ve Kıtlık Hadisesi ile Bunun Sosyo-Ekonomik Sonuçları." *Karadeniz Araştırmaları* (2008): 87–96.

Özdönmez, Metin, and Abdi Ekizoğlu. "Tanzimat ve Meşrutiyet Dönemleri Ormancılığında Katkıları Olan Yabancı Uzmanlar." *Yayın Komisyonuna Sunulduğu Tarih* (1996): 57–68.

Pamuk, Sevket. *A Monetary History of the Ottoman Empire.* Cambridge: Cambridge University Press, 2000.

———. *The Ottoman Empire and European Capitalism, 1820–1913: Trade, Investment, and Production.* Cambridge: Cambridge University Press, 1987.

———. "The Ottoman Empire in the 'Great Depression' of 1873–1896." *Journal of Economic History* 44, no. 1 (1984): 107–18.

Panckoucke, Charles-Joseph. *Encyclopédie méthodique.* Paris: Chez Panckoucke, 1784.

Pardé, J. "Les relations forestières franco-allemandes au XIXe siècle." Spécial: Éléments d'histoire forestière. *Revue Forestière Française* (1977): 153–62.

Parker, Geoffrey, and Lesley M. Smith. *The General Crisis of the Seventeenth Century.* 2nd ed. London: Routledge, 1997.

Peacock, Andrew C. S. *The Frontiers of the Ottoman World.* Oxford: Oxford University Press, 2009.

Pearson, Chris. "'The Age of Wood': Fuel and Fighting in French Forests, 1940–1944." *Environmental History* 11, no. 4 (2006): 775–803.

Pecquet, Antoine. *Loix forestières de France, commentaire historique et raisonné sur l'ordonnance de 1669, les règlemens antérieurs, & ceux qui l'ont suivie: Auquel on a joint une bibliothèque des auteurs qui ont écrit sur les matières d'eaux & forêts, & une notice des coutumes relatives à ces mêmes matières.* Paris: Chez Prault, 1753.

Pekin, E. "Yörük Çuvalları." *Sanat Dünyamız* 2, no. 5 (1975): 14–20.

Perrot, Georges. *Souvenirs d'un voyage en Asie Mineure.* Paris: Michel Lévy frères, 1864.

Pesson, P. *Écologie forestière: La forêt: Son climat, son sol, ses arbres, sa faune.* Paris, 1974.

Pettet, Deirdre. "A Veritable Bedouin: The Chevalier d'Arvieux in the Camp of the Emir Turabey." In *Distant Lands and Diverse Cultures: The French Experience in Asia, 1600–1700*, edited by Glenn J. Ames and Ronald S. Love, 21–46. Westport CT: Praeger, 2003.

Peyriat, Pierre. "La Chambre des eaux et forêts du parlement de Provence au XVIII siècle et son rôle dans la défense des bois." Thesis, Université de Provence, 1951.

Phillips, Carla, and William D. Phillips Jr. *Spain's Golden Fleece: Wool Production and the Wool Trade from the Middle Ages to the Nineteenth Century.* Baltimore: Johns Hopkins University Press, 1997.

Piegay, Herve, and Pierre-Gil Salvador. "Contemporary Floodplain Forest Evolution along the Middle Ubaye River, Southern Alps, France." *Global Ecology and Biogeography Letters* 6, no. 5 (1997): 397–406.

Pina-Cabral, João de. "The Mediterranean as a Category of Regional Comparison: A Critical View." *Current Anthropology* 30 (1989): 399–406.

Pincetl, Stephanie. "Some Origins of French Environmentalism: An Exploration." *Forest and Conservation History* 37, no. 2 (1993): 80–89.

Pinchot, Gifford. "Government Forestry Abroad." *Publications of the American Economic Association* 6, no. 3 (1891): 7–54.

Pisani, Donald J. "Forests and Conservation, 1865–1890." *Journal of American History* 72, no. 2 (1985): 340–59.

Pitcher, Donald E. *An Historical Geography of the Ottoman Empire.* Leiden: E. J. Brill, 1972.

Plaisance, Georges. *Aspects of Mountain Life in Anatolia and Iran.* London: E. Arnold, 1966.

———. *La bibliographie forestière française.* Paris, 1965.

———. *Les formations végétales et paysages ruraux: Lexique et guide bibliographique.* Paris, 1959.

Planhol, Xavier de. "Caractères généraux de la vie montagnarde dans le Proche-Orient et dans l'Afrique du Nord." *Annales de Géographie* 71, no. 384 (1962): 113–30.

———. "Contribution à l'étude géomorphologique du Taurus Occidental et de ses plaines bordières." *Revue de Géographie Alpine* 44, no. 4 (1956): 609–85.

———. *De la plaine pamphylienne aux lacs pisidiens: Nomadisme et vie paysanne.* Paris, 1958.

———. "Expansion et problèmes de l'agriculture turque." *Revue de Géographie de Lyon* 35, no. 1 (1960): 91–103.

———. "Geography, Politics and Nomadism in Anatolia." *International Social Science Journal* 11, no. 4 (1959): 525–31.

———. *An Historical Geography of France.* Cambridge: Cambridge University Press, 1994.

———. "Le boeuf porteur dans le Proche-Orient et l'Afrique du Nord." *Journal of the Economic and Social History of the Orient* 12, no. 3 (1969): 298–321.

———. "Les nomades, la steppe, et la forêt en Anatolie." *Geographische Zeitschrift* 53, no. 2–3 (1965): 101–16.

———. "Les nouveaux villages d'Algérie." *Geografiska Annaler* 43, no. 1–2 (1961): 243–51.

———. "Vie pastorale caucasienne et vie pastorale anatolienne, d'après un livre récent." *Revue de Géographie Alpine* 44, no. 2 (1956): 371–79.

Planhol, Xavier de, and Hamit Inandik. "Études sur la vie de montagne dans le Sud-Ouest de l'Anatolie: Le Yesil gol dag et le Boz dag." *Revue de Géographie Alpine* 47, no. 3 (1959): 375–89.

Poivre, Pierre. *Voyages d'un philosophe, ou observations sur les mœurs et les arts des peuples de l'Afrique, de l'Asie et de l'Amérique.* N.p.: Yverdon, 1768.

Polk, William R., ed. *Beginnings of Modernization in the Middle East: The Nineteenth Century.* Chicago: University of Chicago Press, 1968.

Ponchelet, D. "Le débat autour du déboisement dans le département des Basses-Alpes, France (1819–1849)/The debate surrounding deforestation in the Basses-Alpes department, 1819–1849." *Revue de Géographie Alpine* 83, no. 1 (1995): 53–66.

Popkin, Jeremy K. *A History of Modern France.* 4th ed. London: Pearson, 2013.

Poujoulat, Baptistin. *Voyage dans l'Asie Mineure, en Mésopotamie, a Palmyre, en Syrie, en Palestine et en Egypte.* 1840. Paris: Ducollet, 1864.

Prochaska, David. "Fire on the Mountain: Resisting Colonialism in Algeria." In *Banditry, Rebellion and Social Protest in Africa,* edited by Donald Crummey, 229–52. London: James Currey, 1996.

Pswarayi-Riddihough, Idah. *Forestry in the Middle East and North Africa: An Implementation Review.* Washington DC: World Bank, 2002.

Puyo, Jean-Yves. "La science forestière vue par les géographes français, ou la confrontation de deux sciences 'diagonales' (1870–1914)." *Annales de Géographie* 108, no. 609–10 (1999): 615–34.

Pyne, Stephen J. *Vestal Fire: An Environmental History, Told through Fire, of Europe and Europe's Encounter with the World.* Seattle: University of Washington Press, 1997.

Quataert, Donald. *The Ottoman Empire, 1700–1922.* New York: Cambridge University Press, 2000.

Quenet, Grégory. "Des sociétés sans risques?" In *Les sociétés méditerranéennes face au risque: Disciplines, temps, espaces,* edited by Gérard Chastagnaret. Cairo: Institut Français d'Archéologie Orientale, 2008.

———. *Les tremblements de terre: Aux XVIIe et XVIIIe siècles La naissance d'un risque.* Seyssel: Éditions Champ Vallon, 2005.

Raccagni, Michelle. "The French Economic Interests in the Ottoman Empire." *International Journal of Middle East Studies* 11, no. 3 (1980): 339–76.

Radkau, Joachim. *Nature and Power: A Global History of the Environment.* Cambridge: Cambridge University Press, 2008.

———. *Wood: A History.* Cambridge: Polity Press, 2012.

———. "Wood and Forestry in German History: In Quest of an Environmental Approach." *Environment and History* 2, no. 1 (1996): 63–76.

Rajan, Ravi. "Imperial Environmentalism or Environmental Imperialism? European Forestry, Colonial Foresters and the Agendas of Forest Management in British India 1800–1900." In *Nature and the Orient: The Environmental History of South and Southeast Asia,* edited by Richard H. Grove, Vinita Damodaran, and Satpal Sangwan, 324–71. Delhi: Oxford University Press, 1998.

Rao, Aparna. *The Other Nomads: Peripatetic Minorities in Cross-Cultural Perspective.* Köln: Böhlau, 1987.

Raynal, Guillaume-Thomas-François, abbé, Anthony Strugnell, Rigobert Bonne, and Andrew Brown. *Histoire philosophique et politique des établissements et du commerce des Européens dans les deux Indes.* 1774. Ferney-Voltaire: Centre international d'étude du XVIIIe siècle, 2010.

Recueil des lois composant le code civil: Avec les discours des orateurs du Gouvernement, les rapports de la commission du Tribunat et les opinions émises pendant le cours de la discussion. 9 vols. Paris: Moreau, 1803–5.

Recueil des mémoires de la Société des amis des sciences, des lettres, de l'agriculture et des arts, établie à Aix, département des Bouches-du-Rhône. Aix-en-Provence: Chez Augustin Pontier, Imprimeur du Roi, 1819.

Refik, Ahmet. *Anadolu'da Türk Aşiretleri, 966–1200: Anadolu'da Yaşayan Türk Aşiretleri Hakkinda Divani Hümayun Mühmime Defterlerinde Mukayyet Hükümleri Havidir.* İstanbul: Devlet Matbaasi, 1930.

Reid, James. *Crisis in the Ottoman Empire: Prelude to Collapse.* Stuttgart: F. Steiner, 2000.

Rendu, Christine. "Fouiller des cabanes de bergers: Pour quoi faire?" *Études Rurales* 153–54 (2000): 151–76.

Rey, François. *L'exploitation pastorale dans le département de la Savoie.* Chambéry: Dardel, 1930.

Rey-Goldzeiguer, Annie. *Le royaume arabe: La politique algérienne de Napoléon III, 1861–1870.* Alger: SNED, 1977.

Reynard, J. *Restauration des forêts et des pâturages du sud de l'Algérie.* Alger: Adolphe Jourdan, 1880.

Reynaud, Henri Jean François Edmond Pélissier de. *Mémoires historiques et géographiques sur l'Algérie.* Paris: Imprimerie Royale, 1844.

Ribbe, Charles de. *Des incendies de forêts dans la région des Maures et de l'Estérel (Provence) leurs causes, leur histoire, moyens d'y remédier.* Hyères: Au Siège de la Société Librairie Agricole, 1869.

——— . *La Provence au point de vue des bois, des torrents, et des inondations avant et après 1789.* Paris: Guillaumin, 1857.

Richards, John F., ed. *Land, Property, and the Environment.* Oakland CA: ICS, 2002.

Richez, Jean-Claude. "Science allemande et forestière française: L'expérience de la rive gauche du Rhin." In *Révolution et espaces forestiers*, edited by Denis Woronoff, 232–46. Paris: L'Harmattan, 1988.

Rinschede, Gisbert. *Die Transhumance in den französischen Alpen und in den Pyrenäen.* Münster: Im Selbstverlag der Geographischen Kommission für Westfalen, 1979.

Ritchie, James. *The Influence of Man on Animal Life in Scotland.* Cambridge: Cambridge University Press, 1920.

Rıza, Ali. *Orman ve Mera Kanununun Esbab-ı Mucibe Layıhası.* İstanbul: Mahmud Bey Matbaası, 1910.

Robert, Adolphe, Edgar Bourloton, and Gaston Cougny, eds. *Dictionnaire des parlementaires français.* 5 vols. Paris: Bourloton, 1889.

Roche, Daniel. *La culture équestre de l'Occident, XVIe–XIXe siècle: L'ombre du cheval.* Volume 1, *Le cheval moteur.* Paris: Fayard, 2008.

——— . *Histoire de la culture équestre, XVIe–XIXe siècle.* Volume 2, *La gloire et la puissance.* Paris: Fayard, 2011.

Rogan, E. L. "Aşiret Mekteb: Abdulhamid II's School for Tribes (1892–1907)." *International Journal of Middle East Studies* 28, no. 1 (1996): 83–107.

Rosenthal, Jean-Laurent. *The Fruits of Revolution: Property Rights, Litigation, and French Agriculture, 1700–1860.* Cambridge: Cambridge University Press, 1992.

Roskams, Stephen. "Urban Transition in North Africa: Roman and Medieval Towns of the Maghreb." In *Towns in Transition: Urban Evolution in Late Antiquity* and the Early Middle Ages, edited by S. T. Loseby and N. Christie. Aldershot: Ashgate, 1996.

Rotter, Gernot. "Natural Catastrophes and Their Impact on Political and Economic Life during the Second *Fitna.*" In *Land Tenure and Social Transformation in the Middle East,* edited by Tarif Khalidi. Beirut: American University, 1984.

Rougier de La Bergerie, Jean-Baptiste. *Traité d'agriculture pratique, ou Annuaire de cultivateurs du département de la Creuse et des pays circovoisins.* Paris, 1795.

Rouquette, Pierre. "La transhumance des troupeaux en Provence et en Bas-Languedoc." Thèse, Université de Montpellier, 1913.

Roux, Jean-Paul. "La sédentarisation des nomades yuruks du vilayet d'Antalya." *L'Ethnographie,* n.s. 55 (1961): 64–78.

———. *Les traditions des nomades de la Turquie méridionale.* Paris: Maisonneuve, 1970.

Rowton, M. B. "The Woodlands of Ancient Asia." *Journal of Near Eastern Studies* 26 (1967): 261–77.

Roy, le Comte. *Archives parlementaires de 1787 à 1860: Recueil complet des débats législatifs et politiques des chambres françaises.* Edited by J. Mavidal and E. Laurent. Deuxième série (1800 à 1860). 127 vols. Paris, 1862–1913.

Royer, Jean-Yves. "Les transhumants du Roi Rene." In *Bergers* (Revue éditée par Pays et gens du Verdon). Turin: Imprimerie Mariogros, 1999.

Royle, Trevor. *Crimea: The Great Crimean War, 1854–1856.* New York: Palgrave Macmillan, 2000.

Rubner, Heinrich. *Forstgeschichte im Zeitalter der Industriellen Revolution.* Berlin: Duncker & Humblot, 1967.

Ruedy, John Douglas. *Modern Algeria: The Origins and Development of a Nation.* Bloomington: Indiana University Press, 1992.

Ryder, M. L. *Sheep and Men.* London: Duckworth, 1983.

Şahin, İlhan. *Osmanlı Döneminde Konar-Göçerler/Nomads in the Ottoman Empire.* İstanbul: Eren, 2006.

Sahlins, Peter. *Forest Rites: The War of the Demoiselles in Nineteenth-Century France.* Cambridge MA: Harvard University Press, 1994.

Salzman, Philip. "Political Organization among Nomadic Peoples." *Proceedings of the American Philosophical Society* 111, no. 2 (1967): 115–31.

———, ed. *When Nomads Settle: Processes of Sedentarization as Adaptation and Response.* New York: Praeger, 1980.

Salzmann, Ariel. "An Ancien Régime Revisited: Privatization and Political Economy in the 18th Century Ottoman Empire." *Politics and Society* 21 (1993): 393–423.

———. "Measures of Empire: Tax Farmers and the Ottoman Ancien Régime, 1695–1807." PhD diss., Columbia University, 1995.

Sarre, Friedrich Paul Theodor. *Reise in Kleinasien, Sommer 1895: Forschungen zur seldjukischen Kunst und Geographie des Landes.* Berlin: Geographische Verlagshandlung D. Reimer, 1896.

Sautayra, Édouard. *Législation de l'Algérie: Lois, ordonnances, décrets et arrêtés par ordre alphabétique avec notices, tables (analytique et chronologique) et cartes administratives et judiciaires.* Paris: Maisonneuve, 1878.

Saydam, Abdullah. "The Migrations from Caucasus and Crimea and the Ottoman Settlement Policy." In *The Great Ottoman-Turkish Civilisation*, edited by Kemal Çiçek, 584–93. Ankara: Yeni Türkiye, 2000.

——— . "XIX. Yüzyılın İlk Yarısında Aşiretlerin İskânına Dair Gözlemler." In *Anadolu'da ve Rumeli'de Yörükler ve Türkmenler: Sempozyumu Bildirileri, Tarsus, 14 Mayis 2000*, edited by Tufan Gündüz, 217–29. Ankara: Yörtürk, 2000.

Scheifley, William H. "The Depleted Forests of France." *North American Review* 212, no. 778 (1920): 378–86.

Schiffer, Reinhold. *Turkey Romanticized: Images of the Turks in Early 19th Century English Travel Literature; With an Anthology of Texts.* Bochum: Studienverlag N. Brockmeyer, 1982.

Schneider, Zoë A. *The King's Bench: Bailiwick Magistrates and Local Governance in Normandy, 1670–1740.* Rochester NY: University of Rochester Press, 2008.

Schramm, J. R. "Influence—Past and Present—of François-André Michaux on Forestry and Forest Research in America." *Proceedings of the American Philosophical Society* 101, no. 4 (1957): 336–43.

Schweinitz, H. H. von. *In Kleinasien.* Berlin, 1906.

Sclafert, Thérèse. "À propos du déboisement des Alpes du Sud: Le rôle des troupeaux." *Annales de Géographie* 43, no. 242 (1934): 126–45.

——— . *Cultures en Haute-Provence. Déboisements et pâturages au Moyen âge.* Paris: SEVPEN, 1959.

Scoones, Ian. "Range Management Science and Policy." In *The Lie of the Land: Challenging Received Wisdom on the African Environment*, edited by Melissa Leach and Robin Mearns, 34–53. Oxford: International African Institute in association with James Currey; Portsmouth NH: Heinemann, 1996.

——— . "Sustainability." *Development in Practice* 17, no. 4–5 (2007): 589–96.

——— . "Wetlands in Drylands: Key Resources for Agricultural and Pastoral Production in Africa." *Ambio* 20, no. 8 (1991): 366–71.

Scoones, Ian, and Olivia Graham. "New Directions for Pastoral Development in Africa." *Development in Practice* 4, no. 3 (1994): 188–198.

Scott, James. *Seeing Like a State: How Certain Schemes to Improve the Human Condition Have Failed.* New Haven: Yale University Press, 1998.

Seigue, Alexandre. *La forêt circumméditerranéenne et ses problèmes*. Paris: Maisonneuve et Larose, 1985.

Serres, Olivier de. *Le théâtre d'agriculture et mésnage des champs*. Paris, 1600.

Sessions, Jennifer E. *By Sword and Plow: France and the Conquest of Algeria*. Ithaca NY: Cornell University Press, 2011.

Shaw, Brent D. "Climate, Environment, and History: The Case of Roman North Africa." In *Climate and History: Studies in Past Climates and Their Impact on Man*, edited by T. Wigley, M. Ingram, and G. Farmer, 379–403. Cambridge: Cambridge University Press, 1981.

———. "'Eaters of Flesh, Drinkers of Milk': The Ancient Mediterranean Ideology of the Pastoral Nomad." *Ancient Society* 13–14 (1982–83): 5–31.

Shaw, Stanford J. *History of the Ottoman Empire and Modern Turkey*. 2 vols. Cambridge: Cambridge University Press, 1976.

———. "The Nineteenth-Century Ottoman Tax Reforms and Revenue System." *International Journal of Middle East Studies* 6, no. 4 (1975): 421–59.

Shields, Sarah D. "Sheep, Nomads and Merchants in Nineteenth-Century Mosul: Creating Transformations in an Ottoman Society." *Journal of Social History* 25, no. 4 (1992): 773–89.

Shippers, Thomas. "Le cycle annuel d'un berger transhumant." In *Histoire et actualité de la transhumance en Provence*, edited by Danielle Musset. Mane: Les Alpes de Lumière, 1986.

Simon, Laurent, Vincent Clément, and Pierre Pech. "Forestry Disputes in Provincial France during the Nineteenth Century: The Case of the Montagne de Lure." *Journal of Historical Geography* 33, no. 2 (2007): 335–51.

Sirriyeh, Elizabeth. "The Memoires of a French Gentleman in Syria: Chevalier Laurent d'Arvieux (1635–1702)." *Bulletin (British Society for Middle Eastern Studies)* 11, no. 2 (1984): 125–139.

Sivak, Henry. "Law, Territory, and the Legal Geography of French Rule in Algeria: The Forestry Domain, 1830–1903." PhD diss., University of California, Los Angeles, 2008.

Slane, William de. *Histoire des Berbères et des dynasties musulmans de l'Afrique septentrionale par Ibn Khaldoun*. Alger: Imprimerie du Gouvernement, 1852.

———. *Les prolégomènes d'Ibn Khaldoun*. 3 vols. Paris: Imprimerie impériale, 1863–68. Google Books.

Sluglett, Peter, and Marion Farouk-Sluglett. "The Application of the 1858 Land Code in Greater Syria: Some Preliminary Observations." In *Land Tenure and Social Transformation in the Middle East*, edited by Tarif Khalidi, 409–21. Beirut: American University, 1984.

Smethurst, David. "Mountain Geography." *Geographical Review* 90, no. 1 (2000): 35–56.

Smith, David. *The Practice of Silviculture*. 9th ed. New York: John Wiley & Sons, 1997.

Solakian, Daniel. "De la multiplication des chèvres sous la Révolution." In *Révolution et espaces forestiers: Colloque des 3 et 4 juin 1987*, edited by Denis Woronoff. Paris: Éditions L'Harmattan, 1989.

Şölen, B. Hikmet. *Aydın İli ve Yürükler*. Aydın: G. H. P. Basımevi, 1945.

Somerville, William. "Forestry in Some of Its Economic Aspects." *Journal of the Royal Statistical Society* 72, no. 1 (1909): 40–63.

Spary, E. C. *Utopia's Garden: French Natural History from Old Regime to Revolution*. Chicago: University of Chicago Press, 2000.

Spillmann, Georges. *Napoléon III et le royaume arabe d'Algérie*. Paris: Académie des Sciences d'Outre-Mer, 1975.

Spiridonakis, Basile G. *Empire ottoman: Inventaire des mémoires et documents aux archives du Ministère des affaires étrangères de France*. Thessaloniki: Institute for Balkan Studies, 1973.

Spooner, Brian. "Desert and Sown: A New Look at an Old Relationship." In *Studies in Eighteenth-Century Islamic History*, edited by T. Naff and R. Owen. Carbondale: Southern Illinois University Press, 1977.

Statistique et documents relatifs au Sénatus-consulte sur la propriété arabe, 1863. Paris: Imprimerie Impériale, 1863.

Statistiques des forêts soumises au régime forestier: Forêts domaniales, communales et d'établissements publics. N.p.: n.p., 1892.

Steen, Harold K., Forest History Society, and International Union of Forestry Research Organizations. *History of Sustained-Yield Forestry: A Symposium; Western Forestry Center, Portland, Oregon, October 18–19, 1983*. Santa Cruz CA: Forest History Society, 1984.

Stephenson, Rowland. *Railways in Asiatic Turkey*. London, 1878.

Sümer, Faruk. "Ağaç-Eriler." *Belleten* 26, no. 103 (1962): 521–28.

Surell, Alexandre. *Étude sur les torrents des Hautes-Alpes*. 2 vols. 1841. Paris: Dunod, 1870–1872.

Sutherland, Donald. *France 1789–1815: Revolution and Counterrevolution*. New York: Oxford University Press, 1986.

Swan, Susan L. "Mexico in the Little Ice Age." *Journal of Interdisciplinary History* 11, no. 4 (1981): 633–48.

Sykes, Christopher Simon. *Black Sheep*. New York: Viking Press, 1983.

Tabak, Faruk. *The Waning of the Mediterranean, 1550–1870: A Geohistorical Approach*. Baltimore: Johns Hopkins University Press, 2008.

Takeda, Junko Thérèse. *Between Crown and Commerce: Marseille and the Early Modern Mediterranean*. Baltimore: Johns Hopkins University Press, 2011.

Tankut, Gönül. "Urban Transformation in the Eighteenth-Century Ottoman City." *Journal of the Faculty of Architecture* (Middle East Technical University), 1, no. 2 (1975): 247–62.

Tapper, Richard. "Anthropologists, Historians, and Tribespeople on Tribe and State Formation in the Middle East." In *Tribes and State Formation in the Middle East*, edited by P. Khoury and J. Kostiner, 48–73. Berkeley: University of California Press, 1990.

———. "Black Sheep, White Sheep and Red-Heads: A Historical Sketch of the Shāhsavan of Āzarbāijān." *Iran* 4 (1966): 61–84.

———. *Frontier Nomads of Iran: A Political and Social History of the Shahsevan.* New York: Cambridge University Press, 1997.

———. *Pasture and Politics: Economics, Conflict, and Ritual among Shahsevan Nomads of Northwestern Iran.* London: Academic Press, 1979.

"Tarifs et douanes." *Revue des Eaux et Forêts* 30, no. 5 (1891): 253–60.

Tassy, Louis. *État des forêts en France.* Paris: Octave Doin, 1887.

———. *Études sur l'aménagement des forêts.* Paris: Bureau des Annales Forestières, 1858.

———. *La restauration des montagnes: Étude sur le projet de loi présenté au Sénat.* Paris: J. Rothschild, 1877.

———. *Réorganisation du Service Forestier.* Paris: Typographie A. Hennuyer, 1884.

———. *Service forestier de l'Algérie: Rapport adressé à M. le Gouverneur de l'Algérie (5 août 1872).* Paris: A. Hennuyer, 1882.

Teich, Mikuláš, Roy Porter, and Bo Gustafsson, eds. *Nature and Society in Historical Context.* Cambridge: Cambridge University Press, 1997.

Temple, Samuel S. "The Natures of Nation: State-Building and the Politics of Environmental Marginality in 19th and 20th Century Southern France." PhD diss., University of Michigan, 2010.

Tétreau, Adolphe. *Commentaire de la loi du 4 avril 1882 sur la restauration et la conservation des terrains en montagne.* Paris: P. Dupont, 1883.

Thirgood, J. V. *Cyprus: A Chronicle of Its Forests, Land, and People.* Vancouver: University of British Columbia Press, 1987.

———. *Man and the Mediterranean Forest: A History of Resource Depletion.* London: Academic Press, 1981.

Thobie, Jacques. *Intérêts et impérialisme français dans l'Empire ottoman 1895–1914.* Paris: Publications de la Sorbonne, 1977.

———. *La France, l'Europe et l'Est méditerranéen depuis deux siècles: Économie, finance, diplomatie.* Istanbul: Isis, 2007.

Thornes, John. "Erosional Equilibria under Grazing." In *Conceptual Issues in Environmental Archaeology*, edited by J. L. Bintliff, Donald A. Davidson, and Eric G. Grant. Edinburgh: Bintliff, 1988.

———. "Global Environmental Change and Regional Response: The European Mediterranean." *Transactions of the Institute of British Geographers* 20, no. 3 (1995): 357–67.

———. "Land Degradation." In *The Physical Geography of the Mediterranean*, edited by Jamie C. Woodward, 563–82. Oxford: Oxford University Press, 2009.

Thoreau, Henry David. *Walking*. 1861. Rockville MD: Arc Manor, 2007.

Tocqueville, Alexis de. *The Old Regime and the French Revolution*. Translated by Stuart Gilbert. New York: Doubleday, 1983.

———. *Writings on Empire and Slavery*. Edited and translated by Jennifer Pitts. Baltimore: Johns Hopkins University Press, 2001.

Toksöz, Meltem. *Nomads, Migrants and Cotton in the Eastern Mediterranean: The Making of the Adana-Mersin Region, 1850–1908*. Leiden: Brill, 2010.

Tolley, Cedric. "Qui sont les bergers?" *Bergers* (Revue éditée par Pays et gens du Verdon). Turin: Imprimerie Mariogros, 1999.

Tomaselli, R. "Degradation of the Mediterranean Maquis." In UNESCO, *Mediterranean Forests and Maquis: Ecology, Conservation and Management*. Paris: UNESCO, 1977.

Trabut, Louis, and Auguste Mathieu. *Les Hauts-Plateaux oranais: Rapport de mission*. Alger: Imprimerie Pierre Fontana, 1891.

Trautmann, Wolfgang. "The Nomads of Algeria under French Rule: A Study of Social and Economic Change." *Journal of Historical Geography* 15, no. 2 (1989): 126–38.

Trehonnais, F. Robiou de la. *Rapport: L'agriculture en Algérie*. Alger: Imprimerie Typographique et Ethnographique Bouyer, 1867.

Trivelly, Elise. *Quand les moutons s'en vont: Histoire et représentations sociales du boisement des pelouses sèches du sud-est de la France*. Aix-en-Provence: Publications de l'Université de Provence, 2004.

Trolard, Paulin. "Appel aux Algériens." *Bulletin de la Ligue du Reboisement de l'Algérie* 1 (1882).

———. *Incendies forestiers en Algérie: Leurs causes et les moyens de remédier à ces causes*. Alger: Ligue du Reboisement, 1892.

———. *La question forestière algérienne devant le Senat*. Alger, 1893.

Trottier, François. *Boisement et colonisation: Rôle de l'eucalyptus en Algérie au point de vue des besoins locaux de l'exportation et du développement de la population*. Alger: Imprimerie de l'Association Ouvrière V. Aillaud et Cie, 1876.

Tucker, R. P., and J. F. Richards, eds. *Global Deforestation and the Nineteenth-Century World Economy*. Durham: Duke University Press, 1983.

Tunçdilek, Necdet. *Türkiye İskân Coğrafyası: Kır İskânı (Köy-Altı İskân Şekilleri)*. İstanbul: İstanbul Matbaası, 1967.

————. "Yayla Settlement and Related Activities in Turkey." *Review of the Geographical Institute of the University of Istanbul* 9–10 (1963–64): 58–71.

Türkay, Cevdet. *Başbakanlık Arşivi Belgeleri'ne Göre Osmanlı İmparatorluğu'nda Oymak, Aşiret ve Cemâatlar*. İstanbul: İşaret Yayınları, 2001.

————. *Osmanlı Türklerinde Coğrafya*. İstanbul: Maarif Basımevi, 1959.

Türkiye Diyanet Vakfi and İslâm Ansiklopedisi Genel Müdürlügü. *Türkiye Diyanet Vakfi İslâm Ansiklopedisi (DVIA)*. İstanbul, 1988.

Türkiye Tabiatını Koruma Cemiyeti. *Dağ ve Orman Köylerinin Ekonomik, Sosyal Sorunları ve Çözüm Yolları*. Ankara: Türkiye Tabiatını Koruma Cemiyeti, 1969.

Tzedakis, Chronis. "Cenozoic Climate and Vegetation Change." In *The Physical Geography of the Mediterranean*, edited by Jamie C. Woodward, 89–137. Oxford: Oxford University Press, 2009.

Uluçay, M. Çagatay. *18 ve 19. Yüzyıllarda Saruhan'da Eskiyalık ve Halk Hareketleri*. İstanbul: Berksoy Basimevi, 1955.

Unruh, Jon D. "Integration of Transhumant Pastoralism and Irrigated Agriculture in Semi-Arid East Africa." *Human Ecology* 18, no. 3 (1990): 223–46.

Valbuena-Carabaña, María, et al. "Historical and Recent Changes in the Spanish Forests: A Socio-Economic Process." *Review of Palaeobotany and Palynology* 162, no. 3 (2010): 492–506.

Van Lennep, Henry John. *Travels in Little-Known Parts of Asia Minor, with Illustrations of Biblical Literature and Researches in Archaeology*. London: John Murray, 1870.

Vann, James Allen. *The Making of a State: Württemberg, 1593–1793*. Ithaca NY: Cornell University Press, 1984.

Veinstein, Gilles. *État et société dans l'Empire ottoman, XVIe–XVIIIe siècles: La terre, la guerre, les communautés*. Aldershot: Variorum, 1994.

————. "Sur les sauterelles à Chypre, en Thrace et en Macédoine à époque ottomane." In *Armağan: Festschrift für Andreas Tietze*, edited by I. Baldauf, Suraiya Faroqhi, and R. Vesely, 211–26. Prague: Enigma, 1994.

Velud, Christian, ed. *Les sociétés méditerranéennes face au risque: Espaces et frontières*. Cairo: Institut Français d'Archéologie Orientale, 2012.

Venture, Rémi. "Arles et le Rhône." *Le Rhône à Son Delta* 41–42 (1993).

————. "L'inondation de 1856: Le Rhône, les digues et Napoléon III." Talk presented at Comprendre les inondations, Maison de la Vie Associative—Arles, 9–16 November 2004.

Vernet, Jean-Louis. *L'homme et la forêt méditerranéenne de la préhistoire à nos jours*. Paris: Errance, 1997.

Villeneuve, Christophe de. *Statistique du département des Bouches-du-Rhône*. 5 vols. Marseille, 1821–29.

————. *Voyage dans la vallée de Barcelonette*. Agen: R. Noubel, 1815.

Vincze, Lajos. "Peasant Animal Husbandry: A Dialectic Model of Techno-Environmental Integration in Agro-Pastoral Societies." *Ethnology* 19, no. 4 (1980): 387–403.

Viney, R. "L'œuvre forestière du Second Empire." Spécial: L'histoire forestière, *Revue Forestière Française* (1962): 532–41.

Vita-Finzi, Claudio. *The Mediterranean Valleys: Geological Changes in Historical Times*. Cambridge: Cambridge University Press, 1969.

Vogt, Jean. "Sismicité historique du domaine ottoman." In *Natural Disasters in the Ottoman Empire: Halcyon Days in Crete III: A Symposium Held in Rethymnon 10–12 January 1997*, edited by Elisavet Zachariadou. Rethymnon: Crete University Press, 1999.

Vryonis, Speros. "Nomadization and Islamization in Asia Minor." *Dumbarton Oaks Papers* 29 (1975): 41–71.

Wagstaff, J. Malcolm. *The Evolution of Middle Eastern Landscapes: An Outline to A.D. 1840*. Totowa NJ: Barnes & Noble, 1985.

Wainwright, John. "Weathering, Soils, and Slope Processes." In *The Physical Geography of the Mediterranean*, edited by Jamie C. Woodward, 169–202. Oxford: Oxford University Press, 2009.

Wakefield, Andre. *The Disordered Police State: German Cameralism as Science and Practice*. Chicago: University of Chicago Press, 2009.

Walker, Brett. *Toxic Archipelago: A History of Industrial Disease in Japan*. Seattle: University of Washington Press, 2010.

Warde, Paul. "Fear of Wood Shortage and the Reality of the Woodland in Europe, c. 1450–1850." *History Workshop Journal* 62, no. 1 (2006): 28–57.

Warnier, Auguste-Hubert. *L'Algérie devant l'Empereur, pour faire suite à l'Algérie devant le Sénat, et à l'Algérie devant l'opinion publique*. Paris: Challamel Aîné, 1865.

———. *L'Algérie devant le Sénat*. Paris: Dubuisson, 1863.

———. *L'Algérie et les victimes de la guerre*. Alger: Imprimerie Duclaux, 1871.

Warren, Andrew. "Changing Understandings of African Pastoralism and the Nature of Environmental Paradigms." *Transactions of the Institute of British Geographers* 20, no. 2 (1995): 193–203.

Watkins, Charles. *Ecological Effects of Afforestation: Studies in the History and Ecology of Afforestation in Western Europe*. Wallingford: C.A.B International on behalf of the European Science Foundation, 1993.

Webb, James. *Desert Frontier: Ecological and Economic Change along the Western Sahel, 1600–1850*. Madison: University of Wisconsin Press, 1995.

Weber, Eugen. *Peasants into Frenchmen: The Modernization of Rural France, 1870–1914*. Stanford: Stanford University Press, 1976.

Weil, Benjamin. "Conservation, Exploitation, and Cultural Change in the Indian Forest Service, 1875–1927." *Environmental History* 11, no. 2 (2006): 319–43.

Wenzel, Hermann. *Forschungen in Innerantolien*. Kiel: Buchdruckerei Schmidt & Klaunig, 1937.

Wertime, Theodore A. "The Furnace versus the Goat: The Pyrotechnologic Industries and Mediterranean Deforestation in Antiquity." *Journal of Field Archaeology* 10, no. 4 (1983): 445–52.

White, Frank. *The Vegetation of Africa: A Descriptive Memoir to Accompany the UNESCO/AETFAT/UNSO Vegetation Map of Africa*. Paris: UNESCO, 1983.

White, Sam. *The Climate of Rebellion in the Early Modern Ottoman Empire*. New York: Cambridge University Press, 2011.

———. "Ecology, Climate, and Crisis in the Ottoman Near East." PhD diss., Columbia University, 2008.

Whited, Tamara. "Extinguishing Disaster in Alpine France: The Fate of Reforestation as Technocratic Debacle." *Geojournal* 51 (2000): 263–70.

———. *Forests and Peasant Politics in Modern France*. New Haven: Yale University Press, 2000.

Williams, Michael. *Deforesting the Earth: From Prehistory to Global Crisis*. Chicago: University of Chicago Press, 2003.

Williams, Raymond. *The Country and the City*. Oxford: Oxford University Press, 1975.

Wittek, Paul. "Le rôle des tribus turques dans l'Empire ottoman." In *Mélanges Georges Smets*. Brussels, 1952.

———. "Osmanlı İmparatorluğunda Türk Aşiretlerin Rolü." *İstanbul Üniversitesi Edebiyat Fakültesi Tarih Dergisi* 13 (1963): 257–68.

———. *The Rise of the Ottoman Empire*. London: Royal Asiatic Society, 1938.

Woodhead, Christine. "Letters of Misirli Ibrahim Pasa to Menlikli Ahmed Pasa, Governor of Adana, 1843–46." *Bulletin (British Society for Middle Eastern Studies)* 16, no. 1 (1989): 49–51.

———, ed. *The Ottoman World*. London: Routledge, 2012.

Woodward, Jamie C., ed. *The Physical Geography of the Mediterranean*. Oxford: Oxford University Press, 2009.

Woolsey, Theodore Salisbury. *French Forests and Forestry: Tunisia, Algeria, Corsica, with a Translation of the Algerian Code of 1903*. New York: J. Wiley & Sons, 1917.

Woronoff, Denis. "Histoire des forêts françaises, XVIe–XXe siècles: Résultats de recherche et perspectives." *Les Cahiers du Centre de Recherches Historiques: Archives* 6 (1990).

———. "La crise de la forêt française pendant la révolution et l'Empire: L'indicateur sidérurgique." *Cahiers d'Histoire* (1979): 3–17.

———. "La 'dévastation révolutionnaire' des forêts." In *Révolution et espaces for-estiers: Colloque des 3 et 4 juin 1987*, edited by Denis Woronoff, 44–52. Paris: Éditions L'Harmattan, 1989.

———, ed. *Révolution et espaces forestiers: Colloque des 3 et 4 juin 1987*. Paris: Éditions L'Harmattan, 1989.

Wright, Walter Livingston, ed. *Ottoman Statecraft: The Book of Counsel for Vezirs and Governors (Nasaih ül-vüzera vel-ümera) of Sari Mehmed Pasha, the Defterdar*. Princeton: Princeton University Press, 1935.

Yacono, Xavier. *La colonisation des plaines du Chelif: De Lavigerie au confluent de la Mina*. Alger: Imbert, 1955.

———. *Les Bureaux Arabes et l'évolution des genres de vie indigènes dans l'Ouest du Tell algérois (Dahra, Chélif, Ouarsenis, Sersou)*. Paris: Larose, 1953.

———. "Peut-on evaluer la population de l'Algérie vers 1830?" *Revue Africaine* 98 (1954): 277–307.

Yalçin, B. C. *Sheep and Goats in Turkey*. Rome: Food and Agriculture Organization of the United Nations (FAO), 1986.

Yetişen, Rıza. *Tahtacı Aşiretleri: Adet, Gelenek, ve Görenekleri*. İzmir: Memleket Gazetecilik ve Matbaacılık, 1986.

Yiğitoğlu, Ali Kemal. *Türkiye'de Ormancılığın Temelleri, Şartları ve Kuruluşu*. Ankara: Yüksek Ziraat Enstitüsü, 1936.

Yıldız, M. Cengiz. "Osmanlı'dan Günümüze Demiryolu Politikalarina Genel Bakış." *Ekev Akedemi Dergisi* 18 (2004): 195–208.

Yılmaz, Mehmet. "Policy of Immigrant Settlement of the Ottoman State in the 19th Century." In *The Great Ottoman-Turkish Civilisation*, edited by Kemal Çiçek, 594–608. Ankara: Yeni Türkiye, 2000.

Young, David Bruce. "A Wood Famine? The Question of Deforestation in Old Regime France." *Forestry* 49, no. 1 (1976): 45–46.

Young, George. *Corps de droit ottoman: Recueil des codes, lois, règlements, ordonnances et actes les plus importants du droit intérieur, et d'études sur le droit coutumier de l'Empire ottoman*. Oxford: Clarendon Press, 1905.

Yund, Kerim. "100 Yıllık Türk Ormancılık Öğretimine Bakış." In *Türk Ormancılığı Yüzüncü Tedris Yılına Girerken, 1857–1957*. Ankara: Doğuş Matbaası, 1957.

———. *Türkiye Orman Umum Müdürleri Albümü*. İstanbul: Hüsnütabiat Matbaası, 1959.

Zachariadou, Elisavet, ed. *Natural Disasters in the Ottoman Empire: Halcyon Days in Crete III: A Symposium Held in Rethymnon 10–12 January 1997*. Rethymnon: Crete University Press, 1999.

Zaimeche, S. E. "Change, the State and Deforestation: The Algerian Example." *Geographical Journal* 160, no. 1 (1994): 50–56.

Zerbe, Stefan. "Influence of Historical Land Use on Present-Day Forest Patterns: A Case Study in South-Western Germany." *Scandinavian Journal of Forest Research* 19, no. 3 (2004): 261–73.

Zon, Raphael. "Forests and Human Progress." *Geographical Review* 10, no. 3 (1920): 139–66.

Zürcher, Erik. *Turkey: A Modern History*. 3rd ed. London: I. B. Tauris, 2004.

INDEX

Page numbers in italics indicate illustrations.

Algerians, indigenous: attitudes toward, 51–52, 150, 153; Bureaux Arabes and, 38; cantonnement and, 121, 123; Forest Code (1827) and, 93, 95; as forest guards, 94; forest laws and, 154–56, 172; land ownership law and, 124; revolt (1871) and, 124–25

Algerians, of European descent, 126, 147, 171

Algerian Steppe, 21

Algiers, Algeria, 10, 19, 88

Allauch, France, 114–15, 196–97n42

Allgemeines ökonomisches Forstmagazin, 70

alpages (mountain meadows), 18, 197n64

Les Alpes français (Briot), 165

Alps, 15, 16, 109, 142, 146

Anatolia, xi–xiii, *xii*, xv, xvii–xviii; agricultural expansion in, 176; agricultural production in, 28–29; attitudes toward, 41–42, 52–53; climate change in, 29–30, 40; climate in, 6; commonalities of, with other Mediterranean societies, 31; fires in, 156–57; foreign trade of, 96–97; forest management, French, in, xvii–xviii, 78, 82, 96, 98, 100–106, 176–77, 188–89; forest management, Ottoman, in, 96–98, 177–80; forest management, Turkish, in, 180; forests in, 24–25, 53, 97; French influence on, 39, 41, 99–100, 101, 185; geography of, 12–14; land ownership in, 126–29, 133; natural disasters in, 29–30, 136, 156–60; nomadism in, 4–5, 21–24, 28–29, 39–43, 52, 181; Ottoman state and, 133; resettlement in, 131–32; sedentarization in, 40–41, 130–31, 174–76; technology in, 175–76; transportation in, 175–76. *See also* Ottoman Empire

Annales Forestières, 100, 101

Antalya, Turkey (city), 6, 7, 13, 22, 158, 175

Antalya, Turkey (province), 13, 22, 41, 102, 157, 180

Antalya Plain, 28–29

Anthropocene epoch, 160

antipastoralism, xvi–xvii, 34–35, 54, 60, 105, 184, 188

Arabes des tribus, 124

Arabes des villes, 124

Arabophiles, 38–39

Arabs: attitudes toward, xvii, 35–36, 41, 95; Berbers and, 27, 125; environmental impacts of, perceived, 50–51, 94; fires and, 152; as French subjects, 123–24; narrative about, 36–37, 120; settled, 124; stereotypes of, 153; tribal, 124

Arazi Kanunnamesi. *See* Land Code (Ottoman, 1858)

arch (communal property), 120, 126

Aristotle, 21; *Historia animalium*, 198n88

Arles, France, 10, 29–30, 108, 109, 117, 137–38, 168, 181

arson, 135, 137, 151–54, 156–57

Article 19 (Land Code of 1858), 130

Article 62 (Land Code of 1858), 73

Article 78 (Land Code of 1858), 87

Article 90 (Land Code of 1858), 76–77, 115

Article 110 (Land Code of 1858), 74

Asia Minor. *See* Anatolia

Atlas Mountains, 21, 172–73

Balkans, 97, 176, 201n44

Baltaci, Aristidi, 177

Basse-Alpes, France, 142, 146

Basse-Provence, France, 9–10, 15–16, 109, 137

Beckmann, Johann Gottlieb, *Beyträge zur Verbesserung der Forstwissenschaft*, 70

Bedouin, 31, 41, 42, 173, 201n44

Bent, Thomas, 53

Berbers, 27–28, 36–37, 51, 91–92, 125, 201n144

Bernard, Augustin, 51; *L'évolution du nomadisme in Algérie*, 189

Beylik of the East. *See* Constantine, Algeria (province)

Beyträge zur Verbesserung der Forstwissenschaft (Beckman), 70

biomes, 9

black sheep metaphor, 33

Blanqui, Adolphe-Jérôme, 138–39

Blumenbach, Johann Friedrich, 35

Bonaparte, Louis-Napoléon. *See* Napoleon III, Emperor

Bory de Saint-Vincent, Jean-Baptiste, 49

Bouche, Charles-François, 46–47

Bouches-du-Rhône Department, 9, 60, 67, 82–83, 85–86, 108, 117

Bourbon-Penthièvre family, 55–56

Le bourgeois gentilhomme (Molière), 99

Boyer de Fonscolombe, Étienne Laurent Joseph Hippolyte, 47

Braudel, Fernand, xiv, xv; *Capitalism and Material Life*, 62; *The Mediterranean and the Mediterranean World in the Age of Philip II*, xiii

Bricogne, Louis Adolphe, 102–4, 105, 177

Briot, Félix, *Les Alpes français*, 165–66

Britain and the British, 41, 53, 62–63, 78, 99, 131–32, 157, 158, 159

Buffon, Georges-Louis Leclerc de, 64, 111; *Les époques de la nature*, 45; *Histoire naturelle, gènèrale et particulière*, 34–35, 44, 45; *Mémoire sur la conservation et le rétablissement des forêts*, 63

Bureaux Arabes, 38–39, 123, 124, 126

Burnaby, Frederick, 42, 157, 159

burning, controlled: occupational, 137, 152, 156, 177; for pasture creation and renewal, 25, 53, 97, 102–3, 130, 149, 152; value of, 186. *See also* fires

cabanes (shepherd shelters), 18–19

Camargue, France, 10, 46, 117, 138

camels, xiii, 21, 23, 172, 175

cantonnement (divestment of forest rights), 73, 75, 121–22, 123–24

Capitalism and Material Life (Braudel), 62

carpet manufacture and trade, 158

Catholic Church, 108, 110

cattle, 21, 66, 158, 172

Caucasian race, 35

Celali Rebellions, 40

Çelebi, Evliya, 22, 97

censuses, 66, 124, 128

Cevdet Pasha, Ahmet, 128, 130

Chambre des Eaux et Forêts, 58, 64, 65

charcoal, 52, 157, 176, 177

Charles IX, King, 57

Charles X, King, 71, 88–89, 113

Chateaubriand, François-René, 52

Chelif River Valley, Algeria, 20

cholera, 148, 157

Cilician Gates, xi–xii

Cilician Plain, xi, 13–14, 30

Circassians, 131

Civil Code, 112

civilization: forests important for, 90, 104; levels of, 41–43, 54; pastoralism and, 34, 43–44, 173; theories of, 35–36, 37, 48, 131

civilizing mission, 37, 38, 39, 121, 123

Civil War, American, 167

climate change, 29–31, 44–46, 48, 50, 137, 189

Clutton-Brock, Juliet, 198n85

Cobden-Chevalier Commerce Agreement, 167

Code Napoléon, 87

coefficient of variation (cv), 8

Colbert, Jean-Baptiste, 57, 64
collective tribal responsibility, 135, 155–56
colonialism, xiv, xvii; civilizing mission
of, 37, 123; criticism of, 133, 168–69;
environmental conditions caused
by, 61–63; environmental conditions
justifying, 50–51, 54, 90–91; fire
management and, 149–50, 152; forestry
and, 77–78, 92–96, 170–71, 185; indigenous population and, 169–70, 187–88;
internal, in Provence, 166, 188; islands
changed by, 61–62; land ownership
and, 119–24; pastoralism and, 107,
168–69, 189; theory of progress and,
35; timber supply and, 89–90
La Colonisation, 121
Colonisation de l'Algérie (Enfantin), 37,
38–39
communication development: administration helped by, 54, 130, 134,
164, 167; economic benefits of, 98;
nomads and, 175, 182, 189
conservation: environmental, xvi, 34,
131; ethic of, 61, 64; forest, 71; ideal
of, 68; soil, 145
Constantine, Algeria (city), *6, 7*
Constantine, Algeria (province): agropastoralism in, 28; cantonnement
in, 122–23; climate of, 11; fire bans in,
154; forests in, 93; French presence
in, 120, 179; geography of, 10, 12;
natural disasters in, 148; nomads in,
20; Ottoman presence in, 19
coppices (taillis), 71, 210n89
coppices with standards (*taillis sous
futaie*), coppices with standards, 71
cork trade, 135, 150, 170, 185
The Corrupting Sea (Horden and
Purcell), xiii–xiv
cotton trade, 167
Council of Public Works, 98, 178
coussous (coussouls). *See* pastures

Crau (steppe), 9–10, 15, 109, 117, 118–19,
118, 137
Crimean War (1854–1856), xvii, 41, 98,
99, 131
cv (coefficient of variation), 8

Daudet, Alphonse, 166
Davis, Diana K., *Resurrecting the
Granary of Rome*, xx
declensionist narrative, 44, 104, 151, 165
deforestation, xvi, xviii–xix; in
Algeria, 12, 152; in Anatolia, 102,
103; awareness of, 50, 53–54, 57, 59,
63–64, 187; causes of, 44, 51, 55–56,
64, 67–68, 70, 92, 165; as crime, 172;
effects of, 44–47, 136–37, 138–39,
144; in France, 47, 206n3; laws
regulating, 59, 179–80; mountain,
92, 138–39; in Ottoman empire, xvii;
pastoralism and, 106, 130, 136, 140,
143, 184; reversal of, 163
Demontzey, Prosper, 140–43, *141*; *Étude
sur les travaux de reboisement et de
gazonnement des montagnes*, 142
depopulation, 22, 30, 31, 41, 146
"desert and the sown," xxi, 25–26, 30,
36, 54
desertification, 50–51, 159–60
deserts, 11, 22, 36, 39, 45, 50–51, 52, 160.
See also Sahara Desert
displacement, indigenous. *See* land,
ancestral: dispossession of; land,
ancestral: privatization of
dogs, 17–18
droit d'esplèche (grazing rights), 109
droits d'usage (usage rights), 73, 103, 108
drought, 7–8, 92, 136, 137, 138, 147, 157
Duhamel du Monceau, Henri-Louis, 64;
*Du transport, de la conservation et de
la force des bois*, 64; *Traité complet
des bois et forêts*, 63–64
Durance River, France, 137

using, 66; private, 111–13, 177; products from, 52, 57–58, 177; in Provence, France, 60–61, 82–84; state-owned, 121; terms for, 76, 210n89; Turkish, 100–101, 180; types of, 177

forests, communal, 47, 58, 60, 72–73, 74–75, 82–84, 177. *See also* land, communal

forests, private, 60, 72, 83, 87

forests, public, 83, 87

forests, royal, 72, 83

forests, state, 60, 72, 74, 83, 87

forest school, 70, 78, 105, 177

Forest Service, 90, 93–94, 170–71, 178–79

Foucault, Michel, 56

France, xiv, xv–xvii, xix; Algeria and, 12, 54, 88–89; Anatolia and, 41; environmental awareness of, 62–63; fires and, 152, 154–55; governmentalization of, 56; industrialization in, 68–69; Ministry of the Interior, 46; natural disasters and, 135–36, 138–39; Ottoman Empire and, 99–101; political influence of, 187–89, 191; self-perception of, 44; in war, 65–66

Franco-Prussian War, 124

French Revolution (1789), 48, 55, 56, 59, 61, 64–68, 110–12

French Revolution (1830), 89

French Revolution (1848), 121

French Revolution (1870–1871), 124

French Scientific Congress, 139

La fuite des populations pastorales françaises (Fabre), 165

garrigue (vegetation type), 9, 15, 61, 82

garrigue/maquis (biome), 9

Gaye, Jean-Baptiste Sylvère, 71–72

gazi (holy warriors), 43

gazonnement (pasture regeneration), 142–43

Germany and German influence, xv, 56–57, 69–71, 77, 100, 104, 187

Giono, Jean, 166

goats, xii, xvi, xix; attitudes toward, 47–49, 184; as common pastoral animal, 23, 66, 196–97n42; diet of, 24; domestication of, 21; environmental impacts of, perceived, 34, 47–48, 55, 59–60, 65, 66–67, 102, 188; grazing areas of, 11, 15–16, 83; grazing laws and, 57, 58, 60–61, 74, 87, 93, 114, 171–72; in island colonies, 62; natural disasters and, 158; taxes on, 109; value of, 8, 17, 27, 186

governmentality: definition of, 56; fire control and, 156; forest (term) and, 77; forestry and, 107, 134; French administration and, 78, 91; Ottoman, 130; property rights and, 107, 134; in Provence, 166; web of, 184

grain, 21

grains, 29

"granary of Rome," 50, 205n95

grazing: benefits of, 8, 108, 109, 186; communal, 115, 116–17; environmental impacts of, 95, 186; forest, 116, 170–72, 177–78, 188; illegal, 58, 85, 88, 114, 146, 172; laws regulating, 57, 58, 60–61, 64, 67, 74–75, 83–85, 104, 109–10, 156

Great Britain. *See* Britain and the British

Great Kabylia, Algeria, 10, 28

Great Ordonnance of Colbert, 64

Greek War of Independence, 99

Green Imperialism (Grove), 62

Grove, Alfred, 149

Grove, Richard, *Green Imperialism*, 62

Gülek Pass, xi–xii

Gülhane proclamation (1839), 128

Hamilton, W. J., 52

Hammer, Joseph de, 201n5

North Africa, xiv, 10, 19, 35–36
North Africans, xx, 36, 44, 50–51

The Old Regime and the French Revolution (Tocqueville), 64
Oran, Algeria, 169
ordinance, French (1804), 60
ordinance, French (1811), 60–61
ordinance, Ottoman (1876), 178–79
Ordonnance des Eaux et Foréts (1669), 76
orientalists and orientalism, 35–36, 42, 51, 201n5
Orman Nizamnamesi (law, 1870), 177–78
Osman (tribe), 43
Ottoman Empire, xv–xvi, xvii–xviii; Algeria in, 19; Anatolia in, 13, 21–22, 126–27; environmental degradation in, 160; fires in, 156–57; forest management, French, in, 78, 82, 96, 100–105, 176–77; forest management, Ottoman, in, 97–98, 105–6, 177–80; France and, 98–100, 185, 187; immigrants in, 131–32; land ownership in, 128, 130, 131–32; natural disasters in, 157; natural resources in, 53, 96; nomads, 181; nomads in, 22, 23–24, 31, 39–43, 52–53, 133, 158, 159, 174; peak of, 96–97; problems of, 179–80; sedentarization in, 54, 105–6, 130–31; self-perception of, 42–43; settlement campaigns of, 136; tax collection by, 128. *See also* Anatolia
Ottoman forest administration, 103, 106
Ottoman forest institute, 103, 105, 180
Ottoman period, 22, 25, 136

Pamphylian Plain of Antalya, 30
pastoralism, xiii, xv, xvi–xvii, xix; agriculture and, 25–26; attitudes toward, 33–34, 54, 95–96, 142–43, 169–70; diminishing, 118–19, 163–64, 166–68; economies of, 201n144;

environmental impacts of, 34, 44, 115, 150; fires and, 151–54; Forest Code (1827) and, 74–76; French administration and, 56–57, 58–59, 64; laws regulating, 140; traces of, 3–4
pastoralism, mobile, xvi–xvii, xviii, xviii–xix; advantages of, 8, 21; in Algeria, 19, 50–51; in Anatolia, 21, 172–75; categories of, 4; changing attitude toward, 165–66, 186–87; commonalities among, 14, 183; diminishing, 131, 181; environmental impacts of, perceived, 48–49, 52, 102–3, 105, 130, 149; forests and, 24; French administration and, 78, 82, 183–85; land use of, 124; laws regulating, 126, 128–29; marginalized, 91–92, 108, 161, 164; natural disasters and, 135–37, 158; perceptions of, 36–37, 53–54; popularized in romantic literature, 166; in Provence, France, 14–15; susceptible to governmentality, 107–8
pastoralists, indigenous, 120, 143, 148–49
pastures, xvi; access to, 85–86; in Algeria, 20; communal, 75, 108–9, 111, 114, 116–17, *116*, 133, 172; converted to cultivated land, 107, 117, 176; creation of, 152, 153; degradation of, 173; fallow, 108, 110–11, 112, 116–17; farmland and, 26; forests and, 24–25, 66, 75, 83–84, 96, 102–3, 177–78; loss of, 146; natural disasters and, 137–38, 148, 157, 161; ownership of, 108, 129; poaching of, 114; private, 118; in Provence, France, 9–10; regeneration of, 142–43, 149; rental of, 17, 18, 84–85, 87, 108–9, 113, 129, 181, 188, 197n64; right of pasture and, 172; seasonal, 175; settlements turned into, 39; summer, 23, 24–25, 29, 53, 129, 181; types of, 15; winter,

To order or obtain more information on these or other University
of Nebraska Press titles, visit nebraskapress.unl.edu.

CPSIA information can be obtained
at www.ICGtesting.com
Printed in the USA
LVHW091906230320
650928LV00004B/43

9 780803 290976